DECARBONISING
THE WORLD'S ECONOMY

Assessing the Feasibility of Policies to
Reduce Greenhouse Gas Emissions

DECARBONISING THE WORLD'S ECONOMY

Assessing the Feasibility of Policies to Reduce Greenhouse Gas Emissions

editors

Terry Barker
University of Cambridge, UK

Douglas Crawford-Brown
University of Cambridge, UK

ICP

Imperial College Press

Published by

Imperial College Press
57 Shelton Street
Covent Garden
London WC2H 9HE

Distributed by

World Scientific Publishing Co. Pte. Ltd.
5 Toh Tuck Link, Singapore 596224
USA office: 27 Warren Street, Suite 401-402, Hackensack, NJ 07601
UK office: 57 Shelton Street, Covent Garden, London WC2H 9HE

Library of Congress Cataloging-in-Publication Data
Barker, Terry.
Decarbonising the world's economy / Terry Barker, Douglas Crawford-Brown.
pages cm
Includes bibliographical references and index.
ISBN 978-1-78326-511-4 (hardcover : alk. paper)
1. Carbon dioxide mitigation--Economic aspects. 2. Emissions trading. 3. Carbon offsetting. 4. Environmental policy--Economic aspects. 5. Climatic changes--Economic aspects. 6. Energy policy--Environmental aspects. I. Crawford-Brown, Douglas J. II. Title.
HC79.P55B37 2014
363.738'746--dc23
2014035509

British Library Cataloguing-in-Publication Data
A catalogue record for this book is available from the British Library.

Typeset by Stallion Press
Email: enquiries@stallionpress.com

Printed in Singapore by Mainland Press Pte Ltd.

Contents

Introduction

Terry Barker

The urgency of government action to mitigate climate change has become more apparent since the failure of the Copenhagen meeting in 2009 to agree an internationally coordinated policy to address the problem. The Fifth Assessment Report of the Intergovernmental Panel on Climate Change (IPCC) has confirmed the scientific consensus on the causes, scale and risks of global warming. The risks of unabated change have become obvious through many scientific studies, e.g. on the declining extent and depth of Arctic Sea ice. The gap between existing and agreed government action and what is needed to reduce the risks has been explored by the United Nations Environment Programme (UNEP). In addition, the International Energy Agency (IEA) has explored the risks and costs of delay in action in a series of annual reports on the outlook for world energy. This book, drawing on these studies as motivation, addresses the issue of the feasibility of decarbonisation in terms of requirements and timescales for changes in economic policies and structures, and in energy, transportation, industrial and building systems, by country and sector.

The emphasis in this book is on actual economic and energy policies as implemented by governments rather than the idealised policies (e.g., a global carbon tax) that dominate the academic mitigation literature. Hence the attraction of the book to both general and specialised readers is that it offers a different perspective on the problem from that of the traditional economic literature, e.g. as reviewed in the IPCC reports. We have modelled actual policies and analysed them as a portfolio of policies and measures changing over the future to achieve agreed governmental objectives. We have had to strengthen these actual policies as necessary to reach the objectives of reducing the risks of climate change, with the aim of providing feasible portfolios.

Hopefully future generations are not yet committed to *catastrophic* climate change. The science suggests that there is an increasing chance that they will be, unless we cut and eventually eliminate greenhouse gas (GHG) emissions from burning fossil fuels and tropical forests. In any case we are committed to some change. Mitigation is insurance to reduce the risks of catastrophic change and, given the risks and the uncertainties as well as much lower costs from early action, the sooner the better.

In one way, the problem and solutions are simple to understand. Add more heat to the atmosphere via increasing GHG concentrations and it becomes more agitated. Make emissions more costly and new low-GHG technologies both cheaper and more available, and alternatives to burning coal, oil and gas will emerge. Powerful economic forces, often un-noticed and pervasive, provide incentives for new technologies and new ways of doing things, adding to the more immediate responses of research and development (R&D) in the development of new technologies and ways of saving energy. However, for mitigation to happen governments must take the lead and must agree with each other to act. Cooperation must replace competition between nations in areas such as pricing of oil, gas and coal seen as critical to economic welfare and success. There are precedents — the Montreal Protocol on fluorocarbons is proving a success — but the climate mitigation problem faces stronger opposition from vested interests with deep pockets.

This book is a collaborative effort by members of the Cambridge Centre for Climate Change Mitigation Research (4CMR) to address the issue of decarbonising the global economy by 2050, thus giving a reasonable chance of achieving the internationally agreed target of limiting the rise of global temperatures to 2°C above pre-industrial levels by 2100. The Centre has addressed this problem by developing and applying a large-scale model, and using institutional, economic, energy and engineering literature and data to assess the feasibility of radical reductions in GHG emissions. The model that is described and applied in this book is a 20-region, 41-sector, energy-environment-economy model at the global level, E3MG. The model has developed from a long tradition of disaggregated economic modelling in Cambridge, which started with the Cambridge Growth Project in the 1960s and the development of a dynamic structural econometric model of the UK economy in the 1970s. The global model is designed to explore economic policies for the mitigation, or slowing down, of climate change. It would be better to stop the change, but that is impossible and so practical policies will slow the change to a rate that allows required adaptation to be feasible.

Perhaps surprisingly, we are one of a very few teams anywhere whose energy-economy-environment models are based on an annual series of data. It may seem astonishing, but most other global climate models providing governments with estimates of the costs of climate stabilisation, are nearly all reliant on one year's data, assuming economies in that year to be at full-employment equilibrium. This not only fails to make use of most of the data that might better inform analyses, it weakens the ability to compare results of assessments against historical records. The approach taken in E3MG and in this book removes these weaknesses, providing a sounder base for policy analyses.

Our approach to economics is one of "new economics", so called in contrast to traditional economics. It takes on aspects of the post-Keynesian economics, including an emphasis on institutions, on economic behaviour being specific to its historical, social and political setting, and the critical importance of evolution and technology in addressing long-term problems. The approach follows that of Kenneth Boulding in his 1992 book *Towards a New Economics: Critical Essays on Ecology, Distribution, and Other Themes* (Edward Elgar Publishing Ltd, Cheltenham). Understanding the complexities of the problem and solution requires interdisciplinary research, and an approach which recognises that the evidence is overwhelming that a long-run stable equilibrium is unlikely for the economy. We know from chaos theory and observations of the natural and human systems that change is intrinsic and stability can be short-lived. The modelling on which this book is based reflects this condition of the world.

The book comes from a multidisciplinary team: economics, engineering, mathematics, computing, physics and sociology. The focus of our research is the special intersection of economics and engineering concerned with climate change mitigation — particularly by economic policies inducing low-carbon technological change. This is interdisciplinary research. Traditional economists have until recently assumed that technological change comes from outside the economic system, "like manna from heaven", and engineers have assumed that the economy is not affected by their technologies. Our research programme, which has been largely funded by the Three Guineas Trust, one of the Sainsbury Family Charitable Trusts, has linked micro-engineering studies with macroeconomy studies in terms of the household, transportation and electricity systems at a global level and in the context of how different policy regimes can be designed not only for efficiency and fairness, but also to improve economic performance. A key finding is that environmental quality, energy security and economic stability need not be in competition if policies are well constructed.

This book reports a central outcome of this programme, namely the development of computer-based simulation models that consider the world economy and energy systems both as they have evolved over the last 40 years and as they could change in the future. The book is divided into three parts. We firstly give the background and context to the main analysis. We consider the basic scientific case for decarbonisation, the institutions that have been created, the policies that have been adopted and the methodology of scenario analysis. We then start with the core reference and decarbonisation scenarios and their derivation from recent assessments of the projected course of the world economy and the use of energy. The feasibility of decarbonisation is then considered from four different perspectives in the chapters that follow; namely the economic, the energy-engineering, the sectoral and finally the political feasibility. Finally, we consider the wider impacts on the economy and the environment of the portfolios of policies seen to be feasible.

Chapter 1

The Case for Decarbonisation

Douglas Crawford-Brown and Martin Sewell

1.1. Introduction to the Science of Climate Change Risk

While climate change introduces a wide array of risks related to all aspects of climate, it is driven ultimately by the warming of the atmosphere by sunlight through the greenhouse effect. In that effect, solar radiation is able to pass down through the atmosphere but not back out after being absorbed and re-emitted at the earth's surface as longer wavelengths of (largely infrared) electromagnetic radiation. The increased absorption of the radiation increases the energy stored in the atmosphere and oceans, which appears as increased temperature (a measure of kinetic energy), increased winds and changes in atmospheric water vapour, as well as increased instability of the climate.

Despite its association with global climate change, the greenhouse effect is not inherently problematic. It is a natural process, without which the planet would be too cold to sustain life. Without greenhouse gases (GHGs) such as carbon dioxide, water vapour, nitrogen oxides, ozone, etc., nearly all the heat gained by the earth from sunlight would simply be lost (re-radiated) rapidly and directly back into space.

The past century has seen the composition of the atmosphere as this relates to the solar-driven energy balance change due to the activities of humans, especially from the burning of fossil fuels. As GHGs accumulate in the atmosphere, they trap the sun's heat more effectively, increasing the energy absorbed in the atmosphere, which is in part transferred to the oceans and land masses. This heat is then moved globally through various processes such as the ocean currents (surface and bottom) and the high altitude cells of air that carry energy from the equator to the poles and back. This will, in the long run, result in increased mean global temperatures, although there will be variability across the surface of the earth, with some areas cooling while others heat up. These very long-term processes also mean the earth's atmosphere will not approach thermodynamic equilibrium (its final temperature even if atmospheric GHG concentrations were to remain the same) for several centuries.

The link between human activities and the accumulation of GHGs such as carbon dioxide in the atmosphere has clearly been established by the available scientific data reviewed by the Intergovernmental Panel on Climate Change (IPCC, 2007, 2013). It was not until the Industrial Revolution of the 18th and 19th centuries that atmospheric concentrations of carbon dioxide began to increase appreciably due to human activities (other geological periods have also shown appreciable increases and decreases, but for reasons other than human influence), rising from 280 ppm (parts per million) before the Industrial Revolution to above 390 ppm today. This has been accompanied by an increase in global average surface air temperatures. The increase in average surface temperature across the globe that has occurred over the past 150 years appears to be of the order of 0.6°C ± 0.2°C during the 20th century; best scientific estimates are that this increase could become larger by several degrees celcius by 2100 due to increases in GHGs from human activity. There is, however, significant remaining uncertainty as to the timing and extent of any warming that might take place, and how this compares in magnitude to the natural processes of climate change always at play.

The impacts of climate change will vary between regions of the world, being detrimental in some regions but beneficial in others. Using the UK as an example, the potential effects of climate change include (UKCIP, 2003):

- An increase in mean annual temperatures.
- High summer temperatures, including heat waves, becoming more frequent and very cold winters becoming increasingly rare.
- Winters becoming wetter on average and summers drier on average.

- Average decrease of snowfall amounts.
- Heavy winter precipitation becoming more frequent, as will winter flooding.
- Relative sea levels rising around most of the shoreline.
- Periods of drought becoming longer and deeper.
- Extreme sea level surges (e.g. those capable of beaching existing barriers) being experienced more frequently.
- The possible weakening of the Gulf Stream, further affecting climate in countries such as the UK that are affected strongly by this phenomenon.

By way of an example, the UK Climate Impacts Programme has produced estimates of the change in several key variables of climate over the next century in the UK, summarising these as probability density functions describing uncertainty in those estimates. They do so for three of the IPCC emissions scenarios (low, medium and high). Representative results are shown below in Table 1.1 for change in mean summer temperature by region, and for the medium emission scenario. Their projections extend to the 2080s.

Similar results for projections of increases in winter precipitation are shown in Table 1.2.

Scientific understanding of global warming begins with the study of how carbon moves throughout the environment — described by the carbon cycle shown in Figure 1.1. The figure depicts the flow of carbon regardless of its chemical and physical form, and so values of the flow are usually given in tons of carbon per year. Carbon mass in the atmosphere can be converted to carbon dioxide mass in the atmosphere by multiplying the carbon mass by approximately 44/12 (the ratio of the mass of a carbon dioxide molecule to that of carbon atom).

Note in Figure 1.1 the crucial roles of land use in controlling the flow of carbon into and out of the flora (or vegetation), and of population and societal energy and material use (shown as per capita emissions) in controlling the flow into the atmosphere from the economy (the economy being one of the primary drivers of changes in energy use and technology). Any of these features can be the subject of climate policy. However, risk reduction for climate change need not focus solely on these. One can

Table 1.1. UKCP09 (UK Climate Projections) of increase in mean summer temperature in UK regions by the 2080s.

Change in mean summer temperatures (°C)					
2080s Medium emissions scenario					
	Probability Level				
Region	**10%**	**50%**	**90%**	**Wider Range**	
Channel Islands	1.9	3.6	5.9	1.2	7.3
East Midlands	1.8	3.5	5.8	1.2	7.3
East of England	1.9	3.6	5.9	1.3	7.5
Isle of Man	1.8	3.3	5.2	1.2	6.5
London	2.0	3.9	6.4	1.4	8.1
North East England	2.0	3.7	5.8	1.3	7.3
North West England	2.0	3.7	5.9	1.3	7.3
Northern Ireland	1.7	3.2	5.0	1.1	6.2
Scotland East	1.8	3.5	5.7	1.2	7.0
Scotland North	1.5	3.0	4.9	1.0	6.0
Scotland West	1.8	3.5	5.4	1.2	6.8
South East England	2.0	3.9	6.5	1.4	8.1
South West England	2.1	3.9	6.4	1.3	7.9
Wales	1.9	3.5	5.8	1.2	7.1
West Midlands	2.0	3.7	6.1	1.3	7.5
Yorkshire & Humber	1.7	3.3	5.4	1.2	6.8

instead understand risk reduction as shown in Figure 1.2, in which risk is reduced by:

- Reducing emissions of GHGs, including reducing material and energy demand, improving the efficiency at which remaining demand is met, employing low-carbon materials and energy, or some form of carbon capture and storage at the point of production.
- Sequestering carbon dioxide from the atmosphere through ecosystems or through carbon-scrubbing technologies.

Table 1.2. UKCP09 projections of increase in mean winter precipitation in UK regions by the 2080s.

Change in mean winter precipitation (%)					
2080s Medium emissions scenario					
	Probability Level				
Region	**10%**	**50%**	**90%**	**Wider Range**	
Channel Islands	+3	+21	+48	+3	+62
East Midlands	+3	+19	+41	+3	+54
East of England	+4	+20	+44	+4	+57
Isle of Man	+5	+22	+49	+5	+67
London	+3	+19	+44	+3	+58
North East England	+2	+14	+32	+1	+41
North West England	+3	+16	+34	+3	+50
Northern Ireland	+2	+11	+24	+2	+34
Scotland East	+1	+12	+25	+1	+36
Scotland North	+4	+17	+35	+4	+45
Scotland West	+6	+21	+42	+6	+55
South East England	+4	+22	+51	+4	+67
South West England	+6	+23	+54	+5	+73
Wales	+4	+19	+42	+4	+57
West Midlands	+3	+17	+38	+3	+50
Yorkshire & Humber	+2	+15	+33	+2	+42

- Reducing solar insolation (radiation entering the atmosphere from the sun), for example through injection of particles into the stratosphere to reflect sunlight back into space or otherwise shade the troposphere.
- Reducing the vulnerability of society and ecosystems to a changed climate, either through "hardening" of these systems to withstand greater temperature extremes, storm intensity, etc., or through systems of emergency response that allow rapid evacuation from affected areas when extreme climate events such as flooding take place.

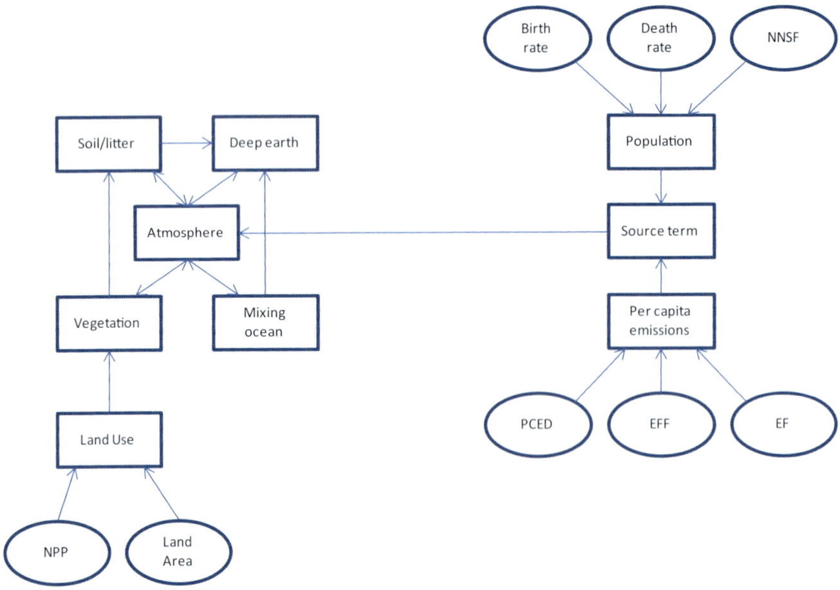

Figure 1.1. A schematic representation of the components of a climate change models suited to policy analyses. (Modified from Crawford-Brown and LaRocca (2006).)

A focus on risk mitigation as a basis for climate change policy requires consideration of the two primary components of risk: (i) the severity of potential effects — losses of value — arising from climate change, and (ii) the probability of each of these effects occurring by any particular time. The central policy task then becomes one of finding a trajectory of decarbonisation (or other form of risk reduction; in this book we are focused on decarbonisation) that can, with reasonable confidence, be said to produce an acceptable level of risk. This level of risk is a function of both the severity of the impacts and the probabilities of them occurring, with both aspects of risk being potential targets of risk reduction. It is common to consider mitigation as changing the probability of extreme climate events leading to impacts, and adaptation as changing the severity of impact from a given extreme event.

The definitive scientific work on climate change risks is summarised by the IPCC (2007, 2013). Their conclusion is that unacceptable levels of

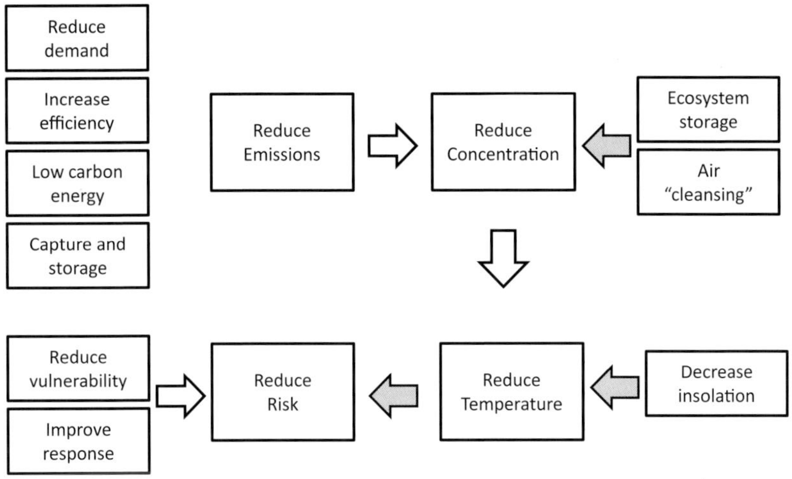

Figure 1.2. A summary of strategies to reduce risks from climate change.

risk to human health, ecosystem integrity (including biodiversity) and physical assets will take place at somewhere above a 2°C rise in mean global temperature above pre-Industrial Revolution levels. While there is some scientific basis for this threshold of effect, related to temperature change in the Holocene, the 2°C value has been largely a choice based on feasibility and consensus of international parties. It is therefore subject to revision in the future, either upwards or downwards as scientific evidence accumulates and as society begins to define the level of risk it is willing to tolerate better.

Avoiding this temperature rise during the 21st century requires preventing a doubling of the pre-Industrial Revolution levels of atmospheric carbon dioxide during that period. As a result, international mitigation targets have tended to focus on reaching an 80% reduction in GHG emissions below recent historical levels as of 2050, with no subsequent increase back up towards current levels of emissions. This is reflected in, for example, the publication by the UK Department of Energy and Climate

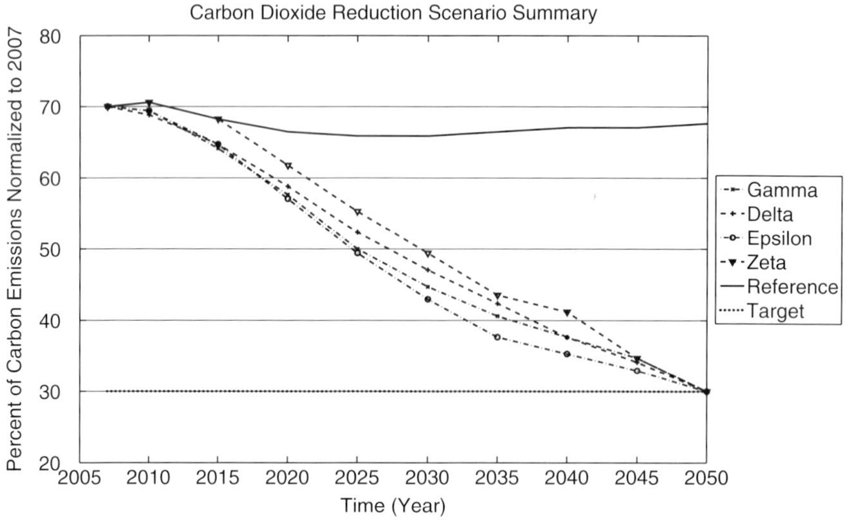

Figure 1.3. DECC trajectories of decarbonisation of the UK economy by 80% up to 2050.

Change (DECC) of scenarios of decarbonisation in the UK economy that meet this reduction target. A summary of these is provided in Figure 1.3, developed from the DECC Energy Pathways report (UK DECC, 2010).

Why does Figure 1.3 not show trajectories that are flat until 2050, or even increasing until then, followed by an immediate drop to the 80% reduction target? Part of the answer lies in the technological, economic and political infeasibility of making such an abrupt decrease in 2050. Nations will already have invested in a high-carbon pathway by that time, with locked-in high-carbon assets, and will be unable to change quickly without massive disruptions to the energy system and the economy. More importantly, the risks of climate change are not only functions of the emissions rate in 2050, but the total amount of carbon dioxide (and other GHGs) emitted between today, the target date for reduction (2050) and beyond. Mathematically, this is equivalent to taking the "area under the curve" for a trajectory, and not simply asking what the emission rate will be once a reduction target has been met.

The necessary area under the curve to prevent unacceptably high risks of climate change, the global carbon budget, depends upon the temperature change deemed acceptable and the atmospheric carbon dioxide concentration producing it. For this book, we assume the policy target remains a temperature rise of 2°C or less relative to the pre-Industrial Revolution level, and that this corresponds to stabilisation of atmospheric carbon dioxide at 350 ppm or less (about 450 ppm or less for GHG measured as carbon-dioxide-equivalent concentrations). Meinshausen *et al.* (2009) examined 18 published climate sensitivity studies both to develop uncertainty distributions on this climate sensitivity and to determine the carbon budget required to stay below the 2°C target with a specified level of confidence.

Figure 1.4 summarises their results for the relationship between the carbon budget and the probability of remaining below the 2°C policy target. In that figure, the carbon budget is given as gigatons (Gt) CO_2 emitted. The central (solid) line shows the central tendency value estimated from the ensemble of models and scenarios considered in their study, and the dashed lines indicate the upper and lower confidence limits on the estimate. Given the nature of their study (Meinshausen *et al.*, 2009), we have chosen not to assign numerical confidence levels to these upper and lower limits, but rather to say more qualitatively that we are "reasonably

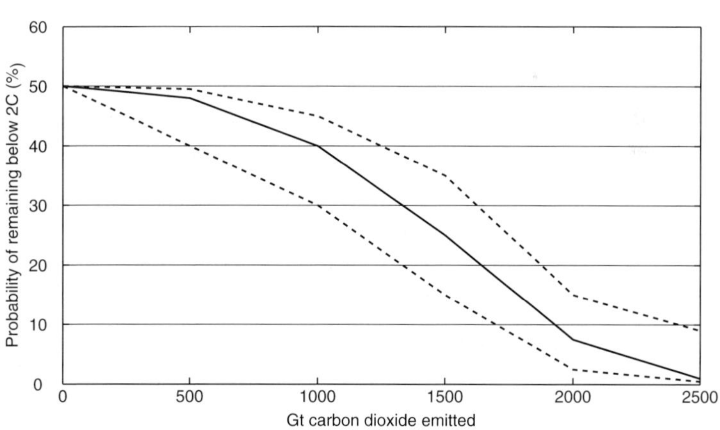

Figure 1.4. The relationship between carbon budget (Gt CO_2 emitted between 2000 and 2050) and probability of that budget producing a 2°C temperature rise or less. Solid line is the central tendency estimate from Meinshausen *et al.* (2009) and the dashed lines are the lower and upper confidence limit.

confident" the true value lies between these limits given uncertainty in which of the existing climate science models is correct, which of the climate sensitivity parameter estimates is correct for use in the models, and which scenario of decarbonisation will occur. Note that a trajectory producing a carbon budget of 1500 Gt yields a 50% confidence that the policy target will be met (and 50% that it will not), if one uses the central tendency estimate of the relationship. If, instead, one uses the lower bound estimates, the policy target is met at 1300 Gt; the upper bound estimates produce the same 50% confidence at 1750 Gt. This issue is reconsidered in Section 1.3 under the precautionary principle. Finally, note that the figure is for emissions of carbon dioxide only, not of all GHGs or all compounds affecting climate.

A more probabilistic interpretation of the above results can be gleaned from the UKCP09 results. Table 1.3 shows the UKCP09 projections of global temperature change under three emissions scenarios (low, medium and high). If one considers the 2080s results as indicative of uncertainties at the end of the 21st century, that uncertainty can be characterised by taking the 10th percentile value divided by the 50th (central tendency) value, and the 90th percentile value divided by this same 50th percentile value. For the former (10%) ratios, the values are 0.8, 0.8 and 0.8 for the high, medium and low-emission scenarios, respectively. For the latter (90%) ratios, the values are 1.2, 1.2 and 1.3, respectively. These results indicate that in estimating global mean temperature resulting from a given emission scenario, the uncertainties in the estimates are approximately

Table 1.3. UKCP09 projections of global mean temperature change under three IPCC scenarios. The results display three percentiles of the probability density function describing uncertainty in these estimates.

Global mean temperature (°C)									
	2020s			**2050s**			**2080s**		
Probability	10%	50%	90%	10%	50%	90%	10%	50%	90%
High emissions	1.0	1.3	1.6	2.1	2.7	3.3	3.4	4.3	5.3
Medium emissions	1.0	1.3	1.6	1.9	2.4	3.0	2.6	3.4	4.2
Low emissions	0.9	1.2	1.6	1.6	2.1	2.6	2.0	2.6	3.4

20% around the central tendency value when those uncertainties are characterised by the 10–90% confidence interval.

This value of 20% is approximately the same as what one finds by examining the upper, lower and central tendency estimates of Figure 1.4, moving horizontally across the graph to determine the range of cumulative emissions associated with any specific level of confidence chosen for staying below a 2°C temperature increase. For example, consider a target confidence of 50% (it is equally likely that the policy limiting cumulative emissions will keep temperature increase below 2°C). From Figure 1.4, the central tendency (solid) line indicates cumulative emissions of 1500 Gt CO_2 will produce this level of confidence. Moving horizontally on that figure, the lower bound is approximately 1200 Gt and the upper bound is approximately 1800 Gt. This is a difference of 300 Gt in either direction from the central tendency value of 1500 Gt, or a 20% difference. As a rough approximation, therefore, one might say that the lower and upper bound values in Figure 1.4 represent 10th and 90th percentile values around the central tendency or 50th percentile value.

If one adopted a probabilistic, risk-based policy approach to decarbonisation, the global target would then be cumulative emissions of 1500 Gt CO_2 up until 2050 (with a continuing rapid decline after that point) if a 50% confidence level were required for remaining below 2°C during the 21st century, or of 1200 Gt CO_2 if 90% confidence were required. Since environmental policies typically have a margin of safety built into them to reflect precaution in the face of uncertainty, it would appear prudent to establish policies of decarbonisation that produce no more than 1200 Gt CO_2 between 2000 and 2050, with continuing significant decline after that period.

1.2. Ethical, Political and Legal Considerations in Decarbonisation

Figure 1.4 displays the relationship between emitted carbon dioxide and the probability of exceeding a 2°C temperature rise. This same figure can be interpreted as summarising the confidence produced in a decision-maker as to the ability of a proposed trajectory of decarbonisation to reach the stated policy. This leaves open the question of the nature of risks it is reasonable to impose on ourselves and others (current and future generations and people in remote places), and how much confidence in estimates of risk is required to meet various ethical, political and legal considerations.

Global climate policy is driven by a concern for both the degree of climate risk across the nations and the ability of those nations to cope with or adapt to that risk. Justifications for global policies therefore focus in part on those nations that are both highly impacted by climate change and do not have the resources to respond when those impacts occur. This is captured in Figure 1.5, where the nations in the bottom right corner of the figure are "vulnerable nations" whose protection by the global community is a significant driver of policies.

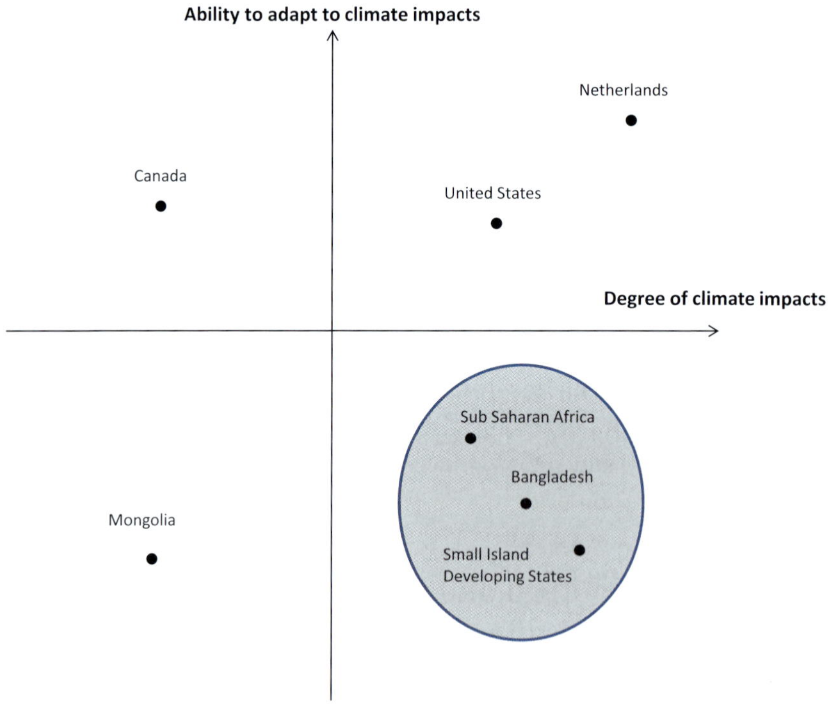

Figure 1.5. The division of nations into those highly impacted by climate change (lower to greater impact or exposure moving from left to right) and ability to respond to impacts (lesser to greater ability moving from the bottom to the top). Most vulnerable nations are in the bottom-right corner, with some examples shown in the blue oval.

Environmental policy traditionally has had three justifications: (i) prudence, or the argument that a lack of sound policy would lead to unacceptable impacts on human life, (ii) rights, or the argument that people and other species have an existence value that must be respected and (iii) contractual duties, or the argument that society involves contracts that control behaviour and impose duties on both parties involved in the contract. Gardiner (2006) has extended this discussion to climate change, making the case for climate change being an especially difficult case of environmental policy because:

- Causes and effects are dispersed throughout society, between countries and between generations.
- There is fragmentation of agency because a wide array of individuals and organisations bear some measure of responsibility for the emissions leading to climate change.
- There is institutional inadequacy because our existing system of laws, regulations and powers has never been called on to control such a complex environmental problem, including a requirement for coordinated international action.

He goes on to call climate change a "perfect moral storm" that both compels action to avert the worst risks but also makes regulation too difficult to implement under existing approaches to policy.

It is useful to consider how climate policy might be harmonised with other environmental policies that are well tested in national and international law, since these other policies can form precedents for legally binding action. A key feature in these other environmental areas is application of the precautionary principle. This principle addresses problems of decisions under significant uncertainty (Deville and Harding, 1997; O'Riordan and Cameron, 1996; Sand, 2000; Sandin, 1999; Stewart, 2002), where a source of risk cannot be well characterised and quantified scientifically. The precautionary principle calls for a framework of decision-making in which damages to property, health and the economy are anticipated before they occur, and provides a measure of protection against this harm even if neither the probability nor severity can be determined accurately by the existing science. It is an evolutionary step in the transition from environmental tort law, where the evidential burden is to demonstrate that a defendant caused the harm claimed; to risk-based environmental policy, where the evidential burden is for the regulator to show that a situation poses some "unacceptable" level of risk to the human race or other species; to recent linkages between environmental protection and social justice in which the evidential burden is on those whose actions (such as the emission of carbon dioxide) might impose risk to demonstrate that this risk is acceptable or justified before carrying out those actions through the environmental commons.

The precautionary principle has become the ground for arguments over:

- Who bears the burden of proof when risks are imposed on society by an actor using the environmental commons (i.e. land, water and air shared by members of a community).
- Standards of the evidential barrier that must be crossed before a risk is allowed or prohibited.
- The formal treatment of uncertainty in quantitative risk assessment, and its relationship to "margins of safety" within decision rules.

The most commonly cited definition, and one that is relevant to climate policy, is from the Rio Conference in 1992: "Where there are threats of serious or irreversible damage, lack of full scientific certainty shall not be used as a reason for postponing cost-effective measures to prevent environmental degradation" (UN, 1992).

The precautionary principle has been formalised in the concept of a "margin of safety" applied in environmental decisions on air and water pollutants. In this approach, formal uncertainty distributions on estimates of harm are produced, as is an "upper bound" estimate of this harm used in choosing a policy remedy, based on the argument that it is better to have over-stated the risk in the face of uncertainty than to have understated it and, hence, have regret for any adverse effects at some future point. In this spirit, there is a well-developed literature on formal uncertainty analysis for climate change predictions (Hall, Fu and Lawry, 2007; Kriegler *et al.*, 2009; Stainforth *et al.*, 2005; Webster *et al.*, 2001; Webster and Sokolov,

2000), and this contributes to understanding how wide the error bars might be on expressions of risk underlying mitigation and adaptation policies, at least as far as they are contingent on existing models. An example was provided in Figure 1.4.

Following calls to apply the precautionary principle in environmental regulations, the EU produced the 2000 European Commission Communication on the Precautionary Principle (hereafter 2000 Communication) (European Commission, 2000; Rogers, 2001; Wiener and Rogers, 2002). It noted that:

> The precautionary principle applies where scientific evidence is insufficient, inconclusive or uncertain and preliminary scientific evaluation indicates that there are reasonable grounds for concern that the potentially dangerous effects on the environment and on human, animal or plant health may be inconsistent with the high level of protection chosen by the EU.

This statement contains an important advance on the 1992 Rio definition by asking for "reasonable grounds" for belief that a threat may be present. It is therefore not simply the existence of uncertainty that triggers a call for application of the precautionary principle, but some degree of evidence that harm may in fact exist at a level "inconsistent with the high level of protection chosen by the EU". "Reasonable" in this context has two meanings: (i) a level of evidence that is neither too weak, allowing all beliefs about risk to pass over the bar and hence justify all proposed regulatory actions, nor too strong, preventing actions in all but a very few cases, and (ii) a means of reasoning about and from that evidence that is in some sense open to scrutiny and rational assessment.

The 2000 Communication further clarifies the precautionary principle by noting six core criteria that should be satisfied for policy choices taken in the name of that principle:

- *Proportional to the chosen level of protection.* The action taken should match the risks posed and the degree to which society should be protected from those risks.
- *Non-discriminatory in their application.* The principle must be applied in the same way to all actions taken in similar situations of risk.

- *Consistent with similar measures already taken.* One should examine past decisions taken to reduce risk, and ensure that new actions — similar in all relevant ways — have the same degree of protectiveness.
- *Based on an examination of the potential benefits and costs of action or lack of action.* Actions taken in the name of the principle may have (perhaps unintended) side effects that increase risk, and this possibility must be considered.
- *Subject to review in the light of new scientific data.* Actions taken in the name of the principle should be subject to periodic review, and changed as the scientific data allow for more accurate estimates of risk.
- *Capable of assigning responsibility for producing the scientific evidence necessary for a more comprehensive risk assessment.* An organisation must be responsible for producing the evidence required to improve the base of quantitative risk assessment so the principle is eventually displaced by a more formal procedure developing a quantified "margin of safety".

The primary lessons brought forward from the 2000 Communication are (i) that climate policy must be understood in a framework that considers the risks and costs, as well as the risk trade-offs, of policies of mitigation and adaptation; (ii) however precaution is applied in climate change, it must cohere with applications of the principle in other realms of environmental protection; (iii) policies developed through invoking the principle must reflect solely a concern for climate change risks, and not act as a disguised route for attaining other goals such as trade protection; and (iv) as knowledge develops of the risks of climate change and the effectiveness of policies at reducing this risk, there will be less of a role for the principle.

What are the implications of the precautionary principle for climate policy in the face of results in Figure 1.4 showing significant uncertainties in the carbon budget required to meet the policy target of 2°C temperature rise or less, and hence to protect even the most vulnerable nations shown in Figure 1.5? What is needed to move climate policy closer to that for tropospheric ozone and related environmental risks, following on from the requirement of the 2000 Communication that precaution in climate change be consistent with similar measures already taken? Here there arise two primary hurdles or challenges: epistemic analysis and margins of safety. The first challenge is to develop more formal and systematic assessments of the epistemic status of model results required to select trajectories of decarbonisation. While there are emerging efforts to do this (an example is the Meinshausen *et al.* (2009) study, which was used in the creation of Figure 1.4), focus has been more on model inter-comparisons than on systematic diagnosis of the reasons for disagreement between models and the formal analysis of uncertainties both within a model and between models, both of which are characteristics of risk analysis in many other areas of environmental policy.

This situation is particularly true in the area of E3 models, which are used by policy makers to understand how policy instruments such as carbon taxes drive investments in low-carbon energy sources or energy efficiency measures, and how these in turn reduce GHG emissions and the concentration of these gases in the atmosphere. Until these models are better tested and vetted, they are unlikely to enjoy the same degree of trust given to the ozone policy models. Increased trust in both climate science models and E3 models will be based on the demonstrated ability of these models to "hindcast" (Forest *et al.*, 2002; Urban and Keller, 2010), or in other words to reproduce and explain data in the past before applying the models to forecasting the impact of future policies.

This reduced epistemic status of model predictions in climate science and the E3 field leads to the second challenge: incorporating margins of safety into climate policy. The "business-as-usual" case of climate policy relies quite heavily on central tendency estimates of the relationships between emissions, atmospheric GHGs, climate sensitivity and risk. This is in distinct contrast to regulation of water and air pollutants (such as ozone) that contain appreciable margins of safety as a routine matter of precaution. Potential steps to developing trajectories of the reduction of emissions needed to provide a margin of safety in the face of uncertainty are:

- Establish a ceiling of acceptable temperature change, representing some upper percentile of the uncertainty distribution of the damage function associated with differing degrees of temperature change. In the current chapter, this is assumed to be 2°C. Insights into how such a probability distribution might be developed using subjective encoding of beliefs are emerging (Hall *et al.*, 2007; Kriegler and Held, 2005; Risbey and Kandlikar, 2002), although have not yet been adopted in climate policy discussions.

- Establish a probability density function (PDF) for the uncertainty associated with estimates of the total carbon dioxide emissions between now and 2050 ("the carbon budget") required to stay just below this ceiling. This is shown in Figure 1.4 based on the Meinshausen *et al.* (2009) analysis.
- Select some upper percentile of this distribution that represents the most rapid trajectory of decarbonisation required, following on from the principles in the 2000 Communication. In order to be consistent with the use of upper bound estimates to provide precaution in other areas of environmental policy, this carbon budget will be somewhere between 1200 and 1500 Gt CO_2, with the latter value providing more than 50% confidence and the former value providing more than 90% confidence of reaching the policy target.
- Identify the policy instruments necessary to bring about this trajectory, and adjust as data on climate change and policy effectiveness flows in over time (i.e. adaptive management).
- Create and maintain a strong community of scientists assessing the reliability of these models, in order to improve the epistemic status of probabilistic risk assessment and the attending probabilistic interpretation of precaution.

With such a series of procedures, climate policy will be harmonised with the application of the precautionary principle in other environmental spheres, where rights of populations to be protected against unreasonable levels of risk have been well established. Presently, however, there is an inadequate scientific basis on which to create formal margins of safety, and hence climate policy targets are the product of less formal processes that are unconnected to the probabilistic approach that underlies many other areas of environmental policy.

1.3. Conclusions

Using the results from Section 1.2, insights can be gained as to the trajectory of global decarbonisation required to reduce the risks of climate change to levels that are currently agreed as necessary by the scientific

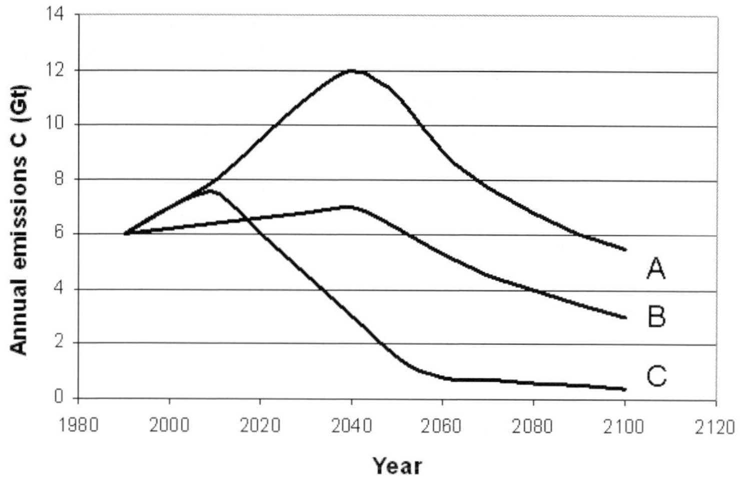

Figure 1.6. Three potential trajectories of decarbonisation of the global economy. As described in the text, curves A and B are from the IPCC SRES analysis, and curve C is required to produce a high level of confidence of meeting the policy target of a 2°C increase or less in mean global temperature.

community. As before, the policy target here is taken to be a 2°C global mean temperature rise or less. Figure 1.6 displays three potential pathways to reduction. Curves A and B are two of the low-emission scenarios from the IPCC. They would produce cumulative emissions of 2050 Gt and 1430 Gt CO_2, respectively, by 2050. Curve C is a hypothetical reduction trajectory produced specifically for this chapter, and which is consistent with the policy targets of the DECC. It would produce cumulative emissions of 1160 Gt CO_2. By contrast, the latest IPCC report (AR5) estimates that the cumulative emissions may need to be as low as 800 Gt when non-CO_2 influences are considered (IPCC, 2013).

Following the analysis in Section 1.2, curve B produces approximately 50% confidence that the policy target (2°C or less) will be met, while curve C produces approximately 90% confidence that the policy target will be met. Under the precautionary principle, it would appear that the more stringent decarbonisation trajectory of curve C is required, with curve B being a candidate default position should such a significant rate of decarbonisation prove infeasible. Curve A produces less than 5% confidence of the policy target being met, and so can probably be rejected as a reasonable basis for the rate of decarbonisation of the global economy.

With what confidence can we make these statements about the necessary rate of reduction of global emissions? Two points concerning uncertainty in the relationship between cumulative emissions and global mean temperature change are relevant:

- The existing science suggests that the value of cumulative global emissions between 2010 and 2050 that is consistent with the global policy goal lies somewhere between 1200 and 1800 Gt. Beyond this, the existing science does not allow for reliable assignments of numerical values of levels of confidence, other than that it is "more likely than not" that the required limit on cumulative emissions is within this range.
- Following a precautionary approach to risk-based policy, it might be reasonable to develop policies that aim at the lower end of this confidence interval (i.e. closer to 1200 Gt than 1800 Gt). However, it is not possible at present to characterise the numerical degree of conservatism or precaution inherent in use of any cumulative emission value.

We also consider the implications of the uncertainty ranges for the rate of decarbonisation required. If the lower bound estimate of 1200 Gt is established as the basis of global policy, this would require a 1.6% reduction per year globally, ending in an approximate 30% decrease in global emissions by 2050 relative to 1990 or 50% relative to 2010. This rate of decarbonisation, however, is a trajectory for the global emissions, not for any currently developed nation such as the UK. Under the principle of contract and converge, which is at least implicit in the Kyoto Protocol, the trajectory of reduction in the economies of the developed world would need to be steeper than this 1.6% per year if currently underdeveloped or developing nations are to be allowed some growth in emissions in at least the short term as a consequence of economic development. Returning to Figure 1.6 and curve C, the DECC trajectories imply a rate of reduction in the UK of 4% per year between 2010 and 2050. This is an already ambitious — in fact unprecedented — rate of decarbonisation for a modern economy. There is, indeed, a significant challenge ahead in reaching the rate of decarbonisation required to remain below a temperature increase of 2°C. Meeting that challenge is the focus of the other chapters in this book.

References

Crawford-Brown, D. and LaRocca, S. (2006). Teaching Systems Principles and Policy Applications using a Reduced-scale Global Warming Model, *Journal of Geoscience Education*, 54, 101–120.

Deville, A. and Harding, R. (1997). *Applying the Precautionary Principle*, The Federation Press, Annandale.

European Commission (2000). *Communication from the Commission on the Precautionary Principle*, Communication (2000)1, Brussels, 2 February.

Forest, C., Stone, P., Sokolov, A., Allen, M. and Webster, M. (2002). Quantifying Uncertainties in Climate System Properties with the use of Recent Climate Observations, *Science*, 295, 113–117.

Gardiner, S. (2006). A Perfect Moral Storm: Climate Change, Intergenerational Ethics and the Problem of Moral Corruption, *Environmental Values*, 15, 397–413.

Hall, J., Fu, G. and Lawry, J. (2007). Imprecise Probabilities of Climate Change: Aggregation of Fuzzy Scenarios and Model Uncertainties, *Climatic Change*, 81, 265–281.

IPCC (2007). *Climate Change 2007: The Physical Science Basis. Contribution of Working Group I to the Fourth Assessment Report of the Intergovernmental Panel on Climate Change*, Solomon, S., Qin, D., Manning, M., Chen, Z., Marquis, M., Averyt, K. B., Tignor, M. and Miller, H. L. (eds), Cambridge University Press, Cambridge.

IPCC (2013). *Climate Change 2013: The Physical Science Basis. Contribution of Working Group I to the Fifth Assessment Report of the Intergovernmental Panel on Climate Change*, Stocker, T. F., Qin, D., Plattner, G.-K., Tignor, M. M. B., Allen, S. K., Boschung, J., Nauels, A., Xia, Y., Bex, V. and Midgley, P. M. (eds), Cambridge University Press, Cambridge.

Kriegler, E. and Held, H. (2005). Utilizing Belief Functions for the Estimation of Future Climate Change, *International Journal of Approximate Reasoning*, 39, 185–209.

Kriegler, E., Hall, J., Held, H., Dawson, R. and Schellnhuber, H. (2009). Imprecise Probability Assessment of Tipping Points in the Climate System, *Proceedings of the National Academy of Science*, 106, 5041–5046.

Meinshausen, M., Meinshausen, N., Hare, W., Raper, S., Frieler, K., Knutti, R., Frame, D. and Allen, M. (2009) Greenhouse-gas Emission Targets for Limiting Global Warming to 2°C, *Nature*, 458, 1158–1163.

O'Riordan, T. and Cameron, J. (1996). *Interpreting the Precautionary Principle*, Earthscan Publishers, London.

Risbey, J. and Kandlikar, M. (2002). Expert Assessment of Uncertainties in Detection and Attribution of Climate Change, *Bulletin of the American Meteorological Society*, 83, 1317–1326.

Rogers, M. (2001). Scientific and Technological Uncertainty, the Precautionary Principle, Scenarios and Risk Management, *Journal of Risk Research*, 4, 1–15.

Sand, P. (2000). The Precautionary Principle: A European Perspective, *Human and Ecological Risk Assessment*, 6, 445–458.

Sandin, P. (1999). Dimensions of the Precautionary Principle, *Human and Ecological Risk Assessment*, 5, 889–908.

Stainforth, D., Aian, T., Christensen, C., Collins, M., Faull, N., Frame, D., Kettleborough, J., Knight, S., Martin, A., Murphet, J., Piani, C., Sexton, D., Smith, L., Spicer, R., Thorpe, A. and Allen, M. (2005). Uncertainty in Predictions of the Climate Response to Rising Levels of Greenhouse Gases, *Nature*, 433, 403–406.

Stewart, R. (2002). Environmental Regulatory Decision-making Under Uncertainty, *Research in Law and Economics*, 10, 71–152.

UK Climate Impacts Programme (2003). *Climate Adaptation: Risk, Uncertainty and Decision-making*, UKCIP, Oxford.

UK Department of Energy and Climate Change (2010). *2050 Pathways Analysis*, DECC, HM Government, UK.

United Nations (1992). Report on the United Nations Conference on Environment and Development, Annex I: Rio Declaration on Environment and Development. Available online at http://www.un.org/documents/ga/conf151/aconf15126-1annex1.htm. Accessed 4 April 2014.

Urban, N. and Keller, K. (2010). Probabilistic Hindcasts and Projections of the Coupled Climate, Carbon Cycle and Atlantic Meridional Overturning Circulation System: A Bayesian Fusion of Century-scale Observations with a Simple Model, *Tellus*, 1–14.

Webster, M. and Sokolov, A. (2000). A Method for Quantifying Uncertainty in Climate Predictions, *Climatic Science*, 46, 417–446.

Webster, M., Forest, C., Reilly, J., Sokolov, A., Stone, P., Jacoby, H. and Prinn, R. (2001). *Uncertainty Analysis of Global Climate Change Projections*, MIT Joint Program on Science and Policy of Climate Change, Report 73, March.

Wiener, J. and Rogers, M. (2002). Comparing Precaution in the United States and Europe, *Journal of Risk Research*, 5, 317–349.

Chapter 2

Policies and Measures for Mitigating Climate Change

Paul Haynes and Yongfu Huang

2.1. Introduction

There are many different strategies with which to address carbon emissions and a range of different policies has been developed to implement such strategies. A combination of these policies forms a climate policy portfolio. Climate change mitigation policies can be divided up in various ways and therefore it will be necessary to assess and combine the different instrument types in a way that prevents contradictory policies from cancelling out each others' strengths and, instead, encourages complementary policies to becoming part of a coherent policy mix. The policy mix in the context of this chapter refers to climate change mitigation policies, which can apply to different sectors of the economy and include many different types of policy instruments. The difference between such policy instruments is often subtle, nuanced and situational. This chapter presents an intuitive and simple way of characterising the policy instruments that exist to address climate change, detailing and exemplifying each category with the policy instruments most relevant to the scope of later chapters (Chapters 4 and 5) in this book.

2.2. Classifying Policy Instruments

Measures are defined by the Intergovernmental Panel on Climate Change (IPCC) as technologies, processes and practices aimed at reducing emissions and/or their effects below anticipated future levels, and include examples such as renewable energy technologies and energy saving strategies. These are contrasted with policies, which are defined as government mandates, often made in conjunction with business and related organisations, to accelerate mitigation and adaptation measures, including examples such as carbon taxes and fuel efficiency standards (see IPCC (2007: p. 19)).

The menu of policy instruments differs over time as some issues emerge as more powerful problems to be addressed, requiring new solutions to be considered and formalised. Policies can be national or regional, or have some degree of international cooperation; be it bilateral, multinational or global in scope. Climate change mitigation is clearly a global problem (Chapter 1), but one which includes issues of specific national consideration such as energy and transport policy, and as such both domestic and international instruments must be included.

There are many methods and procedures which form a part of policy but are not in any way enforced by a policy framework. Awareness campaigns, education and information dissemination, social movements, awards, and climate action naming and shaming might enrol people or organisations to pursue effective action to reduce total carbon emissions, but they are not strictly policy instruments. Nor are other potentially valuable or effective ways of addressing climate change, such as voluntary commitments, schemes or agreements, even where there are timetabled targets, contracts and sanctions for failure to comply with the agreed target. Some of these agreements or schemes might become official policy and made mandatory, but while they remain inconsistently applied across a jurisdiction, they cannot accurately be described as policy instruments. In addition, their voluntary nature means that they are difficult to model in a consistent and appropriate way, which will be a key consideration of the policy assessments discussed in later chapters (Chapters 4 and 5).

Agreements, standards, technical requirements and regulations enforced within a jurisdiction are the most direct types of policy instrument that a government has at its disposal. They are supply-side instruments, which stipulate to producers the terms or limits that they have at their disposal. These types of instruments are often referred to as "command and control" instruments, but are not always inflexible, universal or dictatorial. For example, framework-based instruments enable different types of requirements to be set that will rely upon a degree of interpretation, depending on, for example, local conditions, or local environmental considerations, rather than a single measure. Performance-based regulation will require, instead, quantitative measures or metrics, and will apply universally across the industry and jurisdiction. Similarly, instruments related to technical requirements or standards will also be applied universally and will require compliance with rules determined by regulators and enforced by government in order to use a specific technology, for example any company wishing to sell their cars to Europe has to meet certain CO_2 emission limit per kilometre for their cars. While many of the command and control policy instruments are applied at the national level, some of the key policies on climate change are driven by international agreements including UN directives and EU obligations.

Other types of instrument are economic incentives, but these fall mainly into two very distinct types of instruments, market-based and financial instruments, depending on the nature of these incentives. Economic instruments that directly affect the price or cost of emissions (in the case of CO_2, a carbon price created by carbon trading mechanisms and related market schemes, or carbon tax and charge schemes such as a climate change levy) form one category, while financial instruments, such as feed-in tariffs and subsidies for low-emission research and development or other financial incentives and allowances, comprise the final category of policy instruments to be analysed in this chapter.

Market-based instruments are economic instruments that tackle climate change by acting through price signals to encourage carbon emitters to monitor and reduce their emissions. These types of instruments target either the supply side or demand side, depending upon the beneficiary of the incentive. In reality the costs or incentives are likely to impact on both the supply and demand side as increased costs are often passed on to end consumers. Individual instruments include charge schemes, such as a levy on fossil fuels (or a levy exemption i.e. subsidy for other energy generation) and various trading mechanisms, for example the EU Emissions Trading Scheme (EU ETS), which offer incentives to reduce emissions by setting emission targets to a fixed amount of greenhouse gas (GHG) emitted, and enabling companies that reduce their emissions to sell any remaining credits they no longer need, at the market price, companies that find mitigation too expensive or cannot reduce emissions to meet their target. Removal of subsidies from high carbon-emitting activities and carbon-intensive energy resources can also be seen as a market-based instrument as it increases the price of polluting activities and fuels, i.e. has the same effect to a tax on fossil fuels.

Financial instruments are used to address carbon emissions by making financial resources available for investment in emission-reduction strategies. These include subsidies or tax exemptions for decarbonisation schemes, grants or loans for renewable energy projects, or the availability of funds to tackle emissions directly, for example in enabling the purchase of energy-saving equipment.

While there are some hybrid economic/financial instruments, such as an auction reserve price trading scheme, most incentive-based instruments fall into one class or the other; and, in light of this, the merits of these two types of policy instruments are considered separately in the following sections.

2.3. National versus International Policies

Climate change mitigation is a global issue because GHG emissions diffuse uniformly across the atmosphere. This means that when one country reduces its emissions, the climate benefits are experienced by all countries and there is a "free rider" problem if countries pursue their own self-interest, i.e. they rely on actions made elsewhere. The problem is recognised by governments and a series of international negotiations, for example under the United Nations Framework Convention on Climate Change (UNFCCC), have been only partially successful in coordinating global action. However, individual countries and groups of countries, such as the EU, have introduced unilateral climate change policies to demonstrate that it can be done.

A particular problem associated with unilateral action is that of a potential loss in international competitiveness and relocation of carbon-intensive production to other countries, so-called "carbon leakage". A policy to reduce emissions could become ineffective because the relocation of high-emitting industries to countries without emission constraints could increase overall emissions. Detailed studies, however, conclude that these concerns are exaggerated (Barker *et al.*, 2009 and Barker and Scrieciu, 2010). Although carbon pricing tends to reduce the price competitiveness of carbon-intensive sectors, this may be offset by exchange-rate adjustments or improvements in non-price competitiveness; while the extent of competitiveness impacts will vary with the international exposure of the sector in question (higher-value sectors will tend to have higher international exposure than lower-value industries). However, as later chapters show, actual policies in many countries include special exemptions for polluting sectors after lobbying by those with vested interests, which ultimately weaken the effectiveness of actions. Rather than provide exemptions for sectors threatened by mitigation policies, a better approach is to provide explicit time-limited subsidies to support adjustment to low-carbon outcomes.

2.4. Regulatory Standards: Command and Control Instruments

In the context of policy making and regulation, command and control instruments refer to the imposition of standards on organisations and individuals, which are regulated through an appropriate authority, such as a government-appointed regulator. In addressing the problem of excessive carbon emissions, a command and control approach involves modifying the behaviour of individuals and firms, either by banning or restricting activity deemed responsible for producing unnecessarily high emissions or by introducing standards designed to ensure that ineffective use of resources, such as fuel, are systematically replaced with more efficient practices and processes.

Command and control instruments designed to reduce carbon emissions are generally termed technology-based standards or performance-based standards (Stavins, 1997). The third category of instruments is generally referred to as framework-based standards (Smith and Sorrell, 2001). In addition to these three categories, ambient standards, designating the quality of the outcome to be achieved, are sometimes considered as a separate category (Callan and Thomas, 2007). In the example of climate change mitigation, the level of 450 ppm (parts per million) of carbon dioxide might be an ambient standard. The ambient standard is rarely enforceable but is instead often used as a target or an objective to be addressed by one of the other types of standard. This chapter considers only practical policy instruments, and therefore ambient-type standards or a combination of ambient and technology performance instruments are not examined here.

Technology-based, or design, standards designate the method for achieving a policy objective by stipulating the type of equipment to be used, or the processes and procedures that need to be implemented by specific organisations in order to achieve this objective. These types of instruments are particularly appropriate when there are very limited ways of achieving carbon emission reductions of the type required, and in this context clarification of the options available and specifying the mode of implementation can be cost effective for firms, and can also save on administration and simplify the auditing process. Technology-based regulations and standards include building regulations, predetermined rates of switching to electric cars, and regulations on biofuel shares for vehicles fuels. Future technology-based standards for the energy generation sector could, for example, include compulsory carbon capture and storage for power plants.

Performance-based standards specify the objective to be achieved by organisations but do not specify how they are to be achieved, so allow each organisation to choose how to achieve the objective, on the condition that they meet the minimum statutory requirement specified. Examples are the Energy Performance Certificates required on the sale, rent or construction of a building in the UK (UK DECC, 2011) and the Emissions Performance Standard set at 450 gCO_2/kWh starting in 2013 for new fossil fuel power stations, in the UK's Carbon Plan (UK DECC, 2011). Performance-based standards, although more flexible than technology-based standards, do not necessarily provide incentives to innovate as, once the standard has been met, the firm does not benefit from further developing more effective technologies or practices beyond the ability to make other savings, such as lower energy bills (although there might be disincentives to ceasing development, such as the likelihood that a future round of performance standard targets might be more restrictive). However, empirical research (Lanoie *et al.*, 2011) suggests that performance-based standards have more impact on environmental innovation than their more prescriptive technology-based counterparts. Significant emission-reduction performance-based standards include the European Union Large Combustion Plant Directive and the California Energy Commission's 2008 Building and Energy Efficiency Standards (Lausten, 2008). Future performance-based standards for the energy generation sector might be an international energy emission limit for each participating country.

Framework-based standards set the qualitative performance requirements to be met by the relevant organisations. Framework-based standards generally subject participating organisations to a regulator, but as such standards might involve the implementation of the best available techniques, meeting the requirements of the standard will depend on *ad hoc* judgement or additional reference documents based on revised research, so there is often room to negotiate between the regulator and the regulatee in the course of implementation and assessment. These types of standards are suited to complex regulatory situations that are both too demanding for other types of standards instruments and too difficult to implement in conjunction with the types of emission taxes or tradable permits outlined in earlier sections (Sorrell and Smith, 2002).

Choosing between instruments imposing technology-based, performance-based or framework-based standards depends generally as much on the nature of the market that the firms are operating in as on the nature of the activity or pollutant to be restricted. Once imposed, however, regulator and firms need to establish monitoring, reporting, and verification systems to ensure compliance.

2.5. Market-based Instruments

Market-based instruments are sometimes known as economic or pricing instruments/incentives/measures. This class of policy instruments uses price-related signals to discourage the types of activity deemed to have a negative impact on the environment as a consequence of emissions. The two key examples of market-based instruments are taxation (or the removal of subsidies) and carbon trading.

2.5.1. *Taxation and removal of subsidies*

An indirect or direct tax is levied on good or services and can be used as a policy instrument in order to create an economic disincentive for the production or consumption of a specific good. A carbon tax is normally a tax on fossil fuels, which sets a price on emissions from fuel consumption by either consumers or producers. It is based on the carbon content of fossil fuels in terms of the combustion-related CO_2 emissions and encourages consumers to use less carbon intensive fuels. Because of different levels of carbon content, tax bases for different fossil fuels vary; for example, higher tax bases are usually applied to higher carbon-content fuels, such as coal and oil, while relatively lower tax bases are applied to lower carbon-content fuels, such as natural gas.

In practice, most countries put a carbon tax indirectly on consumers through a tax on the carbon content of fuel consumption. Carbon taxes have been in employed at the national, state and city level. While taxes on fossil fuels have existed for many years, Finland was the first European country to introduce a systematic carbon tax when it introduced its national carbon tax in 1990, followed by several other European countries, including Sweden, Denmark and Switzerland (Andersen and Ekins, 2009). The province of Quebec in Canada enacted a carbon tax in October 2007 and the province of British Columbia began phasing in a carbon tax on all fossil fuels in July 2008 (Richards, 2011).

A related measure is the energy tax, which is levied on the energy content of all energy sources, which could be either fossil fuels or carbon-free energy sources. An energy tax can be considered as an alternative to a carbon tax since it has a direct effect on fossil fuel and therefore CO_2 emissions. On average, for fossil fuels like oil and gas, which have energy content higher than coal, an energy tax represents a greater cost than a carbon tax. In contrast, a carbon tax falls more heavily on coal than an energy tax, because coal releases more CO_2 during combustion in comparison with gas or oil (IPCC, 2006).

Instead of the explicit carbon taxes, many countries, including the UK, have adopted general energy taxes aimed at promoting energy efficiency, energy savings measures and reducing total energy demand as a way of reducing emissions (Pearce, 2006), while other countries use a combination of a carbon tax and an energy tax, for example Finland and Sweden (Andersen and Ekins, 2009).

The removal of fossil fuel subsidies has a similar effect to carbon or energy taxes; it will raise the price of fossil fuel and therefore the price of carbon-intensive products. Fossil fuel subsidies are in use in many developing and transition economies, such as Russia, China, and between Saudi Arabia and Venezuela, etc. and starting from as little as 0.4% in South Korea to 84.4% in Iran of average fossil fuel price in 2010 (IEA, 2010).

2.5.2. *Emissions trading or carbon markets*

Emissions trading is another way of using the price mechanism to achieve reductions in emissions. The scheme acts as an incentive for firms to monitor and reduce their emissions by placing a limit on carbon emissions and developing a system for allowing allowances (permits) for carbon emissions to be traded.

To enable Annex I parties of the Kyoto Protocol (1997) to reduce emissions, the 1997 Kyoto Protocol to the UNFCCC enacted a provision for international emissions trading related to GHG emissions. Via this mechanism, the Annex I parties can acquire units from other parties to meet their emission targets under the Kyoto Protocol. Rather than a single carbon market, several carbon markets operate simultaneously. In addition to the market operating under the Kyoto Protocol, there are several other distinct carbon markets in operation today. There are mandatory and voluntary emissions trading schemes/systems (ETS). After the European ETS, Australia, Canada, New Zealand and the US have all drawn up proposals to create their own domestic mandatory ETS. However, a number of the mandatory ETS at the sub-national level have been created. Voluntary national ETS have also been put in place including the ETS of the UK in 2002–2006, the Japanese voluntary ETS in 2005, and the Chicago Climate Exchange in 2003 (Hood, 2010).

Two main types of ETS are available, depending on different emission targets: the cap-and-trade system and the credit-based system (Tietenberg, 2002). In a cap-and-trade system, the government, or governance body, firstly defines a cap based on the overall maximum amount of carbon emissions over a specified timeframe. This scarcity in the market creates a price for a unit of emissions (which must be covered by an allowance). Every company that is included in the scheme receives a proportion of allowances, which are based, for example, on their historical emissions. The system allows those with lower marginal abatement costs to abate and sell their allowances to those with higher abatement costs. Under a credit-based system, also called "relative cap" system, the government determines a standard of emissions for each source, in either emissions allowed per unit of production or emission intensity. The user who stays under the baseline can have the difference certified as a tradable credit and can sell the credits to these users who did not meet the legal requirement.

There are two key methods used by the regulator to distribute allowances to existing installations: free allocation and/or auctioning. The method of free allocation is based on historical emission levels and is known as "grandfathering". The free distribution of allowances has the potential to reduce the risk of losing competitiveness in energy-intensive and trade-exposed sectors; and it may also be a first step on the path to attracting companies to support an ETS.

Under the second method, auctioning, companies are required to bid for the number of allowances they want to purchase in order to meet their emission targets, as opposed to receiving an initial amount free-of-charge. One of the advantages of auctioning is that it is likely to provide an immediate price signal to the trading market, which enables consumers and suppliers to switch from high-carbon goods and services to low-carbon alternatives for emission reductions. In the third phase of the EU ETS, the amount of auctioning is substantially increased, from less than 4% in Phase II to 100% auctioning in Phase III (Directive 2009/29/EC). Under the Regional Greenhouse Gas Initiative (RGGI), the participating states of the US will auction nearly 100 percent of their annual allowances (RGGI, 2012).

In addition to ETS there are project-based GHG emission-reduction measures defined in the Kyoto Protocol (1997). With project-based transactions, the buyer purchases emission credits from a project that can verifiably demonstrate emission reductions compared with the level of emissions that would have been produced in the absence of the project. The most notable examples of such activities are the clean development mechanism (CDM) and the joint implementation (JI) scheme of the Kyoto Protocol. The CDM is an arrangement under the Kyoto Protocol (Article 12) that allows countries with emission-reduction commitments (Annex I parties) to invest in emission-reducing projects in developing countries as an alternative to making more costly emission reductions in their own countries. JI is a similar arrangement, which allows Annex I parties to invest in emission-reduction projects in other Annex I parties that are in the process of becoming market economies. This leads to investments in emission-reducing projects in countries, such as the Russian Federation, the first- and second-wave accession countries, and other Central and Eastern European states.

While the CDM has demonstrated its ability to drive emission reductions, facilitate essential learning, raise awareness and promote capacity building, it has many limitations, which include low development co-benefits, questionable additionality, i.e. the net difference in emissions resulting from specific interventions, weak governance, inefficient operation, limited scope and uncertain prospects beyond 2012, which is the last year of commitments under the Kyoto Protocol (World Bank, 2010).

2.6. Financial Instruments

Financial instruments represent another policy instrument category commonly used in addressing climate change mitigation. Such instruments are employed by policy makers to tackle emissions by making financial resources available for investment in specific emission-reduction strategies.

The funding required for climate change mitigation, adaptation and technological transfer is substantial, while the efforts to raise funds have been insufficient, with only an estimated 5% of the required global financial resources allocated (World Bank, 2010); however, this section will examine only financial instruments related to climate change mitigation, in particular in the form of policy instruments that enable the development and use of goods and technologies that serve identifiable mitigation goals.

Financial instruments work in one of two ways, which can be characterised as "demand–pull" or "supply–push" depending on their influence.

"Demand–pull" policies are targeted at consumers, designed to induce higher demand for low-emission technologies by reducing their cost for end-users. "Supply–push" policies are targeted at producers, providing production support to entrepreneurs and innovators for the research and development, adoption and deployment of low-emission technologies.

Given the considerable financial costs of such instruments, multiple sources of finance are used in combination in the development of individual policies. Such sources include public climate finance (from national governments, international organisations and dedicated financing mechanisms) and private climate finance.

Public funding is particularly important in terms of enhancing innovation and deployment of new technologies and goods through supporting large-scale projects, developing infrastructure and taking a strategic and coordinated approach to innovation. Such instruments also include new innovative funding mechanisms to create incentives for the private sector to get involved in the climate change mitigation activities and investment, but they are most effective when coupled with developing a supportive environment for multi-sectoral partnerships and international cooperation (Gallagher *et al.*, 2006). Policy instruments that use such financial mechanisms can be divided into domestic and international mechanisms, which can be further subdivided according to the scale in which the instruments are applied.

2.6.1. *Domestic financial mechanisms*

Domestic financial policy instruments are designed to promote the development and deployment of goods and technologies and can be established at national or sub-national level. To facilitate the development of emission-reducing technologies or renewable energy technologies, government funding is made available to encourage scientific research to support new technologies and processes from early development through various stages of innovation and demonstration to the marketing phase of a new product or process. Governmental funding is useful to supplement funding or subsidise the costs of renewable energies or efficient conventional energy sources and helps to offset the learning costs, or the additional costs involved in using the cleaner or efficient technologies by domestic companies or residents of a particular region or the country as a whole.

Three types of financial incentives are used by governments to encourage the development and deployment of low-emission goods and technologies: fiscal measures, price-support measures and investment support. Each of these categories has two or three sub-categories currently used as policy instruments.

Two types of fiscal measures are used to encourage the participation of private sector companies in climate change mitigation efforts. One strategy is to provide tax subsidies, for example, tax exemptions, tax deduction and tax rebates, while the second strategy is to provide tax credits, for example, income tax credits, personal tax credits, corporate tax credits, production tax credits and investment tax credits. Both strategies offer incentives by lowering the burden of tax, but offer different degrees of flexibility (Clement *et al.*, 2005). Such fiscal measures may be either targeted at consumption by rewarding the purchase and installation of certain technologies and goods, or targeted at production by supporting or facilitating investment in the production of low-emission goods and technologies.

Among the price-support measures and mechanisms used to encourage the generation of electricity by means of renewable energy sources, feed-in tariffs and net metering have had the highest profile and are both widely used within OECD (Organisation for Economic Co-operation and Development) countries. A feed-in tariff is a policy mechanism designed to encourage the adoption of renewable energy and to help accelerate the move towards grid parity, by regulating a minimum guaranteed price per kilowatt-hour that an electricity company must pay for renewable energy fed into the national electricity grid by a private producer. The "net metering" strategy is another common measure aimed at reducing costs for an owner of small-scale on-site renewable energy power generation equipment. Under this measure, the owner receives a credit on future energy bills for the amount of power he or she supplies to the national electricity grid minus the amount this consumer takes from the national grid during a certain billing period.

Finally, in the taxonomy of financial incentives, there are two types of investment incentives that have been used by governments to encourage investment in low-carbon technologies and goods: capital grant and favourable lending conditions. The first type of investment support policy is in the form of a capital grant. Capital grants are used to reduce the capital cost of investing in low-carbon technologies and pursuing low-emission practices via the return of a specified percentage of the costs of undertaking emission reduction to the investor, often resulting in significant reductions in the overall cost of such investment. The second type of investment support policy takes the form of favourable lending conditions, or low-cost financing with subsidised interest rates for those investing in emission-reduction technologies or low-emission products. A high-profile example of this type of investment support policy was India's Solar Loan Programme, which, between 2003 and 2007, and with the support of United Nations Environment Programme (UNEP), the Risoe Centre, and two Indian banks Canara Bank and Syndicate Bank, provided low-cost financing for solar energy systems.

2.6.2. *International financial mechanisms*

In addition to domestic financial mechanisms, there is a wide variety of international public funds available for tackling climate change, which includes multilateral initiatives and bilateral initiatives.

There are three key multilateral initiatives, which provide significant support for climate change mitigation strategies. Chief amongst these is the Global Environment Facility (GEF), which is at the centre of the existing multilateral financing programs and projects to protect the global environment. Working with ten agencies, including the World Bank, the GEF has provided primarily grants or concessional funding to protect the global environment in six focal areas: climate change (mitigation and adaptation), biodiversity, international waters, persistent organic pollutants, ozone depletion and land degradation (desertification and deforestation).

The second initiative is the United Nations Collaborative Programme aimed at Reducing Emissions from Deforestation and Forest Degradation (REDD) in Developing Countries, known as the UN-REDD Programme. This Programme was developed in collaboration, between the Food and Agriculture Organisation (FAO), United Nations Development Programme

(UNDP) and UNEP. Its aim is to assist developing countries prepare and implement national REDD+ strategies. There is, however, no guarantee that a global framework for REDD will be put in the place in the future (UNEP, 2011).

The third of the key multilateral initiatives is the World Bank Climate Investment Fund (CIF), which was created in 2008 along with other multilateral development banks as an interim measure to scale up assistance to developing countries and strengthen the knowledge base in the development community. The CIF supports two funding streams relevant for financing climate change initiatives: the Clean Technology Fund and the Strategic Climate Fund.

In addition to these multilateral initiatives, there are a several bilateral initiatives of relevance to climate change mitigation financing. These include the European Commission's Global Climate Change Alliance (GCCA); the UK's Environmental Transformation Fund (ETF); The German International Climate Initiative (ICI); and, the Australian International Forest Carbon Initiative (IFCI). The GCCA helps developing countries to address mitigation, adaptation and poverty reduction, through collaborative partnerships, by providing both technical and financial assistance. The UK's ETF aims to tackle climate change in developing countries by addressing unsustainable deforestation and improving access to clean energy. The German ICI supports sustainable energy systems, adaptation and biodiversity projects related to climate change, stimulating private investment and puts systems in place to ensure that financed projects will strategically support the post-2012 climate change negotiations. Australia's IFCI focuses on reducing emissions from deforestation by providing financial and support incentives to developing countries to address the causes of deforestation.

2.7. The Policy Mix and Interactions between Policies

As the previous sections have shown, there are many instruments available to policy makers with which to mitigate the causes of climate change. The success of individual policy instruments will, however, depend upon the coherence of the policy mix and as such it would be misleading to consider policy instruments in isolation. Some policies are more suitable for encouraging innovation in the development of the more costly technologies that may have the potential for lower costs with widespread use. Other policies may be more appropriate to encourage adoption of lower-cost alternatives to fossil fuels. Johnstone *et al.* (2010), in an empirical study of the effectiveness of different policy instruments for different renewables for innovation using OECD data, find that public policy plays a significant role in determining patent applications. Broad-based policies, such as tradable energy certificates, are more likely to induce innovation on technologies that are close to competitive with fossil fuels. More targeted subsidies, such as feed-in tariffs, are needed to induce innovation on more costly energy technologies, such as solar power.

Governments have a wide range of instruments at their disposal to achieve their climate policy targets (see Table 2.1) (IPCC, 2007: Chapter 13).

Table 2.1. Policies and their constraints and opportunities.

Sector	Policies, measures and instruments	Constraints or opportunities
All sectors	Public RD&D investment in low-emission technologies; standards; carbon prices	Vested interests, environmental tax reform
Energy supply	Reduction of fossil fuel subsidies Taxes or carbon charges on fossil fuels	Resistance from those with vested interests
	Feed-in tariffs for renewable energy technologies; renewable energy obligations	May be appropriate to create markets for low-emission technologies
	Producer subsidies	
Transport	Mandatory fuel economy, biofuel blending and CO_2 standards for road transport	Partial coverage of vehicle fleet may limit effectiveness
	Taxes on vehicle purchase, registration, use and fuels, road and parking pricing	Effectiveness may drop with higher incomes

(Continued)

Table 2.1. (*Continued*)

Sector	Policies, measures and instruments	Constraints or opportunities
	Influence mobility needs through land use regulations and infrastructure planning	Particularly appropriate for countries that are building up their transportation systems
	Investment in attractive public transport facilities and non-motorised transport	
Buildings	Appliance standards and labelling	Periodic revision needed
	Building codes and certification	Enforcement can be difficult
	Demand-side management programmes	Utilities need to profit
	Public sector leadership and procurement	e.g. energy efficient products
	Incentives for energy service companies	Access to third-party financing
Industry	Provision of benchmark information Performance standards Subsidies, tax credits	May be appropriate to stimulate technology uptake. Stability of national policy important
	Tradable permits	Predictable allocation and prices
	Voluntary agreements	Independent monitoring
Agriculture	Financial incentives and regulations for improved land management, maintaining soil carbon content, efficient use of fertilisers and irrigation	Can reducing vulnerability to climate change
Forestry/Forests	Financial incentives (national and international) to increase forest area and to maintain and manage forests; land use regulation and enforcement	Constraints include lack of investment capital and land tenure issues
Waste	Financial incentives for improved waste and wastewater management with renewable energy incentives or obligations	May stimulate technology diffusion

Source of information: Adapted from IPCC (2007).

The IPCC WG3 Report focuses mainly on sectoral options for mitigation and provides a rich source of detail on the economic potential for mitigation at different carbon prices in energy, transport, buildings, industry, agriculture, forestry and waste management. The appropriate policy portfolios for GHG mitigation will, of course, be specific to countries depending on their political systems, available renewable and other energy resources, and the energy efficiency of existing building and equipment stock. Such portfolios will combine policies and measures to produce outcomes that are effective at achieving their main climate objectives, efficient with low costs or even benefits as regards effects on GDP (gross domestic product), and equitable in that vulnerable groups will be most likely to benefit. Importantly for policies to achieve a wide social consensus, they should also address other potential social benefits, such as improvements in air quality, better human health, higher crop productivity, increased comfort from better insulated buildings, or reductions in traffic-related pollution.

The complementarities of different instruments will determine the effectiveness of policy portfolios; in practice policies are often introduced to address problems created by the existing policy landscape, not merely as a means to solve an emerging problem. In order to avoid developing inappropriate climate change mitigation policy portfolios, policy makers and advisors must consider the interaction between different instruments, otherwise polices could be costly to implement, while also increasing the costs to both firms and consumers without contributing greatly to a reduction in carbon emissions. A key challenge will be to assess the interdependencies of different instruments using appropriate metrics, to determine which combinations work effectively and which do not. This assessment is further complicated by the observation that instruments do not always lead to the expected outcomes or instruments that have been effective in one country have been less effective elsewhere (Gunningham and Gabrosky, 1998). In this way, there needs to be an initial coordination between different policy realms so that energy and environmental policy are supported by an appropriate economic framework, and so that the portfolio of policies is appropriately implemented to provide the most effective incentives as a portfolio and ensure that the costs are transparent and that any funding requirements being both planned and adequate.

In addition, there will often be multiple objectives for policy makers to consider in addition to reducing emissions, such as combating energy poverty or providing funds to enrol developing countries into a global decarbonisation programme. There are also trade-offs which need to be considered, such as between reducing transaction costs and encouraging innovation, or ensuring compliance to standards while encouraging participation in trading schemes. There are also potential policy conflicts, as might occur when stringent environmental controls aimed at risk reduction might act as a disincentive to developing technology-driven efficiency by holding the industry to a higher standard of performance each time new technology is introduced to the industry, rather than rewarding innovation in efficiency. The unintended consequences of a policy instrument and the generation of perverse incentives are other factors that could alter the effectiveness of a policy portfolio and require careful consideration (for more details see Smith and Kern (2009)).

An example of complementarity between policies concerns the issue of induced technological change. For example, the US government has generally proposed arguments favouring technological agreements rather than mandatory cap-and-trade schemes such as the EU ETS. It is argued that cap-and-trade will not produce the fundamental new technologies required for a zero- or low-carbon economy (Jaffe *et al.*, 2005). The problem with this argument is that some technological developments (such as carbon capture and storage) require a carbon price to become economic, and their investment prospects depend heavily on the price continuing over their useful life. The most effective policies thus appear to be those that combine the carbon price signal from an ETS with direct incentives to fund low-GHG innovation and research and development, for instance by auctioning emission allowances and using a proportion of the revenues to provide additional incentives to develop new low-carbon technologies.

Another example showing the case for complementarity between policies is that of regulations for energy efficiency leading to the rebound effect. This effect can be offset by a market-based policy leading to an increase in the carbon price. The so-called "rebound effect" (Barker *et al.*, 2009) arises where improvements in energy efficiency reduce the cost of a technology, which then prompts higher use of particular services (for example heat or mobility) that energy helps to provide, so that the energy saving from the innovation is offset by increased energy consumption. Although rebound effects will vary widely in size, any energy efficiency regulations adopted without a carbon price to deter extra carbon use risks higher energy demand overall with only a weak reduction, if any, in emissions, especially at a global level (see Barker *et al.* (2009)). This indicates that a carbon-price signal from market-based instruments is needed to provide a pervasive and long-term basis for investment decisions so that low-GHG options are chosen consistently.

References

Andersen, M. and Ekins, P. (2009). *Carbon Energy Taxation: Lessons from Europe*, Oxford University Press, Oxford.

Barker, T. and Scrieciu, S. (2010) Low Stabilization within a "New Economics" Macro-econometric Framework: Insights from E3MG, *The Energy Journal Special Issue on the Economics of Low Stabilization*, 36, 137–164.

Barker, T., Dagoumas, A. and Rubin, J. (2009). The Macroeconomic Rebound Effect and the World Economy, *Energy Efficiency*, 2, 411–427.

Callan, S. and Thomas, J. (2007). *Environmental Economics and Management: Theory, Policy and Applications*, Thomson South Western, Mason.

Clement, D., Lehman, M., Hamrin, J. and Wiser, R. (2005). *International Tax Incentives for Renewable Energy: Lessons for Public Policy*, Energy Foundation China Sustainable Energy Program, Center for Resource Solutions, San Francisco, California.

De Jonghe, C., Delarue, E., Belmans, R. and D' Haeseleer, W. (2009). Interactions between Measures for the Support of Electricity from Renewable Energy Sources and CO_2 Mitigation, *Energy Policy*, 37, 4743–5472.

Directive 2009/29/EC (2009). European Commission. Available online at: http://eur-lex.europa.eu/legal-content/EN/ALL/;jsessionid=122pTDdXNQy6ct8QGGh2hJwkbsTvGQKspVcYvF16G5ndLt6rJ8KH!2061092372?uri=CELEX:32009L0029. Accessed 1 April 2014.

Gallagher, K., Holdren, J. and Sagar, A. (2006). Energy Technology Innovation, *Annual Review of Environmental Resources*, 31, 193–237.

Gunningham, N. and Gabrosky, P. (1998). *Smart Regulation: Designing Environmental Policy*, Clarendon Press, Oxford.

Hood, C. (2010). *Reviewing Existing and Proposed Emissions Trading Systems*, International Energy Agency, Paris.

IPCC (2006) *IPCC Guidelines for National Greenhouse Gas Inventories*, Cambridge University Press, Cambridge.

IPCC (2007). 'Summary for Policymakers', *Climate Change 2007: Mitigation. Contribution of Working Group III to the Fourth Assessment Report of the Intergovernmental Panel on Climate Change*, Metz, B., Davidson, O. R., Bosch, P. R., Dave, R., Meyer, L. A. (eds), Cambridge University Press, Cambridge, pp. 1–23.

Jaffe, A., Newell, R. and Stavins, R. (2005). A Tale of Two Market Failures: Technology and Environmental Policy, *Ecological Economics*, 54, 164–174.

Johnstone, N., Haščič, I. and Popp, D. (2010). Renewable Energy Policies and Technological Innovation: Evidence Based on Patent Counts, *Environmental & Resource Economics*, 45, 133–155.

Kyoto Protocol (1997). Kyoto Protocol to the United Nations Framework Convention on Climate Change. Available online at: http://unfccc.int/essential_background/kyoto_protocol/background/items/1351.php. Accessed 3 February 2013.

Lanoie, P., Laurent-Lucchetti, J., Johnstone, N. and Ambec, S. (2011). Environmental Policy, Innovation and Performance: New Insights on the Porter Hypothesis, *Journal of Economics & Management Strategy*, 20, 803–842.

Lausten, J. (2008). *Energy Efficiency in Building Codes, Energy Efficiency Policies for New Building*, International Energy Agency, Paris.

Pearce, D. (2006). The Political Economy of an Energy Tax: The United Kingdom's Climate Change Levy, *Energy Economics*, 28, 149–158.

RGGI (2012). Regional Greenhouse Gas Initiative. Available online at: http://rggi.org/design. Accessed 1 April 2014.

Richards, J. (2011). 'Idle Speculation: Prefunding the Boomers: 'Frail Elderly' Care and Legitimizing a Carbon Tax', in Gorbet, F. and Sharpe, A. (eds). *New Directions for Intelligent Government in Canada: Papers in Honour of Ian Stewart*, Centre for the Study of Living Standards, Ottawa, pp. 227–248.

Smith, A. and Kern, F. (2009). The Transitions Storyline in Dutch Environmental Policy, *Environmental Politics*, 18, 78–98.

Smith, A. and Sorrell, S. (2001). Interaction between Environmental Policy Instruments: Carbon Emissions Trading and Integrated Pollution Prevention and Control, *International Journal of Environment and Pollution,* 15, 22–41.

Sorrell, S., and Smith, A. (2002). *Policy Additionality for UK Emissions Trading Projects. A Report for the Department of Trade & Industry*, SPRU, University of Sussex, Brighton.

Stavins, R. (1997). Policy Instruments for Climate Change: How Can National Governments Address Global Problems? *The University of Chicago Legal Forum*, January, 293–329.

Tietenberg, T. (2002). The Tradable Permits Approach to Protecting the Commons: What have we Learned? Fondazione Eni Enrico Mattei, FEEM Working Paper 36.

UK DECC (2011). The Carbon Plan Delivering our Low Carbon Future. UK Department of Energy and Climate Change. Available online at: http://www.decc.gov.uk/en/content/cms/tackling/carbon_plan/carbon_plan.aspx. Accessed 1 February 2013.

UNEP (2011). REDDy-Set-Grow: Opportunities and Roles of Financial Institutions in Forest Carbon Markets, *UNEP Finance Initiative*, Geneva, Switzerland. Available online at: http://www.unepfi.org/fileadmin/documents/reddysetgrow.pdf. Accessed 1 April 2014.

World Bank (2010). *World Development Report 2010: Development and Climate Change*, World Bank Institute Washington, DC.

Chapter 3

Scenario Design for a Global Low-carbon Economy

Jun Li and Aleix Altimiras-Martin

3.1. Introduction to the Scenario Review

3.1.1. *The role of scenarios*

In climate change research, scenarios describe plausible trajectories of climate conditions and other aspects of the future such as demographic changes, technological development, and sociological and economic evolution. Scenarios represent the driving forces, processes, impacts (physical, ecological and economic) and potential responses that are important in informing climate change policy (Moss *et al.*, 2010). According to IPCC (2007c):

> A scenario is a coherent, internally consistent, and plausible description of a possible future state of the world (Carter *et al.*, 1994; Nakicenovic *et al.*, 2000; Raskin *et al.*, 2005). Scenarios are not predictions or forecasts (which indicate outcomes considered most likely), but are alternative images without ascribed likelihoods of how the future might unfold. They may be qualitative, quantitative, or both. An overarching logic often relates several components of a scenario, for example a storyline and/or projections of particular elements of a system.

A set of scenarios is often adopted to reflect the range of uncertainty in projections.

Climate change research focuses specifically on emission, climate and environmental scenarios. Emission scenarios are descriptions of potential future discharges to the atmosphere of substances that affect the earth's radiation balance, such as greenhouse gases (GHGs) and aerosols. It reflects a series of plausible states of the future world based on the research into socio-economic, environmental and technological trends. The emission scenarios are usually based on a set of assumptions about the key relationships and driving forces of change, which are derived from our understanding of both history and the current situation. Climate scenarios are plausible representations of future climate conditions such as temperature, precipitation and other climatic phenomena. On these lines, environmental scenarios focus on change in environmental conditions such as water availability and the rise in sea level. In this chapter only emission scenarios are discussed.

In short, scenarios are developed to understand how human activities can evolve and how those activities interact with the environment. This is a complex exercise which requires identifying and quantifying the causal relationships between human activities themselves and quantifying the effect of those activities on the environment. Indicators on their own, e.g. atmospheric CO_2 concentrations, can accurately describe the state of the environment but do not provide the information on how they affects the environment, i.e. which temperature change does the atmospheric concentration lead to. Neither do they provide information as to what leads to this state of the environment. In order to develop a scenario, it is necessary to identify and organise the drivers causing the release of substances which are a pressure on the environment. It also requires assessing how the changes in the natural state of the environment affect the environment itself.

3.1.2. *A framework for environmental issues*

Methodological frameworks, such as the driver-pressure-state-impact-response (DPSIR) illustrated in Figure 3.1, have been developed to assess the causal relationships between the indicators of human activities and the environment. In this particular framework:

- The **driving forces** are the human activities which cause pressures to the environment (for example driving a car releases CO_2).
- The **pressures** are the direct stresses which are provoked by the driving forces and affect the natural environment, e.g. the CO_2 release.

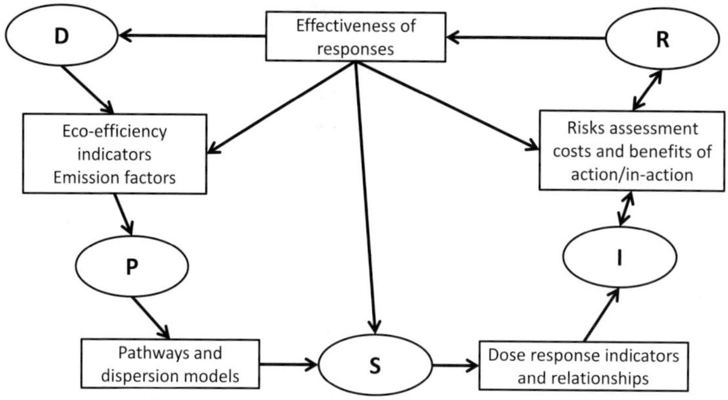

Figure 3.1. Indicators and information linking DPSIR elements.

Source: EEA, 1999.

- The **state** describes the environmental conditions, e.g. atmospheric CO_2 concentration.
- The **impacts** are the environmental effects due to changes in the state of the environment, e.g. an increase in atmospheric CO_2 leads to climate change and ocean acidification.
- The **response** is the strategies developed by humans that aim to reduce the environmental impact of the driving forces. The strategies usually imply policy options which alter the driving forces (e.g. policies which lead to a reduced use of cars, or cars which emit less CO_2). However, some responses may directly affect the other components as well.

In theory, this framework provides an accurate representation of causality, although the understanding and definition of the linking processes between each step is more complicated. Each stage can be accurately described by a set of descriptive indicators (e.g. numbers of cars, concentration of atmospheric CO_2). The relationships between those require an understanding of the underlying science (e.g. how does a car produce CO_2?). In general the relation between drivers and pressures can be characterised by the "eco-efficiency" of the technology and related activities that produce the pressure (e.g. the CO_2 emissions per mile or kilometre). How the pressure affects the state of the environment is characterised by the behaviour of the natural cycles, i.e. the biogeochemical cycles, which are usually summarised by pathways and dispersion models (e.g. how the CO_2 is spread through the atmosphere and absorbed by the oceans).

Linking the change of the environmental state to the corresponding impacts involves understanding the carrying capacities and thresholds for the ecological systems that are affected. The relationship between the impacts and the responses of society is a more subjective analysis since it depends on how these impacts are perceived, evaluated, translated into strategies aiming to modify the drivers, pressures, state or impacts, and how the strategies are executed. At this stage, a full reassessment of the situation — i.e. of the drivers, pressures, state or impacts — is required to understand how the adopted strategy affects the environment and how effective it is. All these steps can be summarised under "effectiveness of the response".

In this way, many different types of strategies aiming at reducing the environmental impact of societies can be explored. For example, a response can be to adopt a tax on CO_2 emissions for main industrial sectors. This will feed back into the affected drivers of CO_2 emissions, which in this case are the power sector and heavy industries (e.g. the metallurgical and cement industries). These industries will take into consideration the new tax and, if well implemented, drive a change in their production technology so that they release less carbon emissions. Chapter 4 of this book give more detail of the actual scenarios we have chosen for the modelling. Chapter 5 uses the model to focus on the economic feasibility of the policies aiming to drive the technological and social change required to mitigate GHG emissions. Chapter 6 and 7 provide a detailed analysis of the technical possibilities to modify the different drivers of GHG emissions.

3.1.3. *Reference and low-carbon scenarios*

A scenario is a simulation of the socio-economic, technological and environmental effects of a specific response. In a DPSIR exercise, suggesting possible responses might be enough. However, in the case of emission scenarios this is just the start of the work. The feedback analysis in the response is actually a key aspect in scenario development. Then scenarios become a useful tool not only to identify the key factors that cause pressure on the environment but to assess the effects of different policies and strategies to mitigate the environmental impact of human activities.

When building scenarios, it is important to first create reference or baseline scenarios. Those are scenarios of "normal" development of human activities which do not include the extra policy or strategy required to address the environmental issues or objectives that are under analysis. In other words, there is no specific response to a specific environmental degradation, although there are different ways for the reference scenario to evolve. That is why there can be different baseline scenarios to illustrate different possible futures. The baseline scenarios are the benchmarks against which the scenarios developed with specific policies and strategies — such as the low-carbon scenarios — can be compared. The effectiveness of the response (i.e. the effectiveness and consequences of the policy or strategy) can then be formally assessed.

3.2. The SRES and IPCC AR5 Scenarios in the Context of Global Low-carbon Scenarios

3.2.1. *Overview*

The Intergovernmental Panel on Climate Change (IPCC) is the organisation charged by governments to review and collate the emission scenarios used for the international scientific exercise of understanding the anthropogenic impact on climate change. The assessment of the subsequent biophysical and socio-economic impacts has been compiled under the various assessment reports. So far, three different sets of scenarios have been used in four different reports on mitigation. The SA90 scenarios have been used in the First Assessment Report (Tegart *et al.*, 1990), the IS92 scenarios have been used in the Second Assessment Report (Watson *et al.*, 1995) and finally the Special Report on Emission Scenarios (SRES) (Nakićenović *et al.*, 2000) has been used in the Third and Fourth Assessment Reports: TAR (IPCC, 2001) and AR4 (IPCC, 2007d), respectively.

The characteristics of the different scenario sets have evolved. Business-as-usual (BAU) scenarios have disappeared in the SRES, which has become instead a series of non-intervention baselines. The implicit storylines have evolved. The SA90 scenarios focused specifically on energy mix and efficiency issues. The IS92 scenarios tackled population, income and resources use issues. Finally, in the SRES, the storylines were explicitly formalised into a set of four and underpinned the four scenario families that the report develops. In the SRES scenarios, the term prediction is changed for projection and the storyline creation is seen as a need to embody the inherent uncertainty of the modelling exercise. Currently, new scenarios have been developed to be used in the Fifth Assessment Report (2013). Section 3.2.6 offers a short review of their new features.

The scenarios have not only evolved as the scientific community improved the assessment tools but also due to policy requirements. Girod *et al.* (2009) have analysed this evolution against saliency, credibility and legitimacy and found that the SRES scenarios gained enhanced credibility through improved scenario construction methods. They also gained legitimacy since the scenario development was transparent with a balanced participation of scientists. However, due to inappropriate classifications and the high number of scenarios, the SRES scenarios have reduced their saliency.

3.2.2. *SRES main features*

Contrary to the IPCC's guidelines, the emission scenarios are original work in themselves rather than an assessment of existing work. The SRES scenarios are built taking into consideration the following driving forces: demography, socio-economic and technical development, resource availability, agriculture and land-use change, and non-climate policies. The starting point is a review of the current literature on each driver or driving force of GHG emissions. The aim of the review is to characterise it by a set of factors, understand its relationships with the other drivers, and quantify those relationships. In most cases, it concludes that a full or at least partial characterisation is possible, but the relationships with the other drivers are sometimes poorly understood and therefore highly uncertain. This procedure is similar to the identification and description of the drivers in a DPSIR framework.

Table 3.1. Drivers of emissions and their corresponding factors.

Drivers (or driving forces) of GHG emissions	Factor
Environmental policies (not related to climate change)	Ozone protection, urban air quality
Land-use change based on food demand and fuel-mix biomass	Population, diets, biomass requirements
Structure of energy systems	Exogenous technological change, different parameter (e.g. efficiencies, costs)
Technological change	Different improvement rates in energy distribution, conversion, extraction
Socio-economic development	Economic growth
Population	Mortality and fertility rates

As the relationships between the driving forces are not always clearly understood, the report assumes that drivers interact following the IPAT model. In this model,

$$\text{Impact} = \text{Population} \times \text{Affluence} \times \text{Technology}$$

and therefore each component of growth is additive. This approach implies a linear growth of the impact, for example of CO_2 emissions. However, this relationship will ultimately rely on the models used to build the scenarios.

It is this uncertainty that actually underpins the adoption of four storylines as the backbone for the scenarios. According to IPCC (2007c)

> Storylines are qualitative, internally consistent narratives of how the future may evolve. They describe the principal trends in socio-political-economic drivers of change and the relationships between these drivers. Storylines may be stand-alone, but more often underpin quantitative projections of future change that, together with the storyline, constitute a scenario.

Another important part of the study is to understand the relations between the drivers and the actors that cause or affect them. The actors can be individuals, local and international communities, local and international institutions, governments, and local and international firms. The storylines also provide the motivations and agency of the various actors, and crucially the relationships between them and their degree of influence.

In the SRES scenarios, the evolution of the driving forces and corresponding factors is articulated according to four storylines in order to create four marker scenarios and their corresponding scenario families. The SRES storylines are built by opposing two sets of trends: globalisation (A) vs. regionalisation (B), and economic focus (1) vs. environmental focus (2) which lead to the four scenario families: A1, A2, B1 and B2.

For example, the storyline A1 depicts a future where economic development dissolves regional disparities, i.e. regional average incomes per capita converge. This convergence is underpinned by advances in transportation of goods and people, improved communication technologies, shifts in national policies on immigration and education, and the development of national and international institutions that enhance industrial productivity and technological diffusion.

Those storylines provide the guidelines and conceptual approach to model the evolution of the emission drivers through the estimation and quantification of the drivers' factors. This process is developed by integrating all the information in a programme specially designed to take into account the relationships and evolution of the different factors — integrated assessment models (IAM). To test different approaches, the different scenarios were developed by a different IAM: the A1 scenario was developed by the AIM model (Jiang *et al.*, 2000), the A2 scenario was developed by the ASF model (Sankovski *et al.*, 2000), the B1 scenario was developed by the IMAGE 2.1 model (De Vries *et al.*, 2000) and the B2 scenario was developed by the MESSAGE model (Riahi and Roehrl, 2000). To complete the scenario developing process, the scenarios are submitted to an open revising process.

The A1 marker storyline is considered the most likely to happen. It is characterised by a balanced energy supply mix and therefore it is called the A1B scenario. It is further disaggregated into three other families: the A1C (characterised by a coal-intensive energy sector), the A1G (characterised by an oil- and gas-intensive energy sector) and the A1T (assuming high technological change which leads to a non-fossil fuel energy sector). The A1C and A1G are combined into the A1FI scenario (fuel-intensive sector). In addition to the original development of each scenario by a different IAM, called the marker scenarios, each family of scenarios is also modelled by the other models, creating a total of 40 scenarios. In order to enable comparison amongst the scenarios, most of them are harmonised, i.e. they follow the same economic growth and population growth paths.

3.2.3. *SRES in the context of new low-carbon scenarios*

A low-carbon scenario (LCS) is an emission scenario leading to very low GHG emissions. The GHG emissions have a cumulative effect and remain in the atmosphere, leading to specific levels of atmospheric GHG concentrations. Typically, the emission scenarios develop futures where CO_2 concentrations are between 450 and 650 ppm (parts per million); a LCS would be of 450 ppm or lower.

Hughes and Strachan (2010) point out that LCS require speculation about future changes through which the technologies used would be very different from the current ones. They are therefore highly uncertain. In those scenarios, decarbonisation can be seen as a driver itself which triggers *even greater* and *more radical* technological change. The social aspects will therefore play a major role, since the adoption of very different technologies implies the necessity of social acceptance and adaptation to the new technology. Therefore, the LCS are more uncertain than normal scenarios, which do not require socio-technical changes of the same magnitude.

Hughes and Strachan (2010) divide the LCS into three categories: trend-driven studies (which are analogous to storylines), technical feasibility studies (which are prominent to low-carbon studies and usually use a backward approach where the technologies are selected according to the predefined emission targets) and modelling studies (which are the most complete kind of studies where the interactions between the different drivers and actors are modelled). The SRES scenarios can be classified as modelling studies since the IAM are used to create them. They are, however, previously based on the storylines and thorough discussion of the available technologies.

The LCS are built by using strong assumptions or exogenous constraints which *force* the scenarios to become low-emission ones. The strong assumptions are required since the technological difference is radical and normal technological development would most probably lead to future scenarios with high emissions. In the case of technical feasibility studies, it is common to set a primary exogenous constraint of emission reduction, e.g. −60% of emissions by 2050. Also, some technological assumptions can be applied such as no nuclear energy by a particular date. Technical feasibility studies do not usually include the economics of the suggested technological changes. In the modelling studies, similar constraints can be applied. Here, the different economic sectors are more integrated, the technological detail of the drivers can vary (they are usually less detailed than technical feasibility studies) and the socio-economic dynamics are modelled with different degrees of detail.

The SRES scenarios are based on four storylines which do not aim to radically change technologies. The SRES scenarios were aimed to provide a basis that could be used to identify and project driving forces of GHG emissions. They are also useful as baseline scenarios to compare with LCS scenarios. For the LCS scenarios, each SRES scenario has to be further developed to include a standard set of policies and measures designed to meet a climate target. The comparison of the results helps to identify a robust set of policies to the projected changing states of the world.

3.2.4. *Methodological issues in the SRES scenarios*

One of the main methodological issues is in the relationships between the main driving forces, which are assumed to follow the IPAT model. The IPAT model is an additive model which does not explicitly explain the relationships between the different driving forces (population, affluence and technology). The IPAT model implies independence between the different drivers. However, it has been demonstrated that, for example, capital and labour productivity (which are related to affluence) are not independent from technological change (Abramovitz, 1993).

Additionally, there are three major inter-related issues with using storylines to create scenarios (Hughes and Strachan, 2010). Firstly, there is the circular logic of trend-based scenarios, secondly the lack of explicit actor characterisation and finally the lack of co-evolution of the different drivers.

The problem of using storylines as a means to articulate and quantify scenarios is the storyline itself. The scenario is justified *a priori* by the storyline, thereby creating a self-justifying cyclical logic. However, there is no clear identification of causes and effects, and therefore the scenario does not provide information about the causes of its development. It is thus not helpful for devising policies since the policy makers cannot identify what is causing the change. The authors of the SRES scenarios acknowledge this, and justify the choice of this cyclical logic by the necessity of representing uncertainty.

Secondly, the SRES characterise the different drivers with quantifiable factors, but they are not associated with specific actors. The storyline vaguely mentions actors such as communities, governments and firms, but does not associate specific responsibilities to each of them. This is a collateral problem to that of the cyclical logic, as the scenario seems justified, but actually it evades the question of who makes the scenario develop as it is planned to happen.

Finally, there is no linkage or effective interaction between the drivers since the "interaction" is based on an IPAT model. In technical terms, all drivers are exogenous, i.e. they do not depend on the evolution of the storyline. A more realistic approach would be to endogenise the drivers by making them evolve according to specific developments of the scenario. However, the literature has not reached consensus regarding most of those interactions. The SRES scenarios acknowledge this knowledge gap during the review of the different drivers. Again, the use of storylines was a practical solution (although not scientifically rigorous) to justify the scenario evolution. So, as a result of the storylines, the SRES scenarios use generic trends for the co-evolution of some drivers. For example, it has been statistically demonstrated that there is a negative correlation between fertility rates and per capita income. Therefore, for example, due to the convergence of economic development between countries in the A1 scenario, the global population grows to some nine billion by 2050 and declines to about seven billion by 2100, implying an increased age average. Although such reasoning seems plausible, there is no specific and explicit scientific reason for this demographic development.

In short, the SRES scenarios are a complex modelling exercise but still have some methodological weaknesses, mainly due to lack of interaction between the drivers of change. The scenarios represent four different socio-economic approaches which oppose economic emphasis to environmental emphasis, and global-based development to local-based development. However, an economic focus does not exclude an environmental one, or *vice versa*. They could even be complementary. The four SRES storylines are therefore a polarised and simplistic way to describe society. It is, however, understandable, given that the aim is to provide a broad accepted basis on which further analysis can be developed.

3.2.5. *Disagreements with specific assumptions of the SRES scenarios*

Building scenarios implies having to describe the world in numbers, and therefore there can be disagreement as to the values or the assumptions that led to those values. Even the SRES scenarios, which have been built with a consensual methodology, have raised criticism regarding the assumptions used.

For instance, Raupach *et al.* (2007) criticise the values of the projected emissions because during the period 2000–2007, real emission grew by 3.5% per year, which is greater than the scenario producing the most emissions — the fuel-intensive one (A1FI). If the trend continues, it would surpass all SRES scenarios. This position contrasts with van Vuuren and O'Neill (2006), who argue that SRES scenarios are consistent with current literature and trends and there is no need for new ones.

The use of market exchange rates (MER) instead of the purchasing power parity (PPP) was denounced by Castles and Henderson (2003). The distribution of the emissions seems to be different depending whether MER or PPP prices are used (Tol, 2004). However, the final difference in temperature has been computed to be minimal despite such differences (Manne *et al.*, 2005).

3.2.6. *The new AR5 scenarios*

The new IPCC Fifth Assessment (AR5) scenarios started to be developed by Moss *et al.* (2008). Van Vuuren *et al.* (2010b) suggest methodological improvements such as "downscaling", which is usually required in IAM exercises to bring the emission variables specified at large spatial scales at a country or grid level. Kriegler *et al.* (2010) suggest the shared socio-economic pathways (SSPs) as a way to systematise the underlying narrative of the scenarios while providing a harmonised base for comparison of socio-economic developments.

Moss *et al.* (2008) address the problem of circular justification of the storylines by devising a new methodology to build scenarios. It uses the concept of representative concentration pathway (RCP), which are a set of selected "benchmark emission scenarios". Those are released with very basic characteristics (emissions, concentrations and land-use change only) to climate modellers which create a climate scenario. Simultaneously, IAMs and vulnerability, impact and adaptation modellers develop new socio-economic variables consistent with the RCP. Finally, all previous work is merged to form the scenario.

This methodology allows a closer interrelation between the three different modelling communities. It therefore creates a more consistent and comparable scenario with more complex links between variables. Additionally, it decouples the RCP from the storyline, i.e. the same atmospheric concentration of GHG can be achieved through many different social, economic and technological paths. Van Vuuren *et al.* (2010a) identify two future challenges of climate change assessments. The first is to model mitigation costs as a function of the assumptions and climate policies. The second is to reproduce the same RCP with different IAMs.

In the AR5 *Summary for Policymakers* from a physical science working group approved by governments in September 2013, the four RCPs adopted for the AR5 are described (IPCC, 2013, in particular Box SPM1). The RCPs are defined by their radiative forcing in the year 2100 compared to 1750 in watts per square metre. Radiative forcing is the change in the global energy balance caused by global warming at the top of the atmosphere. The four are RCP2.6, a mitigation scenario originally from the IMAGE IAM, stabilisation scenarios RCP4.5 from the GCAM model and RCP6.0 from the AIM model and a very-high-emission scenario, RCP8.5 from the MESSAGE model. The RCPs in the report are based on a combination IAMs, simple climate models, atmospheric chemistry and global carbon-cycle models. The AR5 report states that "mitigation actions starting now in the various RCP scenarios do not produce discernibly different climate change outcomes for the next 30 years or so, while long-term climate change after mid-century is appreciable different across the RCPs". This statement justifies the convention that the reference and mitigation scenarios in the literature do not include the effects of climate change, a convention followed in the scenarios in this book.

However, the scenarios underlying each individual RCP do not necessarily provide an internally consistent description of population, income, energy, land use, other relevant driving forces and emissions (van Vuuren *et al.*, 2010a). Since each RCP and scenario originates from a different modelling team and comes from a specific study with its own research questions and assumptions on future developments, there is no reason why the set as a whole would necessarily be representative of the literature for parameters or internal logic, other than that of radiative forcing described in the RCP.

3.3. The IEA Scenarios

3.3.1. *The World Energy Outlook 2008, 2009 and 2010*

This section reviews the scenarios from the World Energy Outlook (WEO) 2008, 2009 and 2010, which form the starting point for the scenarios developed for this book. The simulations and projections of WEO are derived from a large-scale computable model named the World Energy Model. It is designed to replicate how energy markets function. It is used as a major tool to generate the sector-by-sector and fuel-by-fuel projections by region or country for all the scenarios in the outlook series. The model comprises five main modules: final energy demand, power generation, refining and other transformation, investment and CO_2 emissions and fossil fuel and biofuels supply, all of which are modelled on a country or regional basis. It is updated on a yearly basis.

The WEO 2008 considers two types of scenarios, namely, Reference Scenario and Climate Policy Scenarios over the period up to 2030. The Climate Policy Scenarios are further specified as 550 Policy Scenario and 450 Policy Scenario. The Reference Scenario takes into consideration all government policies and measures that were enacted or adopted by the middle of 2008, including those that have not yet been fully implemented. It assumes the implementation of currently adopted energy and environmental policies during the projection period to vary by fuel and by country. Pricing and other market reforms are also assumed to be implemented at different rates. The 550 Policy Scenario and 450 Policy Scenario correspond to long-term stabilisation of GHG concentration at 550 and 450 ppm CO_{2eq} respectively. Each scenario combines three climate policy mechanisms, i.e., cap-and-trade, sectoral agreements, and national policies and measures, while the mix of approaches differs between three groups of countries categorised by the outlook, i.e., "OECD+" (Organisation for Economic Co-operation and Development) countries, "other major economies" and "other countries".

The WEO 2009 considers two scenarios, namely, the Reference Scenario and the 450 Policy Scenario, which are similar to the WEO 2008 Reference Scenario and 450 Policy Scenario. The projection period again runs to 2030. The 550 Policy Scenario was not described in detail, as the Outlook was convinced that the 450 Policy Scenario was gaining widespread support around the world. The Reference Scenario of WEO 2009 describes what would happen if governments were to take no new initiatives bearing on the energy sector beyond those already adopted by mid-2009. The 450 Scenario builds on a "hybrid" climate policy framework introduced in WEO 2008. It reflects a plausible set of policies that could emerge — a realistic combination of a cap-and-trade system, sectoral agreements and national policies tailored to each country's circumstances. The background of scenarios produced is:

1. The global financial crisis and ensuing recession have had a dramatic impact on the outlook for energy markets. World energy demand in aggregate has already plunged.
2. Countries have responded to the financial crisis with prompt and coordinated fiscal and monetary stimuli, which include measures to promote clean energy in many cases.
3. Major economies have adopted new measures in response to and in pursuit of energy and environmental policies, which include new Corporate Average Fuel Economy (CAFE) standards by the US, the EU 20-20-20 package, Japanese photovoltaic (PV) subsidy and feed-in tariffs for households, and China's Golden Sun Programme, feed-in tariffs for wind power plants and the nuclear programme.

The WEO 2010 projections of energy demand and supply, related CO_2 emissions and investment requirements are extended from 2030 to 2035. Three scenarios, namely, the Current Policies Scenario, the New Policies Scenario and the 450 Scenario are presented in the Outlook, differentiated by the underlying assumptions about government policies. The background of WEO 2010 is:

1. The Copenhagen Accord — the UN negotiations on climate change in December 2009 in Copenhagen; to set a non-binding objective of limiting the increase in global temperature to 2°C above pre-industrial levels; to establish a goal for the industrialised countries to mobilise funding for climate mitigation and adaptation in developing countries of US$100 billion per year by 2020; to ensure the industrialised countries (Annex I parties) set emission targets for 2020.
2. The Pittsburgh Commitment — the G20 summit in Pittsburgh in September 2009 for international economic cooperation; to rationalise and phase-out over the medium term inefficient fossil fuel subsidies that encourage wasteful consumption.

The Current Policies Scenario, i.e., the Reference Scenario, took into consideration only those policies that had been formally adopted by mid-2010. The New Policies Scenario took account of the broad policy commitments and plans that have been announced by countries around the world, to tackle either environmental or energy security concerns corresponding to a cautious interpretation and implementation. Specifically, the New Policies Scenario takes into account all policies and measures included in the Current Policies Scenario, as well as:

1. Cautious implementation of the Copenhagen Accord commitments by 2020.
2. Continuation of the EU Emissions Trading Scheme, and introduction of a cap-and-trade system in the rest of the OECD+ after 2020.
3. Phase-out of fossil fuel consumption subsidies in all net-importing regions by 2020 (and in net-exporting regions where specific policies have already been introduced).

4. Extension of nuclear plant lifetimes from five to ten years with respect to the Current Policies Scenario, on a plant-by-plant basis.
5. For 2020–2035, additional measures that maintain the pace of the global decline in carbon intensity.

The 450 Scenario sets out an energy pathway to implement the high end of national pledges and stronger policies after 2020, including the near-universal removal of fossil fuel consumption subsidies. Specifically, the 450 Scenario takes into account all policies and measures included in the New Policies Scenario, plus:

1. Implementation by 2020 of the high end of the range of the Copenhagen Accord commitments.
2. Cap-and-trade systems in the power and industry sectors, from 2013 in OECD+ countries and after 2020 in other major economies (OME).
3. International sectoral agreements for the iron and steel, and the cement industries. International agreements on fuel-economy standards for passenger light-duty vehicles (PLDVs), aviation and shipping.
4. National policies and measures, such as efficiency standards for buildings and labelling of appliances.
5. The complete phase-out of fossil fuel consumption subsidies in all net-importing regions by 2020 (at the latest) and in all net-exporting regions by 2035 (at the latest), except for the Middle East where it is assumed that the average subsidisation rate declines to 20% by 2035.

The WEO 2010 is based on the following five major non-policy assumptions:

1. **Population.** The assumptions about population are the same in each scenario. From WEO 2010, world population is assumed to expand from an estimated 6.7 billion in 2008 to 8.5 billion in 2035, from 1.1% per year in 2008–2020 to 0.7% in 2020–2035.
2. **Economic growth.** The assumptions about economic growth are the same in each scenario. The WEO 2010 assumes that world GDP (gross domestic product) grows on average by 4.4% over the five years to 2015 and 3.1% per year in the period of 2015–2035.

3. **Energy prices.** In the New Policies Scenario of WEO 2010, the IEA crude oil import price is assumed to rise steadily to US$99/barrel in 2020 and US$113 in 2035 in real terms (in nominal terms, prices more than double to US$204/barrel in 2035), while the price rises more rapidly in the Current Policies Scenario (US$110 by 2020 and US$135 by 2035 in real terms) and more slowly in the 450 Scenario (US$90 from 2020 upwards in real terms). Natural gas prices are assumed to remain low relative to oil prices in all scenarios and coal prices rise much less than oil and gas prices, and fall in the 450 Scenario.
4. **CO_2 prices.** In the Current Policies Scenario, carbon pricing is assumed to be limited to EU countries and to New Zealand. The price of CO_2 is projected to reach US$30/tonne in 2020 and US$42/tonne in 2035. Carbon pricing is assumed to be adopted in other regions in the New Policies and 450 Scenarios.
5. **Technology.** In all three scenarios, the performance of currently available categories of technology is assumed to improve, but the pace of improvement varies. It is fastest in the 450 Scenario and slowest in the Current Policies Scenario. A number of major new technologies are assumed to be deployed at various points over the projection period.

3.3.2. *Energy Technology Perspectives 2010*

The IEA's Energy Technology Perspectives 2010 (IEA, 2010) considers two primary scenarios, the Baseline Scenario and the BLUE Map Scenario. In addition, a number of variants of the BLUE Map Scenario are used to assess the impact of individual assumptions. The Baseline Scenario follows the WEO 2009 (WEO, 2009) Reference Scenario for the period 2007–2030 and updates the earlier ETP 2008 analysis for the period 2030–2050. It has the following key features:

- The world economy grows by 3.1% a year on average between 2007 and 2050.
- CO_2 emissions increase from 29 Gt CO_2 in 2007, to 40 Gt CO_2 by 2030, and to 57 Gt CO_2 in 2050 — an average increase of 1.6% a year.

Table 3.2. WEO 2010 scenarios projections in energy-related $CO_{2\text{-eq}}$ emission, global primary energy demand and fossil fuels as proportions of the world primary energy mix.

	Energy-related CO_2 emission	Global primary energy demand	Fossil fuels as proportions of the world primary energy mix
Current Policies Scenario	Energy-related CO_2 emissions rise from 29.3 Gt in 2008 to 42.6 Gt in 2035, an increase of 45%	World primary energy demand increases by 47% from 2008 to 2035 to 18,050 Mtoe — an average annual rate of growth of 1.7%	Fossil fuels account for 79% of the world's primary energy mix in 2035.
New Policies Scenario	Energy-related CO_2 emissions rise from 29.3 Gt in 2008 to 35.4 Gt in 2035, an increase of 21%	World primary energy demand increases by 36% from 2008 to 2035, rising from 12,300 Mtoe to 16,750 Mtoe from an average of 1.4% per year in the period 2008–2020 to 0.9% per year in 2020–2035	Fossil fuels account for 74% of the world's primary energy mix in 2035.
450 Scenario	Energy-related CO_2 emissions reach 31.9 Gt CO_2 in 2020 and fall to 21.7 Gt CO_2 in 2035	World primary energy demand increases by 22% from 2008 to 2035 to 14,920 Mtoe — an average annual rate of growth of 0.8%	Fossil fuels account for 63% of the world's primary energy mix in 2035.

- Primary energy use rises by 84% between 2007 and 2050, with carbon intensity of primary energy increasing by 7%.

The BLUE Map Scenario and its variants were developed to explore what potentially needs to be done, in terms of the introduction of existing and new low-carbon technologies at the lowest possible cost, to reduce global energy-related CO_2 emissions to half their 2005 level by 2050. The BLUE Map Scenario follows the WEO 2009 450 Scenario to 2030 and projects it out further to 2050. The BLUE Scenarios are based on the same macro-economic assumptions as the Baseline Scenario and a set of optimistic but plausible technology assumptions, with the following key features:

- Global CO_2 emissions are reduced to 14 Gt CO_2 in 2050. The emission profile peaks between 2015 and 2020 at just below 31 Gt CO_2, consistent with the long-term global rise in temperatures of 2 to 3°C, assuming similar reductions in other GHG emissions. OECD countries account for just over 30% of the total global emission reduction in 2050 and non-OECD countries reduce their emission by 24%.
- Global emission reduction of 37%, 32%, 17% and 14% by 2050 are accounted for by the transport, power, industry and buildings sectors, respectively. Across all sectors, 38%, 19%, 17% and 6% of global emission reduction is accounted for by end-use efficiency gains, carbon capture and sequestration (CCS), renewable energy and nuclear energy respectively.
- The rate of energy efficiency improvement is 1.5% per year. Electricity demand in 2050 is 13% lower than in the Baseline Scenario. The share of power generation from coal in 2050 declines from 44% in the Baseline Scenario to 12%, while the share from gas declines from 23% to 15%. The share from nuclear and renewables increases from 10% to 24% and from 22% to 48% respectively. In addition, 90% of coal-fired and 30% of gas-fired power stations are fitted with CCS by 2050.
- Results from the BLUE Map Scenario estimate costs of up to US$175/t$CO_2$ saved to reach the target 14 Gt CO_2 by 2050.
- The emission reductions envisaged in the BLUE Map Scenario are driven by increased support for R&D (research and development), demonstration and deployment programmes, CO_2 reduction incentives and policy instruments to overcome other commercialisation barriers.

3.3.3. *Criticisms of the IEA scenarios*

Nelder (2008) criticises the credibility of IEA's WEO 2008 scenarios. He argues that "Adding another 64 mbpd [million barrels per day] of oil production capacity from the existing, very well explored, and very well exploited resource base — a 74% increase over current levels seems quite impossible". Then, Nelder (2009) analyses the IEA's WEO 2009 and argues that the design of the 450 Scenario was politically distorted and not done for the climate change reasons. His analysis of the agency's data shows that uncertainty, specifically about the potential for oil recovery, new discoveries and the extraction of unconventional sources, has repeatedly been used to put an optimistic spin on the reports. He also alleges that there were also a few disturbing discontinuities along the way in which report summaries do not quite match up with the details in the later chapters, and the edges between the solid data and the grey zone gradually became sharper. As for the 450 Scenario, he believes that climate change is a backwards approach to the problem. He wonders how the world can even get to 450 ppm when peak oil, and then peak gas and peak coal, are properly considered. He also criticised that the IEA WEO 2010 makes unconvincing assumptions, publishes data without giving details and projections into the future without fully considering the uncertainty and noise.

Apart from these criticisms, there are some other weaknesses in the modelling reported in the IEA Annual Energy Outlook series, which may impact on the scenarios' credibility.

1. The models used by the outlook are not clearly and fully illustrated.
2. The data associated with the modelling process is not comprehensive and properly provided to the public.
3. The modelling results can't be reproduced and are unjustifiable, partly because of the two reasons illustrated above.

Additionally, the IEA does not model non-energy emissions which accounts for about one-third of total GHG emissions. Several important emission categories such as non-CO_2 emissions, emissions from land-use and land-cover change are exogenous in IEA scenarios. The IEA (and the IPCC) admits that projections of non-energy emission and degree of radioactive forcing of the non-CO_2 GHG are subject to awide range of uncertainty and the availability of the related data in this domain is weak.

3.4. Previous E3MG Scenarios

E3MG is an energy-environment-economy model at the global level (Barker *et al.*, 2006; Barker and Scrieciu, 2010). The model has been designed to assess options for climate and energy policies and to allow for E3 interactions (Barker *et al.*, 2006, 2008) (see Chapter 4 for a description). In the later research (Barker *et al.*, 2012), E3MG is used to assess the policies and measures catalogued from official national and international plans by the IEA in its 2010 WEO (WEO, 2010) as an interpretation of what is adopted by governments to put world GHG emissions by 2020 on a pathway to achieve the long-term climate targets. The E3MG is used to model the effects of strengthening policies to close the gap between those in place and those necessary to achieve the targets. The modelling undertaken in the study adopted three scenarios, namely, the Reference Scenario, the 450 Scenario and Medium 2°C Scenario.

The Reference and the 450 Scenarios were designed mainly based on WEO 2010's Current Policies and 450 Scenarios. The Reference Scenario was intended to represent the outcome of adopting current policies, but without the extra efforts promised in the Cancun Agreement to reduce GHG emissions by 2020. The 450 Scenario includes the policies for reducing GHG emissions assumed in the WEO's 450 Scenario, which can be interpreted as the outcome of the Cancun Agreement at the high end of the range of commitments to reduce emissions. The Medium 2°C Scenario assumes a strengthening of these commitments to reach a level of emissions by 2020 with a medium (50–66%) chance of climate stabilisation. The higher level of investment is designed to accelerate recovery towards full employment in those economies most affected by the Great Recession. In both policy scenarios, the CDM (clean development mechanism) plays a role in transferring funding to developing countries for climate change mitigation.

The 450 and Medium 2°C policy cases are alternative fully dynamic solutions, including the sectoral effects on GHG emissions and energy use, year by year from 2013 to 2020, of all the extra policies and measures introduced in the WEO's 450 Scenario in WEO 2010, strengthened even further in the Medium 2°C Scenario to achieve a reduction in global GHG emissions of 15% below reference case by 2020. Another difference between the WEO's 450 Scenario and the Medium 2°C Scenario is that the climate stabilisation target is reached in 2100 instead of 2150, which is the target year for the IEA's 450 Scenario. The additional annual investment for the post-2012 period 2013–2020 is US$395 bn (2009) in the 450 Scenario and 20% higher at about US$500 bn (2009) prices in the Medium 2°C Scenario respectively. The investment is assumed for every year 2013–2020 and is related to the various policies and measures assumed for the different countries and regions.

To model low climate stabilisation by 2100 with E3MG, Barker and Scrieciu (2010) developed two climate policy stabilisation scenarios, i.e., which refer to a "550 ppm" and a "400 ppm" $CO_{2\text{-eq}}$ stabilisation levels, respectively. The research was done within the EU project ADAM (Adaptation and Mitigation Strategies — Supporting European Climate Policy). As for Baseline Scenario, the ADAM baseline (van Vuuren *et al.*, 2009) has been taken as a starting point for both. A sophisticated method has been developed to calibrate the baseline GDP components (on the demand side) of E3MG over the projection period. An endogenous version of the baseline is then generated, reproducing the calibrated solution to allow for full interaction between the main economic and energy variables. Policies are then applied, which alter relative prices and hence change demand away from the baseline (see Köhler *et al.* (2006) and Barker *et al.* (2006) for a detailed discussion). Both "550 ppm" and "400 ppm" scenarios predominately considered a series of policies on carbon pricing through emissions permit trading and carbon taxation and fiscal incentive through revenue recycling from auctioning permits and implementing carbon taxes. They largely differ in their assumptions on the strength of regulation promoting electric vehicles and the extent of fiscal policies supporting energy efficiency and low-carbon production methods for industry and household end-users. Moreover, with regard to the "400 ppm" scenario, two further sub-scenarios are considered based on the option of its specific policies associated with.

E3MG was also used to assess the feasibility of a 50% reduction in CO_2 emissions by 2050 (Barker *et al.* 2008). Two scenarios, namely carbon price and carbon-plus scenarios, were designed apart from the Baseline Scenario. The carbon price scenario has a global real (year 2000) carbon price of US$2.5/t$CO_2$ in 2011, rising by US$2.5 per year to US$100/tCO_2 in 2050 and assumes that all revenues are recycled through reductions in indirect taxes. The carbon-plus is a scenario for achieving the low climate stabilisation target of a 50% global reduction in CO_2 emissions by 2050 in line with a trajectory for achieving GHG concentrations at a level of approximately 500 ppm $CO_{2\text{-eq}}$. It was constructed from a set of incentives to encourage the deployment of low-carbon technologies in addition to the carbon price. These include additional incentives for electricity, transport and industrial end-use, for efficiency of dwellings and accelerated permit price increase to US$100/t$CO_2$ by 2030 instead of 2050. The scenarios described suggest that achieving a global pathway to a low-carbon economy is feasible at modest carbon prices, but only by applying specific technology deployment policies in addition to the carbon price rising to US$100/t$CO_2$. Those policies include revenues recycling to support investment in low-GHG technologies in energy, manufacturing and transportation, and "no-regrets" options for buildings.

In the E3MG scenarios, CO_2 emissions from land-use changes are incorporated only as an exogenous input with projections taken from another model. The feedbacks from vulnerability, impact and adaptation modelling are not considered in these models' reference and policy scenarios. The IEA and E3MG models are not climate models, so they cannot calculate the feedback effects on climate of the emissions and land-use change they simulate.

3.5. Scenarios Design for Modelling a Low-carbon Economy

3.5.1. *Learning from past scenarios*

In this section, we refer back to the criticisms of the SRES (Section 3.2.4) and explain how our methodology for using scenarios in E3MG has been developed for this book. Drawing on the criticisms in the literature review, the following procedure is suggested for the development of scenarios, with an account of what we have been able to achieve using E3MG:

1. Identify both causes and effects in the climate change domain, i.e., key driving forces and actors (e.g. governments and household), the linkage between the driving forces, and associations between the driving forces and actors, based on which the policy makers are able to identify the drivers causing the climate change and devise policies. Moreover, the scenarios should be developed based on the specific modelling approach and simulation using historical datasets. E3MG includes the main driving forces and main national policy makers. In addition the relationships are based on annual historical data, e.g. levels of population by region and age group affecting consumers expenditure.
2. Address uncertainty issues by using values' spectrum, a range of driving force projections and scenarios. Therefore the results generated by the modelling also should be values' spectrum or probabilities. This is an ambition, not yet fully realised in E3MG applications.
3. A comprehensive emission scenarios development with considering and simulating energy emission, no-energy emissions, CO_2 emissions and other GHG emissions. E3MG is comprehensive enough to include these features.
4. A smooth integration with adaptation models and climate models by simulating the feedback loops between the integrated assessment model and adaptation models, climate models. Again this is at the forefront of research and is only partly realised.

3.5.2. *The driving forces*

The E3MG emission scenarios take into consideration the following driving forces:

- **Population.** Population growth is one of important driving forces in climate change. The estimates are based on the UN projections

(United Nations, 2010). The major differences in the size of populations over the long run are from the assumption variations in fertility, which are constant, low, medium and high; E3MG takes the medium assumption.

- **Economic and social development.** Full measures of economic and social development are difficult to determine, which comprises many dimensions such as politics, economy, social equity and environmental sustainability. GDP has been a dominant measure of national well-being although with many shortcomings. The UN has developed the Human Development Index to address its issues by combining three factors: income as measured by real GDP per capita at PPP, longevity and educational attainment. E3MG treats GDP by region as being solved within the model. However the solution starts from projected pathways for GDP and its components that are consistent with past trends and allow for slower growth of the labour force in the future. The model also adopts the MER measure. We test the robustness of policies for decarbonisation by including an alternative low-growth reference case.
- **Technical development.** Technological innovation, different rates of technology change and the rates of deployment of new technologies have a major impact on decarbonisation and economic growth. However, there also exist large scientific uncertainties, such as possible technological discontinuities. E3MG includes an indicator for technological change as accumulated gross investment enhanced by R&D expenditures, all in constant prices. This indicator is introduced into many relationships in the model, but the more important effects are for more progress to raise export demand and export prices. Chapter 6 considers a number of major new technologies that are approaching commercialisation and are assumed to be deployed at various points over the projection period both for energy supply and end-use. These include carbon capture and storage, renewable energy, electric vehicles and smart grid systems. In contrast to fossil fuel technologies, renewable energy technologies have more potential for considerable cost reductions.

- **Energy and resources.** Total energy demand, final energy intensity, primary energy production, remaining resources in coal, oil and gas, share of renewable in primary energy, energy prices (predominately oil price) and electricity are major determinants in achieving the future GHG emission reduction target. These are modelled in E3MG via energy demand equations, with supplies of energy resources largely projected by assumptions. The oil price assumptions and responses are closely related to those in the IEA's 2010 WEO and are discussed further in Chapters 4 and 5.
- **Agriculture, land use and land cover.** Most changes in land use are induced by the demand for cropland and grassland, which is driven by the demand for food products, the extent of biomass energy use, and policies and practices associated with forest management (IPCC, 2007a). From the literature and various models simulation results, there is a broad agreement that biomass provides the largest potential for emission reduction, and followed by afforestation or reduced deforestation. A subgroup of the IPCC scenarios have taken land-use and land-cover change into account.

The emission pathways for the same temperature target would be different with different assumptions about land use. Rose *et al.* (2011) describe the assumptions and results from four models included in the EMF21 comparison of stabilisation scenarios. In the baselines of the models the general pattern is for continued deforestation in developing regions, so that emissions rise initially, reach a maximum, and finally decline with depletion of forestland that can be cleared because of a drop in population growth and increases in agricultural productivity. In the longer term, land-use changes can become an increasing sink for CO_2 through afforestation. With carbon pricing and schemes such as Reduced Emissions from Deforestation and Forest Degradation (REDD), land-use changes with biomass energy can make a substantial contribution to reducing climate change. In E3MG the contributions are treated as largely exogenous, with a reference table for different carbon prices allowing for prospective effects of REDD by country to 2050.

3.6. E3MG Emission Scenarios for this Book

Development pathway uncertainty (i.e., the fact that future emissions and the evolution of their underlying driving forces are highly uncertain) includes alternative demographic, economic and technological developments that lead to high, intermediary or low emissions of GHG and hence magnitude of future climate change. In climate change research, scenarios are designed to represent major driving forces, processes, impacts and potential responses important for informing climate change policy (Moss *et al.*, 2010).

In this chapter, scenarios are considered with uncertainty and classified into three categories, i.e., Reference Scenarios, Target Scenarios and Policy Scenarios. Reference Scenarios or Baseline Scenarios reflect BAU trends. Target Scenarios set out a pathway to achieve certain targets either in a short or long term such as a reduction in CO_2 emissions or the long-term stabilisation of GHG concentration at 550 and 450 ppm of CO_{2eq} respectively. Policy Scenarios may take account of the broad policy commitments and plans that have been or potentially will be announced by countries around the world to view the development pathways, or take account of a number of variants of policies to assess the impact of individual assumptions such as CCS. The level and types of the policies applied in Target Scenarios and Policy Scenarios are compared with distinct Reference Scenarios (as Figure 3.2 shows below).

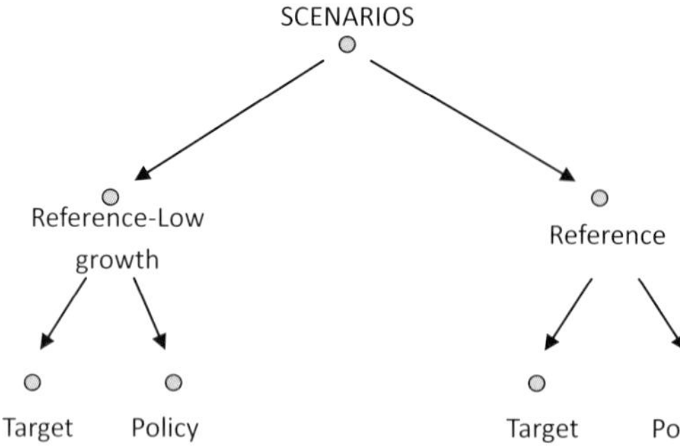

Figure 3.2. An emission scenarios tree for climate change modelling.

3.6.1. *The Reference Scenario*

This is consistent with the New Policies Scenario in WEO 2010. It takes account of the broad policy commitments and plans that have been announced by countries around the world by 2010, to tackle climate change-related issues at a global scale corresponding to a cautious interpretation and implementation, which include the Copenhagen Accord commitments, EU Emissions Trading Scheme, phase-out of fossil fuel consumption subsidies and other commitments.

3.6.2. *The decarbonisation scenario*

This scenario represents an aggressive promotion of low-carbon technologies into the market. It assumes that policies will be implemented to achieve the 2°C internationally agreed target by 2100. This is assumed to be sufficient, if continued, such that the emission levels will return to the pre-industry level by the 22nd century. The pre-industrial level was when the global atmospheric concentration of carbon dioxide, i.e. about 280 ppm (IPCC, 2007b). It develops a pathway toward stabilisation of zero GHG emissions from anthropogenic sources by 2100. Stringent polices are adopted applying a mix of renewable technologies combined with demand-side efficiency changes. For example, to underline the significant potential future role in cutting GHG emissions and powering sustainable development, in one of the IPCC scenarios (IPCC, 2011; Teske *et al.*, 2011) it is assumed that 40% in 2030 and close to 80% of the world's energy supply could be met by renewable by 2050 if backed by the right enabling public policies, with 99% renewable at the end of the century by combining different variable renewable sources. This is the main decarbonisation scenario (to 2050) considered in Chapters 4 and 5.

3.6.3. *The policies evaluation scenarios*

A number of variants of policies or single policy scenarios to assess the impact of individual assumptions, such as universal carbon taxes which may not be feasible, in order to assess the effects and compare them with other results. They may also help to evaluate the impact on climate change in the short or long term of an individual global or regional policy such as financial support for developing countries relating to green growth or implementation of an emissions trading scheme in China.

References

Abramovitz, M. (1993). The Search for the Sources of Growth: Areas of Ignorance, Old and New, *Journal of Economic History*, 52, 217–243.

Barker, T., Foxon, T. and Scrieciu, Ş. (2008). Achieving the G8 50% Target: Modelling Induced and Accelerated Technological Change Using the Macro-econometric Model E3MG, *Climate Policy Special Issue on Modelling Long-term Scenarios for Low-carbon Societies*, 8, S30–S45.

Barker, T., Pan, H., Köhler, J., Warren, R. and Winne, S. (2006). Decarbonising the Global Economy with Induced Technological Change: Scenarios to 2100 using E3MG, *The Energy Journal*, 27, 143–160.

Barker, T. and Scrieciu, S. (2010). Modeling Low Stabilization with E3MG: Towards a "New Economics" Approach to Simulating Energy-Environment-Economy System Dynamics, *Energy Journal Special Issue on The Economics of Low Stabilisation*, 137–164.

Carter, T., Parry, M., Harasawa, H. and Nishioka, S. (1994). *IPCC Technical Guidelines for Assessing Climate Change Impacts and Adaptations with a Summary for Policy Makers and a Technical Summary*, Cambridge University Press, Cambridge.

Castles, I. and Henderson, D. (2003). The IPCC Emission Scenarios: An Economic-statistical Critique, *Energy & Environment*, 14, 159–185.

De Vries, B., Bollen, J., Bouwman, L., den Elzen, M., Janssen, M. and Kreileman, M. (2000). Greenhouse Gas Emissions in an Equity-, Environment- and Service-oriented World: An IMAGE-based Scenario for the Next Century, *Technological Forecasting & Social Change*, 63, 2–3.

EEA (1999). Environmental Indicators: Typology and DPSIR Framework Overview. European Environment Agency, Technical Report 25.

Girod, B., Wiek, A., Mieg, H. and Hulme, M. (2009). The Evolution of the IPCC's Emissions Scenarios, *Environmental Science and Policy*, 12, 103–118.

Hughes, N. and Strachan, N. (2010). Methodological Review of UK and International Low Carbon Scenarios, *Energy Policy*, 38, 6056–6065.

IEA (2010). Scenarios & Strategies to 2050. International Energy Agency, Energy and Technology Perspectives Series, Paris, France.

IPCC (2001). *Climate Change 2001: Mitigation. Contribution of Working Group III to the Third Assessment Report of the Intergovernmental Panel on Climate Change*, Metz, B., Davidson, O., Swart, R. and Pan, J. (eds), Cambridge University Press, Cambridge.

IPCC (2007a). 'Summary for Policymakers', *Climate Change 2007: Mitigation. Contribution of Working Group III to the Fourth Assessment Report of the Intergovernmental Panel on Climate Change*, Metz, B., Davidson, O. R., Bosch, P. R., Dave, R. and Meyer, L. A. (eds), Cambridge University Press, Cambridge, pp. 1–23.

IPCC (2007b). 'Summary for Policymakers', *Climate Change 2007: The Physical Science Basis. Contribution of Working Group I to the Fourth Assessment Report of the Intergovernmental Panel on Climate Change*, Solomon, S., Qin, D., Manning, M., Chen, Z., Marquis, M., Averyt, K., Tignor, M and Miller, H. (eds), Cambridge University Press, Cambridge, pp. 1–18.

IPCC (2007c) 'New Assessment Methods and the Characterisation of Future Conditions', *Climate Change 2007: Impacts, Adaptation and Vulnerability. Contribution of Working Group II to the Fourth Assessment Report of the Intergovernmental Panel on Climate Change*, Parry, M. L., Canziani, O. F., Palutikof, J. P., van der Linden, P. J., Hanson, C. E., Zinyowera, M. and Moss, R. (eds), Cambridge University Press, Cambridge, pp. 131–171.

IPCC (2007d). *Climate Change 2007: Mitigation. Contribution of Working Group III to the Fourth Assessment Report of the Intergovernmental Panel on Climate Change*, Metz, B., Davidson, O., Bosch, P. and Dave, R. (eds), Cambridge University Press, Cambridge.

IPCC (2011). Press Release: Potential of Renewable Energy Outlined in Report, 11th Session of Working Group III. Intergovernmental Panel on Climate Change. Available online at: http://www.ipcc.ch/news_and_events/docs/ipcc33/PRESS%20RELEASE%20Updated%20version%20-%20Potential%20of%20Renewable%20Energy%20Outline.pdf. Accessed 24 February 2014.

IPCC (2013). 'Summary for Policymakers', *Climate Change 2013: The Physical Science Basis. Contribution of Working Group I to the Fifth Assessment Report of the Intergovernmental Panel on Climate Change*, Stocker, T. F., Qin, D., Plattner, G.-K, Tignor, M., Allen, S. K., Boschung, J., Nauels, A., Xia, Y., Bex V. and Midgley P. M. (eds), Cambridge University Press, Cambridge, pp. 3–29.

Jiang, K., Masui, T., Morita, T. and Matsuoka, Y. (2000). Long-term GHG Emission Scenarios of Asia-Pacific and the World, *Technological Forecasting and Social Change*, 63, 2–3.

Köhler, J., Grubb, M., Popp, D. and Edenhofer, O. (2006). The Transition to Endogenous Technical Change in Climate-Economy Models: A Technical Overview to the Innovation Modeling Comparison Project, *The Energy Journal Special Issue on Endogenous Technological Change and the Economics of Atmospheric Stabilization*, 17–55.

Kriegler, E., O'Neill, B., Hallegatte, S., Kram, T., Lempert, R., Moss, R. and Wilbanks, T. (2010). Socio-economic Scenario Development for Climate Change Analysis, CIRED Working Paper DT/WP No 10.

Manne, A., Manne, S., Richels, R. and Edmonds, J. (2005). Market Exchange Rates Or Purchasing Power Parity: Does The Choice Make A Difference To The Climate Debate?, *Climatic Change*, 71, 1–8.

Moss, R., Babiker, M., Brinkman, S., Calvo, E, Carter, T., Edmonds, J. Elgizouli, I., Emori, S., Erda, L., Hibbard, K., Jones, R., Kainuma, M., Kelleher, J. Lamarque, J.F., Manning, M., Matthews, B., Meehl, J., Meyer, L.. Mitchell, J., Nakićenović, N., O'Neill, B., Pichs, R., Riahi, K., Rose, S., Runci, P., Stouffer, R., van Vuuren, D., Weyant, J., Wilbanks, T., van Ypersele, J.P. and Zurek, M. (2008). Towards New Scenarios for Analysis of Emissions, Climate Change, Impacts, and Response Strategies. Technical Summary, Intergovernmental Panel on Climate Change, Geneva, 25 pp. Available online at: http://www.ipcc.ch/pdf/supporting-material/expert-meeting-ts-scenarios.pdf. Accessed 11 April 2014.

Moss, R., Edmonds, J., Hibbard, K., Manning, M., Rose, S., van Vuuren, D., Carter, T., Emori, S., Kainuma, M., Kram, T., Meehl, G., Mitchell, J., Nakićenović, N., Riahi, K., Smith, S., Stouffer, R., Thomson, A., Weyant, J., Wilbanks, T. (2010). The Next Generation of Scenarios for Climate Change Research and Assessment, *Nature*, 463, 747–756.

Nakićenović, N., Alcamo, J., Davis, G., DeVries, B., Fenhann, J., Gaffin, S., Gregory, K., Gruebler, A., Jung, T., Kram, T., La Rovere, E., Michaelis, L., Mori, S., Morita, T., Pepper, W., Pitcher, H., Price, L., Riahi, K., Roehrl, A., Rogner, H., Sankovski, A., Schlesinger, M., Shukla, P., Smith, S., Swart, R., VanRooijen, S., Victor, N. and Dadi, Z. (2000). *Special Report on Emissions Scenarios: A Special Report of Working Group III of the Intergovernmental Panel on Climate Change*, Cambridge University Press, Cambridge, 27 pp.

Nelder, C. (2008). IEA Oil Report: "Time is Running Out" — Nothing Short of an Energy Revolution. Available online at: http://www.getreallist.com/iea-oil-report-time-is-running-out.html. Accessed 24 February 2014.

Nelder, C. (2009). Is the IEA World Energy Outlook Politically Distorted? — Panic and Penitence: The Evolution of the IEA. Available online at: http://www.getreallist.com/is-the-iea-world-energy-outlook-politically-distorted.html. Accessed 24 February 2014.

Raskin, P., Monks, F., Ribeiro, T., van Vuuren, D. and Zurek, M. (2005). 'Global Scenarios in Historical Perspective. Ecosystems and Human Well-Being: Scenarios', in Carpenter, S. R., Pingali, P. L., Bennett, E. M. and Zurek, M. B. (eds). *Findings of the Scenarios Working Group*, Island Press, Washington, DC, pp. 35–44.

Raupach, M., Marland, G., Ciais, P., Leuéré, C., Canadell, J., Klepper, G. and Field C. (2007). Global and Regional Drivers of Accelerating CO_2 Emissions, *Proceedings of the National Academy of Sciences*, 104, 10288–10293.

Riahi, K. and Roehrl, R. A. (2000). Greenhouse Gas Emissions in a Dynamics-as-usual Scenario of Economic and Energy Development, *Technological Forecasting and Social Change*, 63, 231–261.

Rose, S. *et al.* (2011). Land-based Mitigation in Climate Stabilization, *Energy Economics*, 34(1), 365–380.

Sankovski, A., Barbour, W. and Pepper, W. (2000) Quantification of the IS99 Emission Scenario Storylines using the Atmospheric Stabilization Framework (ASF), *Technological Forecasting and Social Change*, 63, 263–287.

Tegart, W., Sheldon, G. and Griffiths, D. (eds) (1990). *Climate Change: The IPCC Impacts Assessment*, Australian Government Publishing Service, Canberra.

Teske, S., Pregger, T., Simon, S., Naegler, T., Graus, W. and Lins, C. (2011). Energy [R]evolution 2010 — a Sustainable World Energy Outlook, *Energy Efficiency*, 4, 409–433

Tol, R. (2004). Exchange Rates and Climate Change: An Application Of FUND. Research Unit in Sustainability and Global Change, Hamburg University. Available online at: http://ideas.repec.org/p/sgc/wpaper/45.html. Accessed 24 February 2014.

United Nations (2011). World Population Prospects (WPP): The 2010 Revision. Department of Economic and Social Affairs, Population Division — UNPD. CD-ROM Edition.

US EIA (2008). *Annual Energy Outlook — Retrospective Review: Evaluation of Projections in Past Editions (1982–2008)*, U. Energy Information Administration, Washington, DC.

US EIA (2010). *Annual Energy Outlook 2010 — With Projections to 2035*, U. Energy Information Administration, Washington, DC.

van Vuuren, D., Isaac, M., Kundzewicz, Z. W., Arnell, N., Barker, T., Criqui, P. , Bauer, N., Berkhout, F., Hilderink, H., Hinkel, J., Hochrainer, S., Hof, A., Kitous, A., Kram, T., Lugeri, N., Mechler, R. and Scrieciu, S. (2009). 'Scenarios as the Basis for Assessment of Mitigation and Adaptation', in Hulme, M. and Neufeldt, H. (eds). *Making Climate Change Work for Us*, Cambridge University Press, Cambridge, pp. 54–86.

van Vuuren, D. and O'Neill, B. (2006). The Consistency of IPCC's SRES Scenarios to 1990–2000 Trends and Recent Projections, *Climatic Change*, 75, 9–46.

van Vuuren, D., Riahi, K., Moss, R., Edmonds, J., Thomson, A. and Nakicenovic, N. (2010a). Developing New Scenarios as a Common Thread for Future Climate Research. Available online at: http://www.ipcc-wg3.de/meetings/expert-meetings-and-workshops/files/Vuuren-et-al-2010-Developing-New-Scenarios-2010-10-20.pdf/at_download/file. Accessed 24 February 2014.

van Vuuren, D., Smith, S. and Riahi, K. (2010b). Downscaling Socioeconomic and Emissions Scenarios for Global Environmental Change Research: A Review, *Wiley Interdisciplinary Reviews: Climate Change*, 1, 393–404.

Watson, R., Zinyowera, M. and Moss, R. (eds) (1995). *Climate Change 1995: Impacts, Adaptations and Mitigation of Climate Change: Scientific-Technical Analyses. Contribution of Working Group II to the Second Assessment Report of the Intergovernmental Panel on Climate Change*, Cambridge University Press, Cambridge.

WEO (2008). World Energy Outlook. Available online at: http://www.worldenergyoutlook.org/media/weowebsite/2008-1994/WEO2008.pdf. International Energy Agency. Accessed 24 February 2014.

WEO (2009). World Energy Outlook. Available online at: http://www.worldenergyoutlook.org/media/weowebsite/2009/WEO2009.pdf. International Energy Agency. Accessed 24 February 2014.

WEO (2010). World Energy Outlook. Available online at: http://www.worldenergyoutlook.org/media/weo2010.pdf. International Energy Agency. Accessed 24 February 2014.

Chapter 4

Modelling Decarbonisation Scenarios

Annela Anger, Terry Barker and Mark Syddall

4.1. Modelling Decarbonisation

The main aim of this book is to analyse the pathways to decarbonisation of the global economy. As shown in Chapter 3, there are various ways to achieve this. One of the ways to build and analyse decarbonisation scenarios is to use a simplified numerical representation of the current and future world — a model. A single model suitable for studying decarbonisation scenarios should at least include representations of (1) the economy, (2) energy technologies and (3) the environment as a module for converting energy use to carbon dioxide and, if possible, then also to other greenhouse gas (GHG) and non-GHG emissions. The models including all three representations are known as energy-environment-economy (E3) models.

Not all such models, however, model all three component systems in full detail. For example, some E3 models do not model the economy in detail, being linked instead to external macroeconomic variables from macroeconomic models to provide constraints on the energy and environmental modules (e.g. MARKAL-MACRO (Strachan *et al.*, 2008b) and MESSAGE-MACRO (Messner and Schrattenholzer, 2000)). MARKAL-MACRO joins a one-sector top–down computable general equilibrium (CGE) model (MACRO) of the economy with a bottom–up energy-technology model MARKAL. The gross domestic product (GDP) from MACRO determines exogenously energy demand in MARKAL. The environmental module of E3 models does not usually include any other environmental impacts/constraints beside atmospheric pollutants.

A difference can be drawn between optimizing and econometric models of the economy. Optimizing macroeconomic global models include the assumption of a "benevolent dictator" maximising global social welfare. These models, included as the economic components of E3 models, dominate the models used for assessment of the impacts of climate change mitigation policies. Most of these optimizing economic models are computable general equilibrium (CGE) models that derive economic behaviour from first-principle theoretical understanding of how actors take economic decisions. CGE models are built on assumptions of the representative agent and theories of rational expectations, profit maximisation and the optimisation of consumer utility. Econometric models, in contrast, use relationships estimated from historical data to describe the future economic behaviour and relate these to energy/emission variables.

4.2. Modelling the Costs and Benefits of Mitigation

One important criterion in assessing and comparing policies for climate change mitigation is their economic costs and benefits in terms of effects on GDP, investment, consumption, employment and inflation. Of course such policies also need to yield environmental benefits, the primary one being reducing climate change through reducing the emissions of GHGs and therefore their concentrations in the atmosphere. An important co-benefit (or secondary benefit) here is a decrease in non-GHG air pollutants, which comes with reductions in the burning of fossil fuels and leads to improved (mostly) local air quality. The scale of this particular co-benefit can match that of the GDP costs of the policies (IPCC, 2007b).

Typically the conventional macroeconomic modelling of mitigation sets aside these primary and secondary environmental benefits and reports the costs in terms of effects on national and global GDP, and we follow this practice, with the qualification that GDP is a measure of marketed output, not welfare, and that it does not take into account the degradation and loss of environmental and resource assets as we explain below. In what follows we focus on macroeconomic effects, with GDP loss as the measure of cost, to compare with results from other models and between different scenarios and solutions from our model.

The economic costs of GHG mitigation are not observable from market prices, since they involve assessment of a complex E3 system as it responds to price signals, regulations and changes in environmental outputs that have no market valuations. The macroeconomic costs estimated for transition to a low-carbon economy are always hypothetical because they involve a comparison of two different states of the E3 system i.e. two scenarios over future years: the "with mitigation" scenario is usually compared with a reference scenario (often termed "business-as-usual" scenario). The risks of future extreme climate events are usually set aside.

4.2.1. *The use of GDP to measure costs*

In the analysis of macroeconomic costs and benefits it is important to distinguish those involving well-being, as opposed to "utility" or those measured by GDP. *Well-being* comes from friends, family, work and comparison with other people after a minimum income is reached. *Utility*, in the specific version used by traditional economists, is aggregated self-interest, setting aside inequality. Utility-maximising integrated assessment models can be deeply flawed, producing misleading if not meaningless conclusions for climate policy because of risks of long-term climate catastrophes. *GDP* is the conventional standard summary measure of macroeconomic effects as it covers all marketed output and is comparable across different countries' national accounts.

The costs and benefits are the differences in GDP in constant prices between two scenario outcomes, and for convenience most of the discussion of costs in this book, unless explicitly stated otherwise, is about the GDP effects, with the caveat that GDP can be a poor indication of well-being, ignoring critical climate, ecosystems and equity effects.

The changes in GDP can be expressed:

(1) *In absolute terms*: The absolute amounts can be misleading if quoted out of context, and they depend on the price base (base year) chosen.
(2) *As percentage of the reference case GDP*: The changes as a percentage of GDP in the reference case show the scale and direction of the costs, avoid the discount rate problem (since the costs and the level of GDP are contemporaneous) and allow easier comparison across years and countries.

(3) *As differences in GDP growth rates*: Differences in GDP growth rates are appropriate for comparing long-term mitigation costs, for example over the years to 2100 and helps to explore accelaration or deceleration of economic growth due to climate change mitigation policies.

As a rule of thumb, a change in GDP is insignificant if it implies a difference from base of less than 1% over a ten-year forecast period. It is not appropriate to use the level of carbon tax or estimated permit price for CO_2 emissions as a summary measure of mitigation costs, because it does not give any indication of the investment costs needed and the costs that indirectly affected industries bear. The carbon tax rate is one of many policy instruments; it is always positive, whereas the economic costs associated with it can be positive or negative; and the carbon price is a partial measure, covering fossil fuel use alone, whereas macroeconomic costs cover the whole economy.

4.3. Literature on Decarbonising the Economy to Achieve the 2°C Target

In the literature, deep mitigation is studied as reductions in GHG for more stringent stabilisation targets that have a probability greater than 50% of reaching stabilisation of the climate at less than 2°C above pre-industrial temperatures, such as 450 ppm (parts per million) CO_{2eq} or lower (den Elzen *et al.*, 2007). Studies which investigate the costs of deep mitigation have been relatively scarce as these targets have been generally considered to be infeasible by some modelling groups, e.g. those in the Stanford Energy Modeling Forum (EMF). The IPCC Fourth Assessment Report (2007a: p. 229) was notable in reporting only six studies with a concentration in the lowest range 445–490 ppm CO_{2eq} out of 177 scenario studies altogether, concluding that there was insufficient evidence to provide a reliable assessment as to costs and feasibility to achieve stabilisation in this range.

This weakness in the literature has been addressed in a series of studies reported after 2007. Many were conducted in the context of comparisons of models and results adopting the same framework of targets and assumptions, although major differences also remained in baselines, model approaches, aggregation and coverage of the models. The main comparative analyses of deep mitigation since the IPCC 2007 Report are summarised in six studies: van Vuuren *et al.* (2007); Strachan *et al.* (2008a); Clarke *et al.* (2009); Edenhofer *et al.* (2010); and Luderer *et al.* (2011). We have reviewed these studies to explore what the costs of deep mitigation are, and how the models have addressed and measured these costs. We draw two sets of conclusions relevant to our own modelling (described in this chapter and in Chapter 5) in this book: one concerns the reasons for the high costs of mitigation, compared to our own results, which find *benefits* in terms of higher GDP when deep mitigation is achieved in the model; the other conclusions concern the feasibility of deep mitigation in this literature and the lessons that can be drawn for climate policy from the studies.

A review of these studies suggests that the reasons for high costs of deep mitigation in the modelling literature are:

- Failure to take account of induced technological change adequately.
- Omission of critical opportunities for substitution (e.g. electric for petrol-powered vehicles).
- Low substitution elasticities for fossil fuel energy use in response to relative price effects.
- Omission of regulatory instruments in the model (e.g. no regulation on standards for energy efficiency).

Below we address these one by one.

4.3.1. *Failure to take account of induced technological change adequately*

The reason why induced technological change (ITC)[1] is a critical issue for climate policy is that it offers a means of substantially reducing the costs

[1] In this book, "induced technological change" is technical change in demand and supply technologies associated with climate change mitigation and brought about by climate policy, specifically the expectation of rising carbon prices. Models for informing climate policy have been developed over the last ten or so years to allow technological change to

of GHG mitigation and increases the rate at which economies can decarbonise. However, some economic models for climate policy do not include ITC at all or include it with restrictive assumptions and in a partial form so that it has only weak effects. In consequence, the results of such models exaggerate, from this perspective, the costs of mitigation and give the impression that stringent mitigation is either infeasible or not possible without serious damage to the economy.

In this section, we address two features of the treatment of ITC in many models, which suggest that the costs of mitigation may be much lower than supposed from the model results. (1) Not all sources of ITC are covered together in the models; if they were all covered, the technological change will be greater and costs lower. (2) Some models assume full "crowding out" by GHG-emission-saving R&D of other R&D and by additional GHG-saving investment by other investment, so that the total of investment including R&D remains unchanged but this assumption has been found to be too restrictive and not justified theoretically or empirically.

4.3.1.1. *ITC in top–down economic models*

A substantial literature in representing ITC and the relevant policies in top-down economic models has been developing (see Köhler *et al.* (2006), IPCC (2007c) and Gillingham *et al.* (2008) for surveys). The studies cover modelling the channels of ITC, such as private sector R&D and learning-by-doing, which may be influenced by climate policy, and climate policies that influence ITC, such as regulation, public spending on R&D, and carbon taxes or trading.

Unfortunately, as Gillingham *et al.* (2008) conclude, there is no agreement as to how ITC should be modelled, although they add:

> Despite these difficulties, with only a few exceptions most studies find that the ramifications and insights elucidated by incorporating [endogenous technological change] are important quantitatively. (p. 2750)

---

be affected by research and development (R&D) spending, energy and carbon prices, or accumulated use of or investment in technologies. In such models technological change is endogenous and if climate policies represented in the models lead to such change, this is defined as induced technological change.

In addition, most of the models have adopted the limiting assumptions of traditional economics about the working of the economy as well as for the representation of technology as noted above. In consequence there is a major problem in interpreting the results for climate policy when these assumptions do not hold, e.g. when economies are not at full employment, when outcomes are uncertain, when there is a widespread lack of information and expertise across consumers and firms, and when non-market considerations are important. A critical issue is the failure of the market to compensate adequately for the potential benefits from innovation, so that insufficient innovation is undertaken because some of the benefits are captured by competitors. The rapid decarbonisation of the energy system implied by the internationally agreed 2°C target calls for a non-marginal economic analysis involving radical adoption of low-GHG technologies in critical sectors.

Clarke *et al.* (2009) provide a framework for classifying the sources of technological change in GHG abatement. They divide the sources into R&D, learning-by-doing and spillovers. These sources can be inside the GHG producing firms or sectors or those supplying the technologies for reducing GHGs, or outside, in other firms and sectors. The spillover effects from other sectors, usually taken as exogenous technological change, can be very important, e.g. from changes in information systems, materials or biotechnology. One of their main conclusions is that few if any of the models in the literature allowing for endogenous technological change include all these sources. The presumption is that the model results will tend to overestimate the costs of abatement by omitting one or more sources of cost-reducing technological change. If R&D effects are combined with learning-by-doing and spillover effects from other sectors, it is also conceivable that the transformed energy system could eventually be lower cost than the fossil fuel-based system.

4.3.1.2. *The crowding out of low-GHG R&D and investment*

One of the channels by which policy can influence technological change is that of R&D and investment in the relevant technologies. There is support in the form of public subsidies to R&D and other innovation in general as a response to the market failure that insufficient innovation is undertaken by private firms because they cannot capture all the benefits. In other words, the social returns to R&D are much higher than the private returns. However the literature (e.g. Newell, 2010: p. 258) emphasises that, unless the R&D is primarily demand driven by higher GHG emission prices, it is unlikely to lead to lower GHG emissions.

Another critical issue is the extent of crowding out of other R&D and investment by policy-induced R&D and investment. Gerlagh (2008) has directly modelled the effects of including such change using a traditional economic approach and found that the carbon price is halved to achieve the same levels of mitigation. With induced technological change and R&D allowed to increase above base, consumption was less than 0.5% below base by 2100 for the 450 ppm scenario, compared to just over 6% below base without. Interestingly, consumption is slightly higher than base in 2050 in the 450 Scenario with induced technological change. This is in contrast with the conclusion of Nordhaus (2002), who found a very small effect of R&D as a component of climate policy, due to the crowding out of non-GHG R&D. The critical difference in Gerlagh's model that brings about the cost reduction comes from allowing higher carbon prices to switch R&D out of high-carbon R&D and into low-carbon R&D. A study of US patents 1973–1997 (Popp and Newell, 2012) is consistent with Gerlagh's assumptions in that it finds no significant evidence of crowding out of non-energy R&D by energy R&D and some evidence of crowding out of carbon-intensive energy R&D by low-carbon energy R&D.

4.3.2. *Omission of critical opportunities for substitution (e.g. electric for petrol-powered vehicles)*

As a contribution towards the literature reviewed by the IPCC for its Fourth Assessment Report (2007a), the CCSP[2] commissioned a study

[2]The Climate Change Science Program (CCSP) is the organisation responsible for bringing together the information to inform US climate policy, which in 2009 became the US Global Change Research Program (USGCRP).

of the costs of global decarbonisation by three modelling groups in the US:

- The Integrated Global Systems Model (IGSM) of the Massachusetts Institute of Technology's (MIT's) Joint Program on the Science and Policy of Global Change.
- The MiniCAM Model of the Joint Global Change Research Institute, which is a partnership between the Pacific Northwest National Laboratory and the University of Maryland.
- The Model for Evaluating the Regional and Global Effects (MERGE) of GHG reduction policies developed jointly at Stanford University and the Electric Power Research Institute.

These models were selected because they could provide estimates of the long-term macroeconomic costs of mitigating climate change. The results appear prominently in the IPCC (2007a) estimates of costs. However, one model produced exceptionally high costs. The CCSP–IGSM model gave GDP costs of about 5% by 2050 for the most stringent scenario of 450 ppm CO_{2eq} by 2100, compared to about 1% from the other two models, with the costs rising to 16% of GDP by 2100. The carbon price for this model rises to 230 US\$/tCO_2 by 2050 and 1651 US\$/tCO_2 by 2100. This high price is partly due to the assumption of the limited substitution of fossil fuels by electricity as an energy source for transportation: "In the IGSM scenarios, fuel demand for transportation, *where electricity is not an option* and for which biofuels supply is insufficient, continues to be a substantial source of emissions" (US CCSP, 2006: pp. 4–21, emphasis added). It appears that the MIT model did not allow for electric vehicles to replace gasoline vehicles, even when the carbon prices were extremely high. These high-cost results are clearly outliers in the IPCC report (see IPCC, 2007a: p. 205).

Another example, from many model results, is that of the omission of the substitution of fossil fuel electricity by biomass electricity-generation with CCS (BECS). In the Energy Modelling Forum study 22 (EMF 22) comparison (Clarke *et al.,* 2009), models that included BECS showed lower costs. In case of using MESSAGE model the present values of total global costs over the century was US$1.5 trillion (2005US$) lower when BECS was included in the modelling.

Another example is the study by Azar *et al.* (2010) using a global energy-economy model (GET 5.0), globally aggregated and including three end sectors and ten primary energy options. Estimates of the costs of stabilising atmospheric CO_2 concentrations at roughly 450 ppm CO_{2eq} by 2100 are presented, both with and without CCS technologies applied to fossil fuels and biomass. Results show that costs are significantly reduced by 80%, from US$26 to 6 trillion, with BECS. GHG emissions become negative after 2070, reaching −4 Gt CO_{2eq} by 2100, with a reduction below base of 1.21% GDP.

In summary, one reason why some models show higher costs is because they do not include all the technological options for substituting low-carbon for high-carbon energy supply.

4.3.3. *Low substitution elasticities for fossil fuel energy use in response to relative price effects*

One feature of the models is that they use estimates of low responses of energy demand to increases in the relative price of energy, generally drawn from the econometric literature. These are usually from time-series analysis, but cross-section analysis, in which different countries' use of energy is related to the relative price of energy in each country, yields much higher elasticities (e.g. Liddle, 2012). These low price elasticities are likely to be too low when the relative price increase is very long term, as expected for effective climate policy.

4.3.4. *Omission of regulatory and other instruments in the model (e.g. no regulation on standards for energy efficiency)*

The models used for assessing decarbonisation may not include all the policy instruments that affect ITC and GHGs. The traditional approach in modelling climate policies adopts a single instrument (carbon price) to reach a single target (CO_2 reduction). This is the case with the modelling studies in the EMF 22 (Clarke *et al.*, 2009) focused on scenarios for climate stabilisation. However, in practice, governments adopt many instruments in a portfolio to achieve multiple objectives. Regulation of standards can create new markets for GHG technologies, allowing for greater production of standardised products and economies of scale. An approach that assumes the world maximises global welfare, with one instrument for the global warming externality and no instrument for the low-GHG innovation externality, will exaggerate costs, especially when there are inefficiencies in energy use, and unemployed resources.

4.3.4.1. *Modelling portfolios of policies*

The issue for policy is how to design a portfolio of economic and regulatory instruments that induces rapid technological change to reduce costs over the long term. Addressing the problem of rapid reduction in GHG emissions becomes an interdisciplinary task involving, for instance, natural science, engineering and economics.

There are three main ways by which technological change can be influenced by climate policy: regulation to increase energy efficiency or reduce permitted GHG emission rates, e.g. from vehicles or power stations; innovation policies to increase R&D spending in order to improve the emission performance of existing technologies or develop new low-GHG technologies; and the effects of taxes or emissions trading schemes (ETS) in increasing fossil fuel and carbon prices, which in turn directly increase the demand for and take-up of low-GHG technology. All three ways can lead to acceleration of technological change especially if they focus on technologies with more potential for learning-by-doing.

Regulation of standards in various forms is often adopted by governments to reduce energy use and GHG emissions (e.g. fuel and/or emission standards for the road vehicle fleets in Europe and the US and the renewable energy directive of the EU; see Directive 2009/28/EC). Traditional economic analysis for climate policy assumes that the economy is working at full efficiency. However, it takes time for new or improved technologies to be adopted. Actual equipment and practices can be far from the most efficient available. Regulation of markets to require higher standards of efficiency or reduced GHG emissions can induce technological change and shift the emission characteristics of energy use. One example is the regulations imposed to improve fuel efficiency in the US automobile industry (Lee *et al.*, 2010).

Most traditional economic models assume that regulation will reduce economic welfare, but this may not always be the case. Just as in some cases health and safety regulations can restrict choice but improve welfare, so its extension to regulation for climate effects may be similarly beneficial. The critical issue for the costs of regulation is the lifetime of the capital stock affected. The period of transition to a low-carbon economy is some 40 years from 2010 if the 2°C target has a good chance of being achieved. This comparatively long time period means that technologies have time to be developed, diffused and scaled to mass markets. Technological change via low-cost regulation, such as GHG-emission standards on new vehicles and electricity plant, has time to make a major contribution to reducing GHG emissions. In addition, if the regulation leads to substantial increases in investment, and there are unemployed resources, then the wider economy will benefit and unemployment will be reduced. If in turn long-term economic growth is demand-driven then the growth rate may even increase.

There is a notable contrast between the policies considered in the modelling literature and the portfolios of policies introduced and proposed by governments, including the International Energy Agency in its projections of the 450 Scenario (IEA WEO, 2010–2012) and the German government (Jochem *et al.*, 2008). The modelling literature tends to focus on highly stylised carbon price assumptions via taxes or emissions trading, with some studies including either R&D or learning-by-doing (Gillingham *et al.*, 2008), but only a few (e.g. Magné *et al.*, 2010) including both (so called "two-factor" endogenous technological change). In contrast, the government scenarios include all three ways for policy to affect ITC, in varying mixes depending on sector and country-specific institutions.

One reason for a portfolio approach comes from recognising the feedback effects in the E3 system. One feedback is the rebound effect associated with improvements in energy efficiency. The improvement will lower the real price of energy and hence this will lead to an increase in the energy demand, if the demand is not at saturation levels. The inclusion of increases in carbon and energy prices through taxes and trading alongside the efficiency regulation will offset this effect (Barker *et al.*, 2009).

However the main reason for including an increase in real carbon prices is that the modelling strongly suggests that without the price incentive the technological change will not be implemented and the costs will not fall. Newell (2010) addresses the issue directly and emphasises the importance of a GHG emission price to encourage low-GHG innovation from both the demand and supply sides.

Fischer and Newell (2008) develop a ranking of policy instruments for a portfolio to reduce GHG emissions from the US electricity sector. They find that emission price is most cost-effective, then emission performance standard, a fossil fuel tax, a renewables share requirement, a renewables subsidy and finally R&D subsidy. The combination of instruments is expected to reduce emissions at significantly lower cost. Gerlagh and van der Zwaan (2006) comparing a similar list of instruments find that the emission performance standard, achieved through carbon taxes recycled as subsidies to non-fossil-fuel use, is the most cost-effective option. Barker *et al.* (2008) and Barker and Scrieciu (2010) develop and use a hybrid energy-environment-economy model on a global scale (E3MG) allowing for all the channels for ITC. They bring together several instruments of policy (carbon trading, carbon taxes, R&D subsidies, regulation) in a portfolio designed to accelerate technological change over the period to 2050 towards low-GHG technologies. They report that stringent abatement targets can be achieved with carbon prices rising to high levels, but with macroeconomic benefits. The carbon prices are much lower if regulation is included in the policy portfolio. The benefits come mostly from induced investment and lower unemployment, but also from higher technological change.

However, in order for climate policy portfolios to induce more rapid technological change and reduce costs over the long term, it is necessary that current institutional factors constraining their implementation or adoption are well understood and effectively addressed.

4.3.5. *Conclusions from the literature on stringent decarbonisation*

1. The assumptions made by the studies available on the overall costs of meeting more stringent stabilisation targets are very important in

determining the results. If the models assume that the reference case is optimal, as many do, then any mitigation policies will reduce welfare, and GDP if that is used as the measure of welfare.

2. The studies also highlight that if more stringent targets are to be achieved then portfolios of complementary policies will be necessary to achieve the required transformation. Policies to raise carbon prices, such as carbon taxes, will have to be combined with policies on regulation to drive technological development and energy-efficiency technologies.
3. Low-cost stringent mitigation can be achieved by a wide range of technological options in the electricity sector. Different studies come up with a diverse set of mixes for low-carbon generation, ranging from high renewables, high nuclear to high BECS (Edenhofer *et al.*, 2009). The different mixes depend on the assumed scope for learning-by-doing in the different technologies, the initial conditions assumed for the projections and the political choices of governments in leading economies.
4. The carbon intensity of the reference case is important for the feasibility and cost of deep mitigation. With high oil prices, as expected from limited supply in the face of growing demand, the switch from gasoline to electric vehicles will also be a feature of the reference case. And the long-term historical trend of improving energy efficiencies can be expected to continue in all the scenarios.
5. Although global net present value costs of meeting stringent targets are estimated to be in trillions US$ over the century and annual costs are as high as several percent of annual GDP, these mitigation costs are relatively modest compared to the projected levels of GDP from the economic growth assumptions in the scenarios.

4.4. The Modelling Approach used for this Study

4.4.1. *A short description of E3MG*

In this book we have developed the scenarios for global decarbonisation using E3MG (see, for example Barker *et al.* (2012)). In contrast with traditional models, economic growth in E3MG is demand-driven and supply constrained, with no assumption of the economy being in full-employment equilibrium. The world economy is treated as an open system of interacting economies with different levels of unemployment and financial imbalances. In order to implement the institutional aspect of economics, namely that economic activity is highly specific to location and timing, the model is disaggregated into 20 world regions and 41 industrial sectors in each region. Although there are regional macroeconomic variables derived by aggregation, such as GDP, and global variables, such as the world oil price, many economic activities, such as output, investment, employment, exports and imports, and associated prices, are treated at the industrial or product level. The assumption is that each industry has its own institutional rules and procedures. The model is further disaggregated in its treatment of consumers' and government expenditures, in energy demand and supply, and in emissions of pollutants into the atmosphere.

A crucial feature of the model is that it represents observed behaviour. It is a simulation econometric model with the version we use based on annual data from 1970–2006 and input–output data for the year 2000. The economic data are organised around the social accounting matrix with supply–demand balances for 20 regions and 41 products. The model includes 22 sets of stochastic equations estimated with short- and long-term components by instrumental variables for aggregate consumption, consumers' expenditure shares, investment, employment, exports, imports, prices and wage rates, energy demands, fuel shares, and labour participation rates. Population, exchange rates and interest rates, government spending, tax and other fiscal policies, and the availability of natural resources are taken as exogenous.

E3MG is a non-equilibrium model with an open structure such that labour, foreign exchange and public financial markets are not necessarily closed. In this approach, fiscal policy is paramount in managing the economy, with monetary policy responsible for maintaining financial stability, interest rates and exchange rates. In the modelling, industrial output is determined as the sum of intermediate demands from other industries and the final demands of households, government, investment and net international trade. Employment is derived from output, allowing for varying returns to scale across industries and over time. The estimated equations show that, in the long run, employment rises less than output in most industries, i.e. there are economies of scale and specialisation. Employment can increase through changes in industrial structure with economic growth, if employment-intensive sectors such as health and education increase relative to manufacturing, or through fiscal policy changing the level or structure of demand over time. Unemployment is the difference between employment and the total labour force, given the participation of the working population. Participation in turn depends on the level of unemployment (the "discouraged worker effect" with higher unemployment).

There is an extensive dynamic treatment of prices, wages and costs and the effect of changes in relative prices on the real economy, making the model much more complex than a Leontief input–output model. Prices depend on unit costs, expected world inflation represented by world oil prices, and the ratio of actual to expected output, with exchange rates, taxes and subsidies affecting costs. The industrial wage equations for each sector are based on a theory of real wage bargaining, allowing for levels of unemployment, consumer price inflation and wage rates in other sectors and other countries. Since the equations are estimated from historical data for each sector in each region, the prices and wage rates vary by sector as to how much costs are passed on to prices.

Considering economic theory, the model is based upon a post-Keynesian economic view of the long run. In other words, in modelling long-run economic growth and technological change we have adopted the "history" approach of cumulative causation and demand-led growth (Kaldor, 1957, 1972, 1985; Setterfield, 2002), focusing on gross investment (Scott, 1989) and trade, and incorporating technological progress as an indicator based on accumulated gross investment enhanced by R&D expenditures. Other post-Keynesian features of the model (see Holt (2007) for a discussion of such features) include: varying returns to scale (that are derived from estimation), non-equilibrium, not assuming full employment, with varying degrees of competition, and the feature that industries act as social groups and not as a group of individual firms (i.e. no optimisation is assumed but bounded rationality is implied). The grouping of countries and regions has been based on political criteria, the main one being the division into UNFCCC (UN Framework Convention on Climate Change) Annex 1 and non-Annex 1 parties. At the global level, accounting conventions are imposed so that the expenditure components of GDP add up to total GDP and total exports equal total imports at a sectoral level allowing for imbalances in the data.

4.5. The Ability of E3MG to Fit the Past

One critical limitation of the traditional economic components of large-scale integrated assessment models (most of those reviewed in the IPCC 2007 Report) is that the economic components are based on one year's data, which means that there is no observational basis within the models for representing dynamic behaviour. This implies that the equilibrium models that have been used to report costs of stabilisation (IPCC, 2007a) are nearly all inadequate for assessing the costs and benefits of accelerated transitions to deep mitigation because all the dynamics have to be assumed.

In contrast, E3MG goes one step further than nearly all the other global economic models used for stabilisation research in that some 30 years of data are used to estimate the dynamics and economic behaviour in response to GHG reduction policies.

Given that the model is estimated on past data, E3MG should be suitable for simulating the past economic behaviour. However there are major problems in simulating the past:

1. It is a challenging intellectual task, basically because the global economy is a complex system in which there are major interdependencies between countries and variables, as well as over time. The most advanced macroeconomic model to tackle the problem, Global Vector AutoRegressive (GVAR) (Garratt *et al.*, 2006), treats all the economic variables, including oil prices as endogenous in the system, so that the solution is driven by the starting point and the internal dynamics of the model, but with a very limited role for policy instruments, and with no coverage of carbon taxes, ETS, R&D or even government expenditure.
2. In addition, the dynamics become more relevant and interesting for climate policy when technological diffusion is treated as responding to climate policies in the models. However, this has been a relatively recent development in integrated modelling. Previously, as reported in the IPCC 2001 Report, technological change was treated as

exogenous in most of the results reported, and any diffusion was assumed to be incorporated in the baseline.

3. The potential for international climate policy to yield substantial benefits in terms of higher global GDP via the adoption of specific portfolios of instruments (international emissions trading with auctioning and use of revenues for technological incentives and/or environmental tax reform) has only recently been modelled in stabilisation scenarios (Barker *et al.*, 2006). Moreover it has taken until the IPCC in 2007 (see IPCC (2007b), in particular Box SPM4 on page 28) to recognise that the GDP benefits from induced technological change may be even larger at more stringent levels of mitigation.

This section reports how we have applied E3MG to simulate the year-by-year stochastic dynamics of greenhouse gas emissions, including effects of changes in oil prices and in existing policies, such as those affecting fossil fuel taxes, over the period 1972–2006, with technological diffusion treated as responding to changes in relative prices. Data on energy and the economy are available on a reasonably consistent basis for the years 1972–2006 from international sources to estimate the model.

We have developed an endogenous solution 1973–2006, similar to the solution 2008–2050. In doing so, we have used the existing estimated equations to provide the first estimate of the solution for annual GHG emissions, given observations on global oil prices, exchange rates, economic policies and population. However, we have had to restrict the dynamic solutions for following reasons:

1. There are missing data for former Soviet Union regions to 1992, so these regions have been solved for the later years only, from 1997.
2. The model solution for Brazil is unstable in early years, so this region is solved from 1994 onwards.
3. Sector data is occasionally missing in early years (e.g. US data on electronics output is near-zero until 1987), so these data have been left exogenous at historical values.
4. Due to lack of data, the simulation modelling of the electricity system was restricted to the period after 1995. The emissions from power

generation were held exogenous up to 1995 in the endogenous solution 1973–2006.

Although there are major problems of model instability, fully recursive solutions can be found, given these assumptions and systematically removing or respecifying unstable equations. These procedures have been a substantial research task involving solutions of each region's equations, then all the equations for each endogenous variable.

The exercise revealed limitations in the treatment of some industries and regions — e.g. simulation of CO_2 from manufacturing processes (cement, metal refining) — and problems in feedback. Therefore, CO_2 emissions in the historical solution revealed less fit with the actual data than it was observed by estimation of macroeconomic variables such as GDP, consumption and investment (see Figure 4.1).

4.6. Developing the Reference and Decarbonisation Scenarios

Any modelling that forecasts the impacts of GHG emissions reduction policies has to make assumptions about the future. Usually there are at least two scenarios specified: a business-as-usual/baseline/reference scenario that is the basis for comparison and a policy scenario that includes the climate change policies under observation and is compared against the business-as-usual/baseline/reference scenario.

The modelling undertaken in this study required the specification of scenarios to reflect the set of policies for GHG mitigation — see Chapter 3. More than 20 different policy scenarios were modelled including the two main scenarios: the Reference and the Decarbonisation Scenario that achieves the emission reduction target.

The **Reference Scenario** is based on "current policies" as covered by the IEA (International Energy Agency) 2010 report. It is intended to represent the outcome for the global economy 2013–2050 of adopting policies without the extra efforts promised in the Cancun Agreement[3] or any

[3] See http://cancun.unfccc.int/ for more information.

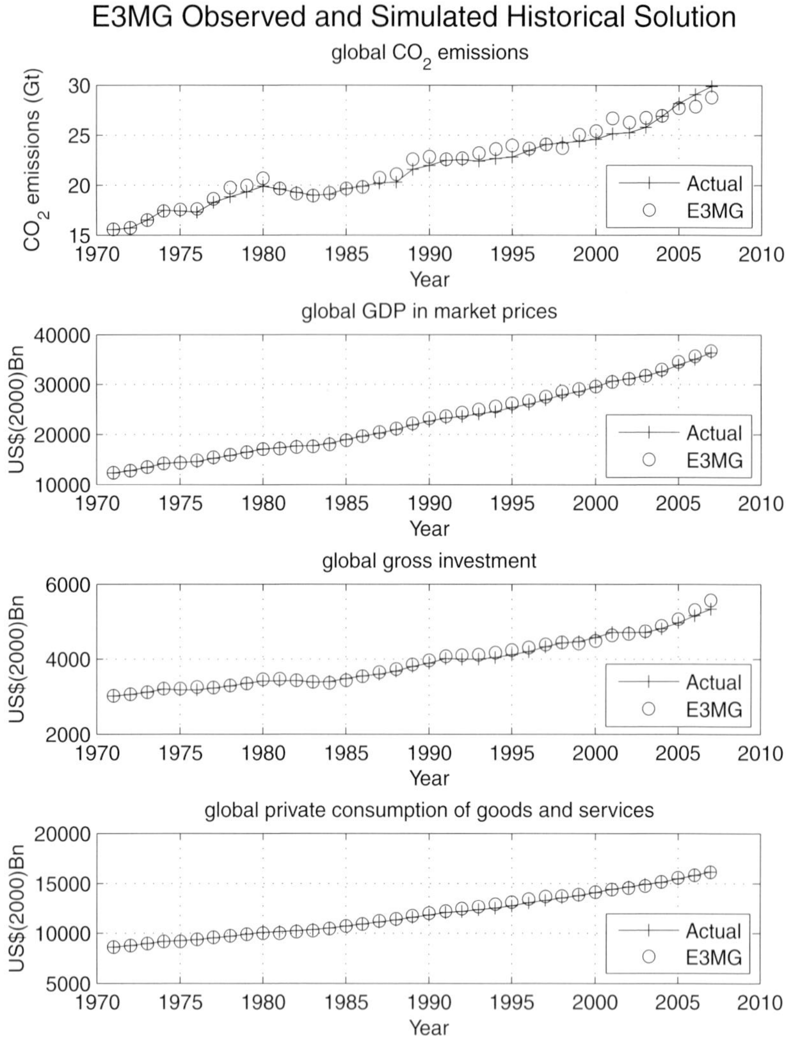

Figure 4.1. Comparison of actual and simulated variables in E3MG in the historical solution.

other additional measures to reduce GHG emissions by 2050. This scenario is mainly based on the IEA's WEO (World Energy Outlook) (2010) current policies scenario using our model to provide a fully dynamic solution over the period, with results initially calibrated to match those in

WEO (2010) to 2035, but extended to 2050 and revised for the different profile of oil prices in WEO (2013). This reference case is compared with scenarios that include policies to increase investment in low-carbon technologies and reduce GHG emissions.

The **Decarbonisation Scenario** is intended to achieve the agreed 2°C target and includes the policies for reducing GHG emissions assumed in the IEA's 450 Scenario (IEA WEO, 2010). These policies can be interpreted as the outcome of the Cancun Agreement at the high end of the range of commitments to reduce emissions. However our modelling with E3MG (Barker *et al.,* 2012) assumes a strengthening of these commitments to reach a level of emissions by 2020, which is 15% lower than those expected from the Cancun Agreement.

This further reduction is required to give a medium (50 to 66%) chance of achieving climate stabilisation at 2°C as summarised in Table 4.1. The higher level of investment is designed to accelerate recovery towards full employment in those economies most affected by the Great Recession. In both policy scenarios, the CDM plays a role in transferring funding to developing countries for climate change mitigation.

4.6.1. *Using E3MG to model emissions trading schemes for the energy sector*

A major component of the policies we have modelled for decarbonisation in E3MG is a set of ETS covering the energy sector, in particular electricity generation, which are assumed to start in 2015. We have chosen the ETS as the market-based instrument to establish a carbon price because the international mechanism has been a component in the Kyoto Protocol since 1997 and has evolved in practical application in several countries. Experience of developing and using ETS schemes has been gained in the EU, in the western and eastern groups of states in the USA, in New Zealand and Japan.

Our assumption is that following the Durban Agreement of December 2011, countries will prepare for a legally binding set of reductions in emissions to start by 2020. One component of the preparations is assumed to be an ETS covering industrialised countries and another for China, both starting in 2015 at relatively low rates. We have assumed that the schemes

Table 4.1. Some details of the E3MG Reference and Decarbonisation Scenarios.

E3MG Scenario	**Reference**	**Decarbonisation**
Description	Current policies	Medium chance (50 to 66%) of reaching 2°C target in 21st century
Source	IEA WEO 2010 extended until 2050	Described in this chapter
Emissions trading scheme	EU only from 2005	EU from 2005 Other OECD from 2015 China from 2015 Mexico from 2015 Other Annex 1 and selected non-Annex I parties from 2030 Rest of the world from 2040
Average carbon price 2020 (2012US$/tCO$_2$)	1	5
Average carbon price 2050 (2012US$/tCO$_2$)	1	65
Annual additional investment into low-carbon technologies % of investment	n/a	2.9% by 2020 rising to 4.5% by 2030
Other policies	No new climate change mitigation and energy policies in addition to those formally adopted by mid-2010	Removal of fossil fuel subsidies in net oil importers by 2020 Regulation for electric cars and housing Clean development mechanism (CDM) used to fund extra investment in least-developed countries

are restricted to the energy-intensive sectors, partly because the transaction costs are relatively low, since the number of entities covered is very much smaller than in the service and household sectors, and partly because the institutions and experience necessary for smooth operation have already been established in the energy sector in many countries, with the European ETS being the most advanced.

We assume that the energy-sector ETS starts in a group of industrialised countries and China at rates shown in Table 4.2. We have included China here because it is obvious that without major action to decarbonise the country's energy system it will be impossible to reduce emissions to about 12Gt CO_2 by 2050 consistent with the 2°C target. The level of emissions in China in 2011 was about 10 Gt CO_2 and in the Reference Scenario this grows to over 20 Gt CO_2. Further schemes are introduced in other countries or regions, with the permit (allowance) price starting at low levels and then rising towards a common global price, although this process is not completed by 2050.

The coverage of the regional schemes has been chosen depending on existing smaller-scale schemes, at the national or local levels, e.g. the cap-and-trade schemes for California, although scaled up to the national levels. The table also shows the world average carbon price assuming these permit prices for the energy sectors and no other carbon taxes. All the remaining reductions in emissions in the Decarbonisation Scenario are achieved through the effects of the ETS schemes by increasing electricity prices, reducing energy subsidies and imposing regulation.

The allowance prices in Table 4.2 are the result of assumed caps on GHG emissions to reach the 2°C target. These schemes assume that all allowances are freely allocated to industry and that the extra costs (including opportunity costs) are passed on as higher final product and electricity prices. An important decision for governments is *how to set caps* on GHG emissions in the schemes and *how to allocate and distribute* emission allowances to trading companies. The overall reductions in GHG emissions are assumed to be set by governments on the basis of international agreements on the acceptable extent of global warming and the implications for the levels of GHG emissions. The allocation of a global reduction across countries and regions is highly contentious and we have adopted a stylised approach, taking the policies as included in the IEA's 450 Scenario (2010) and strengthening them to achieve a stronger target in terms of the reduction in global CO_2 emissions by 2050 to half the emissions of 1990. The set of national policies is assumed to include an ETS for the energy sectors, with regional caps, which yield gradually rising permit prices with the prices for selected years shown in the table. The permit prices that emerge from this process depend on national circumstances, the overall CO_2 reduction target and the effects of the other GHG mitigation policies on all sectors.

4.6.2. *The world oil price, decarbonisation and the global energy market*

Table 4.2 and Figure 4.2 show the assumptions for the world oil price (adapted from IEA WEO, 2013) in the Reference Scenario and the assumed lower oil price when policies for decarbonisation are implemented. As the world economy shifts away from the use of fossil fuels, the

Table 4.2. Carbon[4] and oil prices in the modelling scenarios.

	2012	2015	2020	2030	2040	2050
The Reference Scenario assumptions						
Crude oil price (2012US$/bbl)	112	103	107	123	139	155
US ETS price (2012US$/t$CO_2$)	0.0	0.0	0.0	0.0	0.0	0.0
EU ETS price (2012US$/t$CO_2$)	8	13	26	28	27	25
China ETS price (2012US$/t$CO_2$)	0	0	0	0	0	0
Global average carbon price (2012US$/t$CO_2$)	1	1	1	1	1	1
The Decarbonisation Scenario						
Crude oil price (2012US$/bbl)	112	103	108	119	124	123
US ETS price (2012US$/t$CO_2$)	0	21	47	110	143	187
EU ETS price (2012US$/t$CO_2$)	8	21	47	110	143	187
China ETS price (2012US$/t$CO_2$)	0	2	8	110	143	187
Global average carbon price (2012US$/t$CO_2$)	1	1	5	29	41	65

Sources: *E3MG*.

[4]The carbon price shown in the table is the ETS allowance prices for energy-intensive sectors in the countries shown. Other countries and regions are also assumed to implement an ETS. See text. The global average carbon price is the average across all sectors and countries.

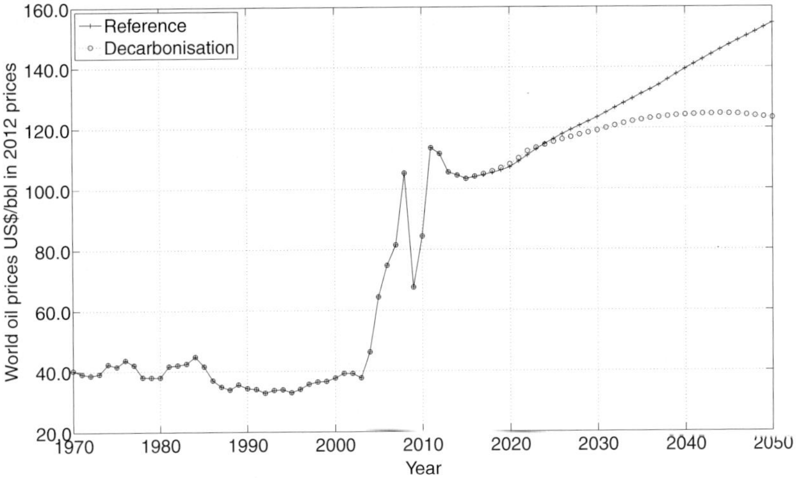

Figure 4.2. The assumed world oil price in the Reference and Decarbonisation Scenarios.

Source: IEA (2013) and E3MG.

prices of all the main types of fossil fuel, i.e. oil, gas and coal are liable to fall, unless the reduction in demand is accompanied by a corresponding reduction in supply.

However, the future supply–demand balance is different for oil, gas and coal. The supply of *conventional oil* appears likely to peak over the next 20 years (Sorrell *et al.*, 2010) so that demand is likely to be constrained by supply, and oil prices will rise unless (1) GHG mitigation or energy security policies are effective in reducing oil demand or (2) unconventional sources of oil become much lower in cost due to new technologies. Any constraints on the supply of *natural gas* seem more uncertain with the development of fracking techniques to extract gas from shale deposits. The potential reserves of gas have been increasing dramatically from 2011, with the demonstration of this new technology in increasing supplies of gas in the US. The main limitations to supply of shale gas appear to be the need for water and the potential for pollution and gas leakage. *Coal* reserves are plentiful so that the supply of coal is potentially much higher that likely demand. The supply of coal is restricted by costs of extraction and transportation.

The availability of natural gas and the substitutability of coal for gas in power generation mean that there will be a shift away from coal to gas in the Reference Scenario. With an ETS covering power generators, the shift to gas will be accelerated depending on the price of carbon allowances. For all fossil fuels, when GHG mitigation policies lead to a reduction in the global demand for fuel, the prices in the Decarbonisation Scenario are expected to rise more slowly than the prices in the reference-case projections.

Because oil supplies are most restricted in the likely growth of supplies, the world price of oil is expected to rise over the projection period as shown in Figure 4.2. In the Reference Scenario, which has been developed following the profile of the projections in the IEA WEO (2013) current policies scenario, oil prices reach about US$150/bbl in 2012 prices by 2050.

4.6.3. *Additional investment and regulation*

The required additional annual investment for the post-2012 period 2013–2020 is presented for 2020 in Table 4.3.

The investment structure is based on the IEA's 450 Scenario (WEO, 2010: Figures 13.14 and 13.15 therein, pp. 401) and is 22% higher for each of the sectors in order to be consistent with the 2°C target. The scale of the additional investment is some 0.6% of world GDP by 2020. The investment is assumed for every year 2013–2050 and is related to the various policies and measures assumed for the different countries and regions.

Table 4.3. Annual additional and total investment in 2020 in the Decarbonisation Scenario (billion 2000 US$).

	Additional investment	Total investment
US	82	2623
EU	75	3151
China	70	2577
OPEC	16	600
Rest of the world	119	4501
World total	362	13452

Source: E3MG (2012) and IEA WEO (2010).

The levels start low in 2013 and reach the full 22% above IEA levels by 2016; they then rise to reach a maximum by 2035 then fall thereafter. The investment is associated with additional regulation requiring gradually more stringent standards for lower-carbon intensity in power generation, equipment, buildings and vehicles (see Table 4.4 for additional transport policies) and is assumed to be funded by increases in the price of electricity or the costs of buildings, equipment and vehicles.

The average additional global investment in years in 2020 is about 2.7% higher compared to the Reference Scenario, rising to 4.5% higher in 2030, reaching a peak in the mid-2030s then declining to 3.4% higher in 2040 and 1.6% higher in 2050. The energy savings associated with this investment are cumulative and continue to rise over the whole period. The extra investment amounts to about 1.1% of global GDP in 2030.

4.7. Impacts of Decarbonisation on the Global Economy

The differences between the Decarbonisation and Reference Scenarios give estimates of the impact of these policies on the global economy. Table 4.5 shows average growth rates for two 40-year periods, 1970–2010 and 2010–2050 for some key indicators. The growth in final energy consumption and CO_2 emissions already slows in the Reference Scenario as governments, industry and consumers respond to changes in policy in the pipeline, with examples as given in the IEA's current policies scenario. Growth in GDP is lower in the future projections and growth in employment is much lower, partly because the growth of urbanisation in China falls over most of the period. Inflation is also lower than in the past, with a decarbonised economy benefiting from lower projected oil prices. These projections are subject to great uncertainty, and we explore the effects of substantially lower growth on the scenarios in the next section of the chapter.

The projections for CO_2 and GHG are illustrated in Figure 4.3. The Decarbonisation Scenario shows a reduction in CO_2 emissions by 2050 to about 12 Gt CO_2, about half the level in 1990. All major countries have substantial reductions in their emissions from a wide range of policies. In particular, China, with strong regulation and a rapidly rising price of CO_2 allowances, transforms its power sector away from coal from 2015 to 2030.

Table 4.4. Examples of transport sector policies and measures included in the Decarbonisation Scenario in key regions.

New Policies Scenario		**450 Scenario**			
United States	Renewable Fuel Standard Support to natural gas in road freight traffic Increase of ethanol blending mandates		OECD+	OME	OC
Japan	Target shares of new car sales according to Next Generation Vehicle Strategy 2010:	Extended emission target for passenger light-duty vehicles (95 gCO_2/km by 2035) Light commercial vehicles	75	85	105
		2020	2030		Full technology spillover from passenger light-duty vehicles
	Conventional ICE vehicles	50–80%	30–50%	Medium- and heavy-freight traffic	5% more efficient by 2035 than in CPS
	Hybrid vehicles	20–30%	30–40%	Aviation	Sectoral target of 45% efficiency improvements by 2050 and support to use of biofuels
	Electric vehicles and plug-in hybrids	15–20%	20–30%	Other sectors such as maritime and rail	National policies and measures

(*Continued*)

Table 4.4. *(Continued)*

New Policies Scenario		**450 Scenario**			
	Fuel cell vehicles	0–1%	0–3%	Fuels	Retail fuel prices kept at a level similar to current policies scenario
	Clean diesel vehicles	0–5%	5–10%	Alternative clean fuels	Enhanced support to alternative fuels
European Union	Extended emission target for passenger light-duty vehicles (95 gCO_2/km by 2020) Emission target for light commercial vehicles (135 gCO_2/km by 2020) Enhanced support to alternative fuels Several national EV targets, subsidy extension Aviation and international maritime shipping in EU ETS as of 2013				
China	Vehicle fuel economy standard 7l/100 km by 2015 Extended subsides on the purchase of alternative vehicles				
India	Increased utilization of natural gas in road transport				
Brazil	Increase of ethanol blending mandates				

Table 4.5. Average annual world growth rates[5] of CO_2 emissions and key macroeconomic variables, Reference and Decarbonisation scenario, % per annum.

	1970–2010	2010–2050	
Scenario	**Reference**	**Reference**	**Decarbonisation**
CO_2 Emissions	1.9	1.5	−2.8
Final energy demand	2.4	1.0	0.4
GDP	3.0	2.6	2.7
Price index consumers' expenditure	4.3	2.7	2.5
Employment	1.6	0.7	0.7

Source: E3MG modelling.

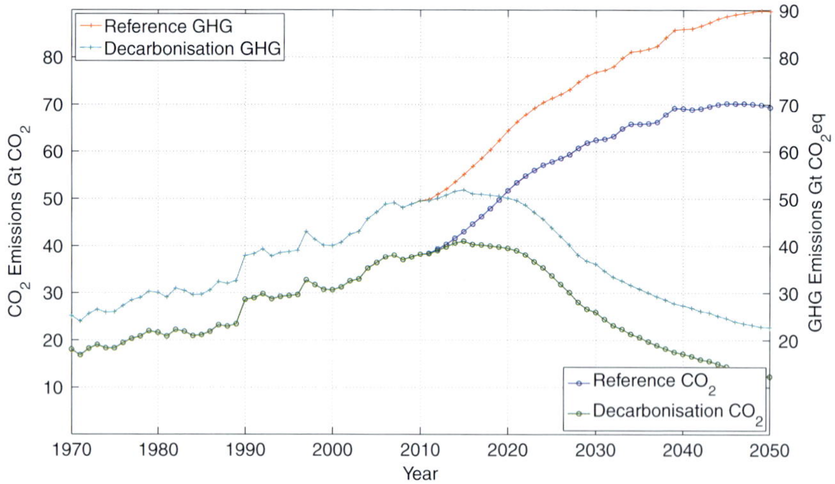

Figure 4.3. Global CO_2 and GHG emissions in the Reference and Decarbonisation Scenarios.

Source: E3MG modelling.

Note: There are discontinuities in the historical data.

[5]The projections are not forecasts. The significant figures given in this and later tables should not be taken to indicate reliability.

The pathway gives an accumulated reduction from 2010 to 2050 of 1036 Gt, which is above the reduction in the AR5 Representative Concentration Pathways (RCPs) of RCP2.6, a mitigation scenario that easily achieves the 2°C target by 2100 (van Vuuren *et al.*, 2011). However this is below the reduction for the RCP4.5, which misses the target. The Reference Scenario is between the RCP4.5 and RCP6.0, implying substantial global warming.

The projections for world GDP and investment are illustrated in Figure 4.4, which shows both the Reference and Decarbonisation Scenarios. Both GDP and investment are higher when the global economy is decarbonised, although by barely noticable differences in the graph. GDP is 3.3% and investment is 3.7% above the Reference Scenario by 2050.

Figure 4.5 shows the exogenous investment associated with the additional regulations, excluding those for the power sector. It also shows how the total additional investment (additional investment plus induced investment) is higher after the mid-2020s compared with the reference case. GDP is also higher, with E3MG providing a muliplier of 1.6 in 2020, i.e. the GDP increase is 60% above the injected exogenous increase in investment. The induced investment is higher when the effects of the regulations on oil demand begin to affect world oil prices. With oil prices rising more

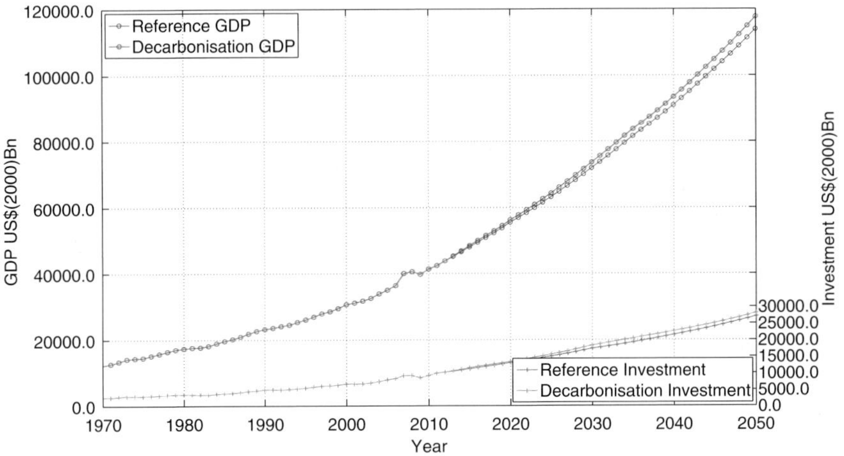

Figure 4.4. Global GDP and investment in the Reference and Decarbonisation Scenarios.

Source: E3MG modelling.

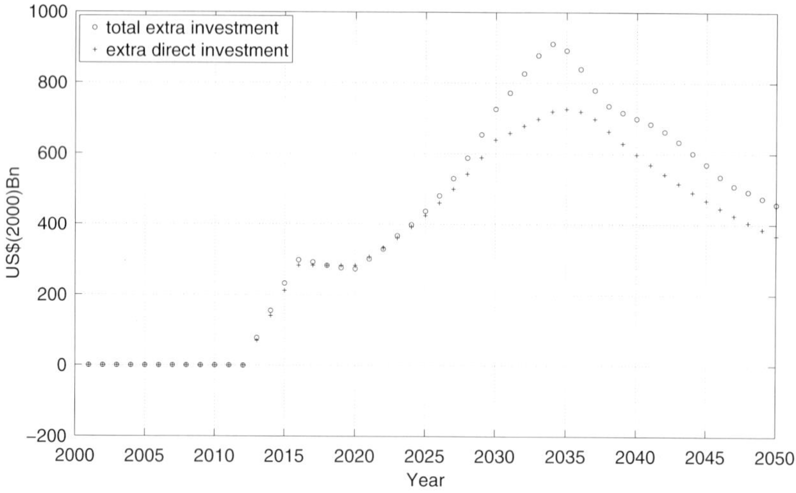

Figure 4.5. Additional investment associated with extra regulations for building, industry (non-power) and transport, direct and total (assumed plus induced).

Source: E3MG modelling and IEA WEO, 2010.

slowly than in the reference case, world inflation is slightly lower and real incomes slightly higher.

The Decarbonisation Scenario also shows higher global employment throughout the projection, rising to 0.9% or about 40 million above the Reference Scenario by 2050. Increases in employment are likely to be concentrated in the construction and engineering sectors, as these produce much of the equipment that will be necessary to reduce carbon emissions to the required level. Companies that feature in these supply chains, such as the metals and minerals sectors, are also likely to benefit, even though they can be quite energy intensive themselves. Employment in the coal and oil industries will be lower.

The low-carbon sectors often have their own specific skills requirements and this is a key issue in meeting the challenges of decarbonisation. This is an issue that the modelling is not able to address and requirements could be specific to time and location. However, if skilled workers are not available, the results for employment and GDP could be less positive than suggested in this chapter. Therefore it is clear that careful planning is required to manage the transition.

4.8. The Resilience of the Policy Mix

There are particular uncertainties about world growth and the world oil price that may affect the policy mix required for decarbonisation. The effects of a lower growth rate are discussed in this section.

4.8.1. *The effects of higher oil prices*

There is a major uncertainty about the future long-term change in oil prices. This section discusses how the scenarios change with higher oil prices sustained over the period to 2050.

Higher oil prices tend to depress world GDP growth. This will tend to reduce the growth of energy demand, but the effect on CO_2 emissions will depend on how much substitution is possible between oil, gas and coal. Higher oil and gas prices in the Reference Scenario will reduce the relative price of coal and encourage the use of coal to generate electricity. CO_2 emissions will tend to rise even as energy demand grows more slowly. However, if regulations are in place to reduce the carbon intensity of power generation and ETS force up the price of allowances when coal becomes relatively cheaper, then substitution toward coal will be restricted. In the Decarbonisation Scenario coal is prevented from re-entering the electricity market by regulation and the higher oil prices lead to faster decarbonisation of transport via the switch to hybrid and all-electric vehicles.

4.8.2. *The effects of lower world growth*

We have tested the effects of lower growth on the policy mix by developing a set of reference scenarios and corresponding decarbonisation scenarios with higher savings rate, thereby reducing the growth of consumption and GDP. Figure 4.6 shows these projections, with the red lines showing reference and the blue lines showing the decarbonisation outcomes. In all cases the effect of decarbonisation on the lower-growth reference scenarios is to raise the growth rate, with the relative increase being higher at the lowest-growth reference scenario. This effect is due to the impact of the absolute levels of investment required for decarbonisation being higher in relation to GDP.

Figure 4.7 shows the effects of low growth on global CO_2 emissions. Higher savings and less consumption in decarbonisation scenarios also

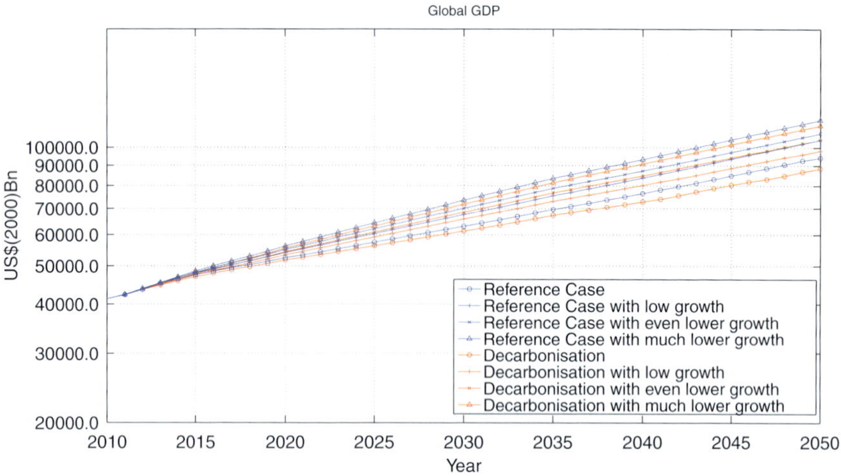

Figure 4.6. The effects of lower world growth on GDP with and without decarbonisation.

Source: E3MG modelling.

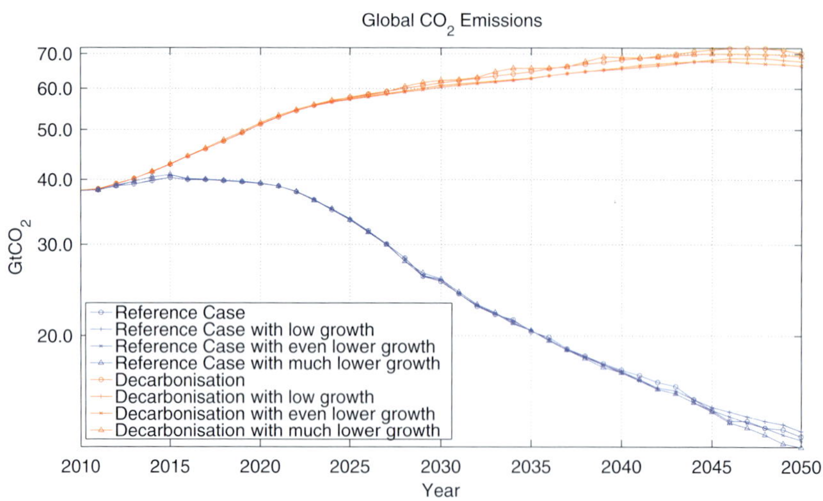

Figure 4.7. The effects of lower world growth on CO_2 emissions.

means lower investment into low-carbon technologies and that results in somewhat higher CO_2 emissions in decarbonisation scenarios. For the reference scenario, only the lowest-growth results in higher CO_2 emissions and this is due to switching to higher levels coal consumption to satisfy the global energy demand. The two less extreme growth reductions do bring about some reductions in CO_2 emissions, but these are very small reductions compared to those brought about by targeted climate policy.

These results have implications for the debate on "green growth" and "de-growth". Green growth appears feasible, given policies that ensure the recycling of revenues from auctions of emission permits and extra regulations to encourage energy saving and switching away from fossil fuels. However, policies that aim to lower consumption *per se* may result in more CO_2 emissions by reducing the rate of investment and the rate of technological change, so reducing the switch away from coal. Reducing consumption in the expectation that this will lead to lower GHG growth seems to be misplaced. Policies directed at decarbonisation through technological change enables the investments needed to develop low-carbon pathways.

References

Azar, C., Lindgren, K., Obersteiner, M., Riahi, K., van Vuuren, D., den Elzen, K., Möllersten, K. and Larson, E. (2010). The Feasibility of Low CO_2 Concentration Targets and the Role of Bio-energy Carbon-capture and Storage, *Climatic Change*, 100, 195–202.

Barker, T. and Scrieciu, Ş. (2010). Modeling Low Stabilization with E3MG: Towards a "New Economics" Approach to Simulating Energy-Environment-Economy System Dynamics, *Energy Journal Special Issue on The Economics of Low Stabilisation*, 31, 137–164

Barker, T., Anger, A. Chewpreecha, U. and Pollitt, H. (2012). A New Economics Approach to Modelling Policies to Achieve Global 2020 Targets for Climate Stabilisation, *International Review of Applied Economics*, Special Issue on Economic Policies of the New Thinking in Economics, 26, 205–211.

Barker, T., Dagoumas, A. and Rubin, J. (2009). The Macroeconomic Rebound Effect and the World Economy, *Energy Efficiency*, 2, 411–427.

Barker, T., Foxon, T. and Scrieciu, Ş. (2008). Achieving the G8 50% Target: Modelling Induced and Accelerated Technological Change Using the Macro-econometric Model E3MG, *Climate Policy Special Issue on Modelling Long-term Scenarios for Low-carbon Societies*, 8, S30–S45.

Barker, T., Pan, H., Köhler, J., Warren, R. and Winne, S. (2006). Decarbonizing the Global Economy with Induced Technological Change: Scenarios to 2100 using E3MG, *The Energy Journal*, 27, 241–258.

Clarke, L., Edmonds, J., Krey, V., Richels, R., Rose, S. and Tavoni, M. (2009). International Climate Policy Architectures: Overview of the EMF 22 International Scenarios, *Energy Economics*, 31, S64–S81.

den Elzen, M., Meinshausen, M. and van Vuuren, D. (2007). Multi-gas Emission Envelopes to Meet Greenhouse Gas Concentration Targets: Costs versus Certainty of Limiting Temperature Increase, *Global Environmental Change*, 17, 260–280.

Edenhofer, O., Knopf, B., Barker, T., Baumstark, L., Bellevrat, E., Château, B., Criqui, P., Isaac, M., Kitous, A., Kypreos, S., Leimbach, M., Lessmann, K., Magné, B., Scrieciu, S., Turton, H. and van Vuuren, D. (2010). The Economics of Low Stabilization: Model Comparison of Mitigation Strategies and Costs, *Energy Journal*, 31, 11–48.

Fischer, C. and Newell, R. (2008). Environmental and Technology Policies for Climate Mitigation, *Journal of Environmental Economics and Management*, 55, 142–62.

Garratt, A., Lee, K., Pesaran, M. and Shin, Y. (2006). *Global and National Macro-econometric Modeling: A Long-Run Structural Approach*, Oxford University Press, Oxford.

Gerlagh, R. (2008). A Climate-change Policy Induced Shift from Innovations in Carbon-energy Production to Carbon-energy Savings, *Energy Economics*, 30, 425–48.

Gerlagh, R. and van der Zwaan, B. (2006). Options and Instruments for a Deep Cut in CO_2 Emissions: Carbon Capture or Renewables, Taxes or Subsidies? *Energy Journal*, 27, 25–48.

Gillingham, K., Newell, R. and Pizer, W. (2008). Modeling Endogenous Technological Change for Climate Policy Analysis, *Energy Economics*, 30, 2734–2753.

Holt, R. (2007). 'What is Post Keynesian economics?' in Forstater, M., Mongiovi, G. and Pressman, S. (eds). *Post Keynesian Macroeconomics: Essays in Honour of Ingrid Rima*, Routledge, London, pp. 89–107.

IEA (2010). *World Energy Outlook*, IEA, Paris.

IEA (2011). *World Energy Outlook*, IEA, Paris.

IEA (2012). *World Energy Outlook*, IEA, Paris.

IEA (2013). *World Energy Outlook*, IEA, Paris.

IPCC (2001). *Climate Change 2001: Mitigation. Contribution of Working Group III to the Third Assessment Report of the Intergovernmental Panel on Climate Change*, Metz, B., Davidson, O. R., Swart, R. and Pan, J. (eds), Cambridge University Press, Cambridge.

IPCC (2007a). *Climate Change 2007: Mitigation. Contribution of Working Group III to the Fourth Assessment Report of the Intergovernmental Panel on*

Climate Change, Metz, B., Davidson, O. R., Bosch, P. R., Dave, R. and Meyer, L. A. (eds), Cambridge University Press, Cambridge.

IPCC (2007b). 'Summary for Policymakers', *Climate Change 2007: Mitigation. Contribution of Working Group III to the Fourth Assessment Report of the Intergovernmental Panel on Climate Change*, Metz, B., Davidson, O. R., Bosch, P. R., Dave, R. and Meyer, L. A. (eds), Cambridge University Press, Cambridge, pp. 1–24.

IPCC (2007c). 'Mitigation from a Cross-sectoral Perspective', *Climate Change 2007: Mitigation. Contribution of Working Group III to the Fourth Assessment Report of the Intergovernmental Panel on Climate Change*, Metz, B., Davidson, O. R., Bosch, P. R., Dave, R., Meyer, L. A. (eds), Cambridge University Press, Cambridge, pp. 619–690.

Jochem, E., Jaeger, C., Battaglini, A., Bradke, H., Cremer, C., Eichhammer, W., Förster, H., Haas, A., Henning, E., Idrissova, F., Kasper, B., Köhler, J., Köwener, D., Krause, J., Lass, W., Lilliestam, J., Mannsbart, W., Müller, M., Meißner, F., Pflüger, B., Radgen, P., Ragwitz, M., Rauschen, M., Reitze, F., Riffeser, L., Saure, K., Schade, W., Sensfuß, F., Toro, F., Walz, R. and Wietschel, M. (2008). *Investitionen für ein klimafreundliches Deutschland*, Studie im Auftrag des Bundesministeriums für Umwelt, Naturschutz und Reaktorsicherheit. BSR-Sustainability, European Climate Forum (ECF), Fraunhofer-Institut für System- und Innovationsforschung (ISI), Öko-Zentrum NRW, Potsdam-Institut für Klimafolgenforschung (PIK).

Kaldor, N. (1957). A Model of Economic Growth, *Economic Journal*, 67, 591–624.

Kaldor, N. (1972). The Irrelevance of Equilibrium Economics, *Economic Journal*, 52, 1237–1255.

Kaldor, N. (1985). *Economics without Equilibrium*, Cardiff Press, Cardiff.

Köhler, J., Grubb, M., Popp, D. and Edenhofer, O. (2006). The Transition to Endogenous Technical Change in Climate-Economy Models: A Technical Overview to the Innovation Modeling Comparison Project, *Energy Journal*, Endogenous Technological Change and the Economics of Atmospheric Stabilisation, Special Issue #1, 17–56.

Lee, J., Veloso, F., Hounshell, D. and Rubin, E. (2010). Forcing Technological Change: A Case of Automobile Emissions Control Technology Development in the US, *Technovation*, 30, 249–264.

Liddle, B. (2012). The Systemic, Long-run Relation Among Gasoline Demand, Gasoline Price, Income, and Vehicle Ownership in OECD Countries: Evidence from Panel Cointegration and Causality Modeling, *Transportation Research Part D: Transport and Environment*, 17, 327–331.

Luderer, G., Bosetti V., Jakob M., Leimbach M. and Edenhofer O. (2011). The Economics of Decarbonizing the Energy System: Results and Insights from the RECIPE Model Intercomparison, *Climatic Change*, 114(1), 9–37.

Magné, B., Kypreos, S. and Turton, H. (2010). Technology Options for Low Stabilization Pathways with MERGE, *Energy Journal* Special Issue #1, 31, 83–107.

Messner, S. and Schrattenholzer, L. (2000). MESSAGE-MACRO: Linking an Energy Supply Model with a Macroeconomic Module and Solving it Iteratively, *Energy Journal*, 25, 267–282.

Newell, R. (2010). The Role of Markets and Policies in Delivering Innovation for Climate Change Mitigation, *Oxford Review of Economic Policy*, 26, 253–269.

Nordhaus, W. (2002). 'Modeling Induced Innovation in Climate Change Policy', in Grubler, A., Nakicenovic, N. and Nordhaus, W. (eds). *Technological Change and the Environment*, Resources for the Future Press, Washington, DC.

Popp, D. and Newell, R. (2012). Where Does Energy R&D Come From? Examining Crowding Out From Energy R&D, *Energy Economics*, 34, 980–991.

Scott, M. (1989). *A New View of Economic Growth*, Clarendon Press, Oxford.

Setterfield, M. (ed.) (2002). *The Economics of Demand-led Growth — Challenging the Supply-side Vision of the Long Run*, Edward Elgar, Cheltenham.

Sorrell, S., Speirs, J., Bentley, R., Brandt, A. and Miller, R. (2010). Global Oil Depletion: A Review of the Evidence, *Energy Policy*, 38, 5290–5295.

Strachan, N., Foxon, T. and Fujino, J. (2008a). Policy Implications from the Low-Carbon Society (LCS) Modeling Project, *Climate Policy*, 8, S17–S29.

Strachan, N., Kannan, R. and Pye, S. (2008b). Scenarios and Sensitivities on Long-term UK Carbon Reductions using the UK MARKAL and MARKAL-Macro Energy System Models, UK Energy Research Council Research Report 2. Available online at: www.ukerc.ac.uk/support/tiki-download_file.php?fileId=200. Accessed 25 February 2014.

US CCSP (2006). Scenarios of Greenhouse Gas Emissions and Atmospheric Concentrations, Report by the U.S. Climate Change Science Program and approved by the Climate Change Science Program Product Development Advisory Committee.

van Vuuren, D., den Elzen, M., Lucas, P., Eickhout, B., Strengers, B., van Ruijven, B., Wonink, S. and van Houdt, R. (2007). Stabilizing Greenhouse Gas Concentrations at Low Levels: An Assessment of Reduction Strategies and Costs, *Climatic Change*, 81, 119–159.

van Vuuren, D., Stehfest, E., den Elzen, M., Kram, T., van Vliet, J., Deetman, S., Isaac, M., Goldewijk, K., Hof, A., Beltran, A., Oostenrijk, R. and van Ruijven, B. (2011). RCP2.6: Exploring the Possibility to Keep Global Mean Temperature Increase Below 2°C, *Climatic Change*, 109, 95–116.

Chapter 5

The Economic Feasibility of Policies for Decarbonisation

Terry Barker, Annela Anger and Hector Pollitt

5.1. Introduction

This chapter is concerned with the *economic* feasibility of the rapid decarbonisation of both national and the global economies described in the previous chapter. At the global level, we are interpreting decarbonisation in terms of the likelihood of achieving the 2°C target in terms of both the accumulated global anthropogenic CO_2 emissions 2000–2050 and the reduction of annual emissions to 50% of 1990 levels. Chapters 6 and 7 consider feasibility from the point of view of the engineering. What technical limits are there in replacing the use of fossil fuels by low-GHG (greenhouse gas) alternatives? And how far can processes and products be re-engineered to reduce the emissions of GHGs? These technical limits must be respected in assessing the economic feasibility of the policies and measures.

This chapter examines economic feasibility in terms of how high any explicit carbon price has to be set, how many sectors and countries it has to cover, what additional regulations are required and how the decarbonisation can be achieved at low economic cost or even benefit in macroeconomic performance (e.g. GDP) and employment through the portfolio of policies and measures we have discussed in Chapter 4. See Knopf *et al.* (2010) for a more general view of feasibility.

Policy instruments are considered here in groups, with some rooted in economic signals, such as carbon prices, and others relying on regulation through standards for energy use or for GHG emissions from buildings, equipment and vehicles. The different economic instruments are considered with a focus on how large these signals and forces need to be to drive decarbonisation. The chapter considers the evidence, using the energy-environment-economy model at the global level (E3MG), that these economic instruments are effective at driving such large changes in material and energy supply and demand. It assesses the interaction between the main groups of instruments, for example how regulation can support and strengthen the effect of carbon-price incentives on the switch to renewable electricity generation or electric vehicles. This helps in designing the mix of instruments and their strengths needed to achieve the overall GHG reduction target discussed in Chapter 4. This mix includes the kinds of regulatory instruments that must accompany the economic instruments.

Chapter 3 discussed and presented the scenario approach to modelling and assessing climate change mitigation. Chapter 4 reviewed the modelling of E3 systems, with details of the model we apply in this chapter in comparison with other modelling approaches. This chapter also reports the Reference and Decarbonisation Scenarios judged necessary to achieve the 2°C target and lower-growth alternative scenarios both for the reference and decarbonisation cases. Chapters 6 and 7 cover the engineering feasibility of decarbonisation from technological and sectoral perspectives, respectively. The engineering approach tells us how much improvement in energy efficiency we can expect from a given investment in buildings, equipment or vehicles. However it does not tell us how much investment will be undertaken, since this depends on the level of demand for the services concerned (comfort, light, mobility and power), the regulations and the various prices and costs. In addition the engineering approach does not tell us about what behavioural changes are needed, why apparently cost-effective mitigation is not actually taken up by business or households. The regulations cover standards for efficiency. The prices and costs include the price of electricity, the cost of fuels, the effects of any carbon tax or emission permit requirements, the prices of the assets, as well as any subsidies.

The current chapter covers the feasibility of policies, in particular the issue of whether a carbon tax is necessary in the portfolio of policies. It also covers the implications of these policy portfolios and trajectories of emission reduction for macroeconomic performance in many different categories of nations with differing economies. The goal is to determine whether some of the scenarios that pass the engineering tests might prove more or less desirable in purely economic terms, including equitable distribution of wealth and access to energy.

An important feature of economic feasibility is the opportunity of a reduction in costs of new technologies as the scale of production increases. There have been major reductions in unit costs of solar and wind technologies for power generation as governments have provided incentives for adoption of the technologies. Learning-by-doing is discussed in Chapter 6, but the general feature of induced technological change from a carbon price is discussed in this chapter.

However one problem for economic feasibility is that as the speed and scale of the investment in low-GHG technologies increases, the producers reach capacity and are limited in how fast they can increase their own investment. The prices of the low-GHG equipment and materials can then rise as a result of shortages of capacity. The increase in prices and the shortages in supply may then reduce the growth in demand, creating a boom–bust cycle leading to bankruptcies and waste of resources in the equipment industries and damaging the investment prospects.

5.2. The Price of Carbon and Decarbonisation

5.2.1. *The carbon price as representing a global economic externality*

The price of carbon is a social cost, achieved explicitly through carbon taxes and emissions trading schemes (ETS) and implicitly through regulation. If the carbon tax is applied comprehensively on all use of fossil fuel that leads to CO_2 emissions into the atmosphere, then it will provide a pervasive signal throughout the economy leading to reduced demand for goods and services according to their contribution to the CO_2 emissions.

Past anthropogenic GHG emissions are an externality that affects the global economy through the accumulation of GHG in the atmosphere, which leads to climate change in the present and, due to the long lags in responses, in the future. If the emissions are going to increase, then their impacts are going to be more intensive and extensive on the global economy through increasing the risks of extreme weather events, rising temperatures and sea levels, leading to damages to human life and health, damages to ecosystems and biodiversity (IPCC, 2007).

In traditional economic theory, this externality is converted into a social cost by putting a monetary valuation on the damages of climate change over the coming centuries. This allows for a cost–benefit analysis to account for the full costs and benefits of climate change, adaptation and mitigation to achieve the optimal amount of GHG emissions and the appropriate carbon tax in the market, thereby avoiding some, but not all, of the consequences of climate change.

However, estimation of current day and future external costs of climate change are subject to high uncertainties. These uncertainties stem from, for example, insufficient scientific knowledge about the extent and nature of climate change damages, as well from a limited understanding of future economic systems. For example, accounting for costs of low-probabilistic extreme weather events involves the valuation of loss of life and property on a huge scale across generations and countries. This makes it problematic and controversial to use the one-dimensional (using money as the measure) cost–benefit analysis to assess the effects of a climate change mitigation policy (Barker, 2008; Weitzman, 2009). These challenges in calculating external costs can make it meaningless if not impossible to find the optimal cost of carbon. The alternative is to adopt a target reduction in GHG emissions, and assess the cost effectiveness and other attributes of policies and measures to reach the target. This is the alternative adopted in this book and addressed in this chapter.

For the carbon price to exert the effects expected in theory, it needs to be reliable. The more reliable the carbon price, the higher the investments in CO_2-reducing equipment and technologies and the greater the benefits from decarbonising the economy. In addition there is a second economic externality associated with innovation in GHG-reducing technologies (Jaffe *et al.*, 2005). Markets are not as good as they should be at driving innovation, because innovators are not able to capture all the benefits that accrue from innovation, even with the patent system. This is why policies for decarbonising also require, according to traditional theory, other policies to address the inadequate level of innovation, such as public investment, subsidy and regulation in addition to the carbon price. For example, carbon tax revenues and auctioning revenues from carbon markets could be used to invest in R&D (research and development) in low-carbon technologies and reduce or offset the financial risk of implementing new technologies.

The price of carbon needs to be pervasive covering all countries and economic sectors. Only then can carbon leakage be avoided and emission reductions be met at a lowest possible cost. Carbon leakage is an increase of carbon dioxide emissions in a country/region/industry that are not covered with a carbon reduction measure as a response of carbon reduction measure imposed on another country/region/industry. The other effect related to carbon reduction measures is the so-called rebound effect. A rebound effect is likely to occur if the amount of energy required per a unit of physical output is reduced and therefore producing a unit of output becomes cheaper and/or more energy becomes available, which results in smaller reductions in energy use and related emissions than expected. Rebound effects can be direct and indirect. Direct rebound occurs in the same industry where energy efficiency was improved. Indirect rebound effect affects other industries through changes in relative prices or through availability of excess energy.

5.2.2. *Policy instruments to generate a price of carbon*

The most straightforward way of imposing the social cost of global warming is by a universal carbon tax. However, in practice, no government has used a carbon tax as the sole instrument for reducing CO_2 emissions. The most frequently used policy is regulation in the form of energy efficiency standards, i.e. insulation standards for dwellings. Market-based instruments, such as carbon and energy taxes and emissions trading, are also widely in use, but at low rates and with very limited coverage. Voluntary carbon reduction measures such as offsetting and voluntary carbon trading are growing but are uncertain in their effects and so far seem unlikely to achieve substantial reductions in emissions. Economists prefer market-based instruments, such as emissions trading, to command-and-control measures that can ensure the same results, because they have the ability to achieve the same results at a lower cost (Hansjürgens, 2005). In addition to this, market-based instruments are often promoted because of their potential ability to promote innovation (Fisher, 2005). Hence, emissions trading should promote sustainable development by encouraging companies to use greener technologies and practices in order to comply with the trading scheme. However, there are others that argue that market-based instruments, compared with command-and-control measures, gives companies a "licence to pollute" (Kelman, 1981), and therefore, the "polluter pays principle" is not seen as completely ethical. The main argument here is that if a company can and is happy to pay for its pollution, then it can carry on polluting and not care about the damages caused to others by its activities.

5.2.3. *Achieving decarbonisation by price alone*

One important question is how high any carbon price must rise to achieve the 2°C limit for average temperature increase to 2100 above pre-industrial levels. This depends on when the price takes effect, how many countries impose the carbon price, which sectors are covered, how the price is imposed, whether by an ETS or carbon taxes, and whether the carbon price increase is anticipated, so that those who have to pay can adjust their consumption and production patterns in advance.

We have experimented with E3MG to assess how high the carbon price has to be to achieve the decarbonisation targets without regulation. A global carbon price scenario is the typical instrument assumed in most modelling of GHG mitigation (IPCC, 2007: Chapter 3), although we have phased in the increases starting at different rates in different groups of countries, reflecting political feasibility. In E3MG we have imposed the carbon price as a mixed ETS and carbon tax regime, with the trading scheme being introduced region-by-region in the energy-intensive sectors, mainly power generation, and the carbon tax throughout the rest of the economy. For each region, the carbon prices increase year by year after they are introduced. The effect of introducing the carbon price in the industrialised countries first, including China, is that the average global price increases faster through the period as more and more countries are covered.

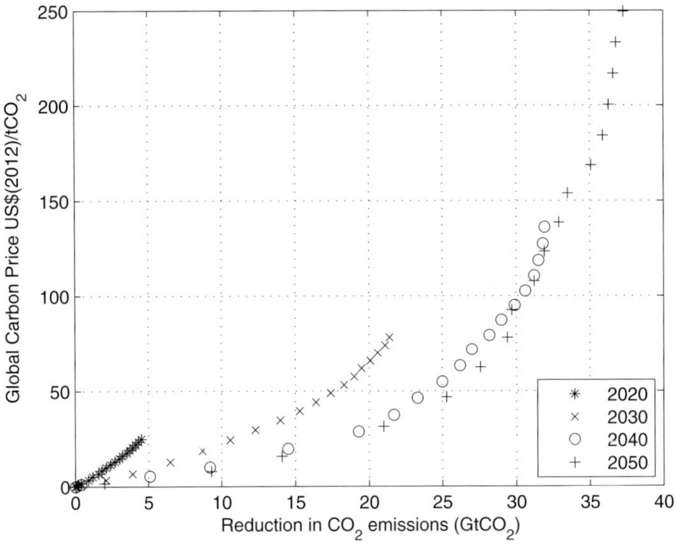

Figure 5.1. Reduction in global CO_2 emissions for a range of average carbon prices for periods ending with the years 2020, 2030, 2040 and 2050.
Source: E3MG calculations.

We have generated 19 dynamic solutions of E3MG annually 2009–2050, with the carbon prices set at the same profile over the period as those of the ETS price in the main Decarbonisation Scenario, but with the prices in the profile being scaled by different amounts. Some of the results for the carbon prices for selected years and the consequent reductions in global CO_2 are shown in Figure 5.1. The figure shows a series of weighted average global prices in terms of US\$/tCO_2 in 2012 prices for the years 2020, 2030, 2040 and 2050, and the reductions in CO_2 for these years that are associated with the price trajectories. Note that in each region the carbon prices rise each year after they are introduced and the emission reductions shown in the figure are partly the results of previous years' carbon prices.

Figure 5.1 illustrates four features in the responses to carbon prices modelled in this version of E3MG. The first is that as the prices rise the reductions in CO_2 are fairly regular, non-linear but monotonic in the expected direction, i.e. a rise in the carbon price leads to a reduction in emissions. At the global level, the solutions are stable. The second feature is that as more time is allowed for the responses, the sizes of the responses are higher. For any of the carbon prices shown above, the later the year, the higher the reduction. This is partly because over the period more regions introduce carbon prices so that the opportunities for low cost reductions increase, but it is also because the model allows for lagged responses. The third feature is that the responses are much larger for low prices and became almost zero at very high prices. It is apparent from Figure 5.1 that as the carbon price rises it has a smaller and smaller effect on CO_2 emissions. The easy options in reducing CO_2 become increasingly exhausted. A carbon price rising to US\$91/tCO_2 (2012 prices) is estimated to reduce CO_2 emissions by 50% below the reference case. However, by 2050 the highest carbon price in the solution of nearly US\$250/tCO_2 reduces CO_2 emissions by 55% below the reference case, some 33 Gt CO_2, well below the 46 Gt CO_2 reductions required to meet the 2°C target. The fourth feature of Figure 5.1 is that by the last decade of the projection, the reductions in CO_2 at lower CO_2 prices are largely exhausted as the high prices cover most regions of the world and further reductions become increasingly expensive.

It is clear that we have estimated and modelled some GHG emissions as being rather unresponsive to relative prices. Of course, it is impossible for any ETS or carbon tax to cover all emissions. There will always be leakage and evasion, especially with high carbon prices. The rates of leakage and evasion will also depend on the fines and other costs of non-compliance. There is also a risk that in some countries the use of coal begins to increase again as demand for energy continues to rise and alternative energy sources are either unavailable or more expensive, even allowing for a high carbon price. We reach the conclusion that a carbon price alone will not achieve the necessary reductions in GHG emissions without covering all countries and sectors and rising to very high levels that seem politically impossible.

5.3. Emissions Trading Schemes and Other Policies

5.3.1. *Allocation of ETS allowances by benchmarking or auctioning*

The allocation of emission allowances to the entities (e.g. companies) under the trading scheme could be set on the basis of their historical emissions, i.e. by "grandfathering" (e.g. on average emissions of the last three years) or using a type of "benchmarking" (e.g. output-based method that builds on best practices and/or best available environment friendly technologies). "Grandfathering", especially if the allocated amount of allowances is updated after a certain time period, is argued to create perverse incentives to energy-intensive industries to maintain their high emissions in order to get a higher industry-wide cap for the next trading period (Neuhoff *et al.*, 2006). The benchmarking method is argued to cause less distortion on the market — i.e. it has less impact on competitiveness and causes less leakage (Demailly and Quirion, 2006).

It is argued that free allocation of emission allowances can provide an incentive for pollution sources to identify themselves and report their emissions in order to obtain their allowances (Stavins, 2005).

Revenue-raising instruments such as auctioned emission allowances and carbon taxes, with revenue recycling that reduces other taxes, reduce emissions at a lower cost for society than free emission allowances or other market-based instruments with no revenue recycling and at a lower cost than command-and-control methods (Stavins, 2005; Sterner and Hammar, 2005).

The benefits of revenue recycling are argued to be higher where cuts in employers' payroll taxes increase labour demand, especially where there is unemployment (Ekins and Barker, 2001). The same benefits could be reached if the revenues from selling free allowances owned by state-owned companies are redistributed, i.e. invested in retraining and promoting entrepreneurship in areas with unemployment. In other words, if a company is owned by the state, then the state can use the revenue from excess allowances in the same way as it uses the revenue from carbon taxes or auctioned carbon allowances. It has been stressed that evaluation of the effects of carbon taxes is incomplete without analysing how the tax revenues are used (Ekins and Barker, 2001). The same holds for a carbon trading scheme in the case of auctioning initial emission allowances or selling excess allowances of a state-owned company.

From a microeconomic perspective, there is no difference whether the initial allowances are auctioned or distributed free-of-charge for firms that are covered by an ETS, provided that the amount of allocated allowances is independent of their current behaviour (Demailly and

Quirion, 2006). Freely distributed allowances imply an opportunity cost that firms add to their marginal production cost. Therefore, under profit maximisation, whether the allowances are auctioned or freely distributed does not impact firm's production or emissions, but affects their profits and stock value (Demailly and Quirion, 2006; Keppler and Cruciani, 2010). Tietenberg (2002) argues that the allowance distribution method used does not affect the cost-effectiveness (efficiency) of the ETS, since the transferability of allowances "allows them to ultimately flow to their highest-valued users".

The European Union Emissions Trading Scheme (EU ETS) (Directive 2003/87/EC, 2003) and GHG emissions trading under the Kyoto Protocol (Kyoto Protocol, 1997) are examples of ETS. The EU ETS is currently the world's largest ETS and the first that crosses country borders. The EU ETS includes CO_2 emissions from energy-intensive industries in the EU (Directive 2003/87/EC, 2003). Companies, that belong to these industries (currently about 10,000), trade among themselves. It came into operation on 1 January 2005 and is the centrepiece of the current European climate change policy. Its second phase started in 2008. Phase 1, from 2005 to 2007, failed to deliver a substantial carbon price – the price of a European Union Allowance Unit (EUA) was close to zero in 2007 and prices in Phase II (2008–2012) also became very low due to the global financial crisis and recession in Europe; this has left an "overhang" (excess) of allowances in Phase III (2013–2020) that again is depressing carbon prices. The design of the EU ETS post-2020 is currently under review.

In the E3MG modelling of the various ETS, we have assumed that all the allowances are auctioned and the revenues used to reduce employment and other taxes. However we have also assessed the outcome if allowances are benchmarked.

5.3.2. *The effects on GDP of recycling revenues from carbon taxes and permits*

One issue in the GHG mitigation literature has been whether the use of revenues from carbon taxes or the auctioning of emission permits or allowances could lead to an increase in employment and GDP (gross domestic product). The issue is part of a wider one, that of environmental tax reform (ETR) covering all environmental taxes, the reform being to shift the tax burden from the "goods" of employment creation to the "bads" of pollution-creating activities, such as the burning of fossil fuels. The debate has given rise to the "double-dividend hypothesis", the two dividends being (1) the environmental one from reduction in environmental pollution, e.g. GHGs from the increase in carbon price, and (2) the increase in economic activity and employment from the reduction in employment taxation, if that was used as the recycling channel.

Ekins and Speck (2011) review the literature and provide studies on the modelling of ETR in the EU. Earlier surveys are by Bosquet (2000) and Patuelli *et al.* (2006). The main findings from the many national and regional studies are that ETR usually leads to substantial reductions in environmental pollution, and if the revenues raised are used to reduce employment taxes (usually social security contributions by employers) then employment increases both in the short term and long term, but by rather small percentages. The effects on GDP reported in the studies are very small and mixed. The revenues raised are not sufficient to generate substantial increases in employment.

In the context of GHG mitigation, the recycling issue becomes one of how the revenues from the auctions of emission allowances or carbon taxes are used and what effect the way they are used has on the growth of GDP and employment. If the CO_2 emission allowances are assumed to be auctioned, there is a stream of new government revenues. For example, we have assumed in the Decarbonisation Scenario that in Annex 1 parties (most industrialised countries including economies in transition such as Russia) the revenues are used to reduce employers' contributions to social security, and in the rest of the world the revenues are used to reduce direct tax rates. The results of using E3MG to assess this effect confirm those in the literature. The effect of recycling the revenues from auctioning of allowances for the energy sector is to raise GDP and employment in the regions and periods covered by the ETS: by 2050, global GDP is 0.8% higher than the reference case and employment is 0.2% higher. The effects of recycling revenues are higher if a carbon tax is imposed on all sectors of the economy as well as those covered by the ETS, since the energy sector is responsible for some 30–40% of CO_2 emissions from the burning of fossil fuels, although this percentage rises in the Reference Scenario as developing countries increase their use of coal for electricity generation.

Figure 5.2 shows the reference case and two low-carbon scenarios from E3MG to illustrate the effects of recycling revenues from government receipts from auctions in ETS for the energy sectors and from carbon taxes for all the other sectors. In both low-carbon scenarios the carbon price covers all sectors in the regions when it is introduced, and with that price rising to a global average (2012 prices) of US\$154/tCO_2. When no revenues are recycled, the revenues from allowance auctions and carbon taxes are used to reduce governments' financial deficits. With recycling all revenues are used to reduce employers' taxes in the industrialised countries (Annex 1 parties) and to reduce direct taxes in other countries.

The main feature of Figure 5.2 is that the uses of revenues have very small effects on CO_2 emissions or on GDP. CO_2 emissions fall to about 55% below the reference case by 2050, with the absolute levels falling by less when revenues are recycled because the higher GDP leads to more

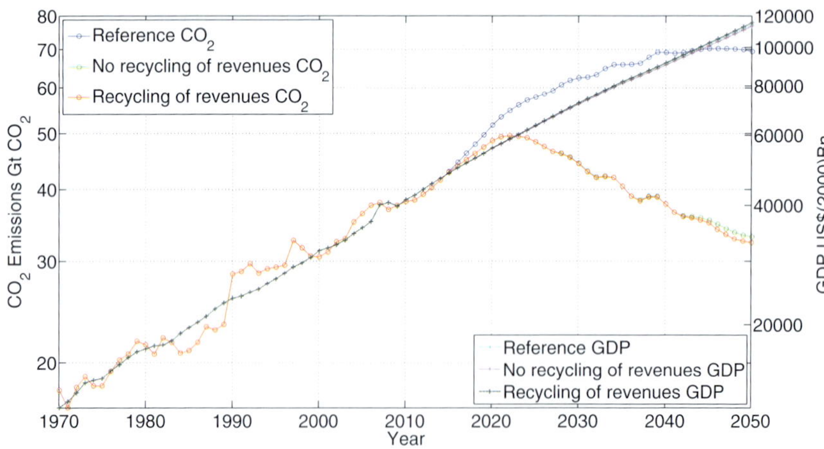

Figure 5.2. The effects of recycling of revenues from carbon prices on levels of global GDP and CO_2 emissions 2010–2050.

Source: E3MG calculations.

Note: There are discontinuities in the historical series.

use of energy and fossil fuels. GDP is close to the reference case projection throughout the period.

The effects of the recycling assumption on GDP show up more clearly in Figure 5.3, which presents the same results as Figure 5.2 but as percentage differences from the reference case for each year. Global GDP is 1% below the reference case by 2050 with no recycling of revenues but about 1.5% above the reference case with recycling. When revenues are recycled, GDP moves immediately above the reference case, because the extra spending from increased employment and consumer incomes adds to the stimulation of the world economy. The response of GDP shows a clear profile of an increasing response through the 2020s. This is the effect of the extra investment associated with the increasing regulations for energy efficiency and low-carbon equipment and vehicles over this period, with the growth in the extra investment slowing after 2030, then the investment declining. The effect here excludes any response of reduced oil prices as world oil demand is reduced below the reference case. This effect is discussed in the next section.

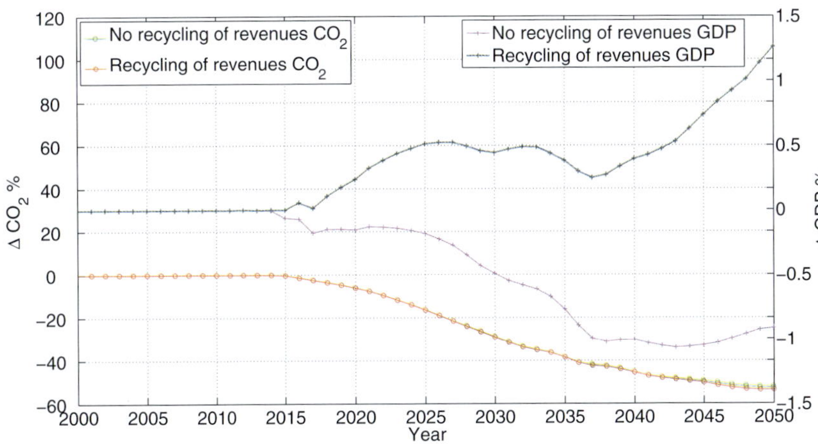

Figure 5.3. Percentage difference from reference case for the effects of recycling of revenues from carbon prices on global GDP and CO_2 emissions 2010–2050.

Source: E3MG calculations.

It should be emphasised that these are very small differences since they are spread over the whole projection period: global GDP growth 2010–2050 is between 2.53% pa (per annum) and 2.59% pa depending on the recycling of revenues, with the reference case showing growth of 2.55% pa. The effect of a carbon price leading to lower growth of GDP corresponds to the estimated effects of mitigation found in many results in the literature that do not include recycling of revenues via reductions in other taxes (IPCC, 2007).

5.3.3. *World oil prices and decarbonisation*

Figure 5.4 shows the projections of oil prices in the reference case, the Decarbonisation Scenario and a scenario in which energy-security policies lead to a major shift in transportation away from use of oil to use of electricity. The oil price in the Decarbonisation Scenario is less than US$123 per barrel (/bbl) (2012) prices by 2050, compared to an average of US$155/bbl in the Reference Scenario in 2050 and US$112/bbl in 2012. The lower oil price when regulations are introduced to reduce transport demand for fossil fuels shows how much these contribute to the overall slower growth in oil prices when the world economy is decarbonised.

This lower trajectory for increasing oil prices has a major effect on both CO_2 emissions and the world economy. This can be shown by adopting the same assumptions for oil prices in the Decarbonisation Scenario as the reference case. If oil prices are much higher, by about 40% above those in the central Decarbonisation Scenario, then the reduction in oil demand will lead to CO_2 emissions 30% lower by 2050, from an already low level. Other energy prices are also higher in this high-oil-price Decarbonisation Scenario, although not nearly as much as those for oil. The main way in which oil and other energy prices might be the same as in the reference case comes through a response to the decarbonisation policies by those investing in oil production, to essentially stop further investment in new facilities. The higher oil prices are accompanied by 1.5% lower world GDP by 2050. However, it seems unlikely that oil and other fossil fuel prices will not be lower when demand falls and the results are reported as an indication of the role played by world fossil fuel prices in offsetting carbon prices at the global level.

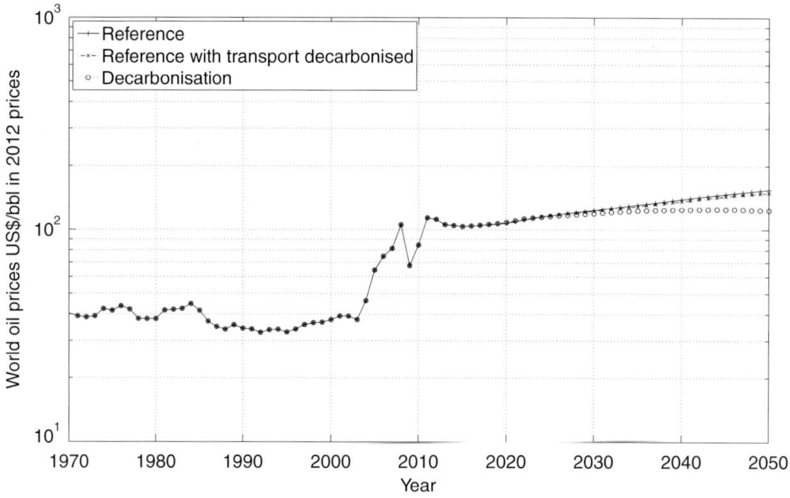

Figure 5.4. World oil price in selected scenarios.

Source: E3MG calculations.

5.4. Regulations for Decarbonisation of Buildings, Industry and Transport

As discussed in Chapter 4, additional regulation in the form of higher standards for energy efficiency, GHG emissions, and performance, is a critical complement to the introduction of carbon prices in achieving effective and efficient reduction in GHGs. We have already discussed above how carbon prices alone are infeasible for reaching the GHG reduction targets. Chapter 4 has covered the details of the regulations and how they are included in the reference case; this chapter provides some analysis.

The regulations are modelled through three effects introduced exogenously into E3MG, sector by sector and region by region. The first is through an increase in investment in energy-saving materials, equipment and vehicles. The second effect is the increase in prices paid by those using the more energy and carbon-efficient investment, e.g. the increase in the capital cost of electric vehicles compared to gasoline-powered vehicles. The third effect is the reduction in coal, oil and gas and the increase

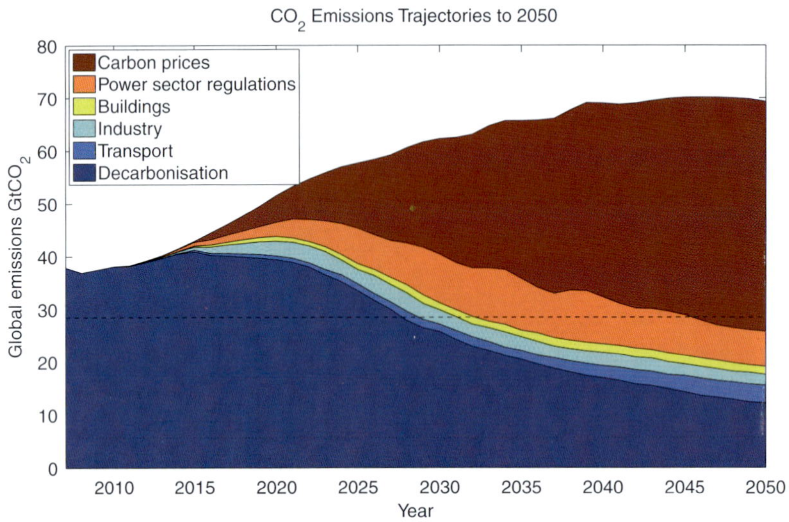

Figure 5.5. Contributions of carbon prices and regulations for the power sector, building, industry and transport to the Decarbonisation Scenario.

Source: E3MG calculations.

in electricity used by each sector. These effects are introduced throughout the period and work though the world economy, affecting output, trade, consumption, investment and employment.

Figure 5.5 shows the contributions of these regulations, calculated by removing them cumulatively, sector by sector, from the Decarbonisation Scenario and resolving E3MG each time. The regulations for the power sector to reduce the carbon intensity of electricity generation successively over time are combined with extra incentives for renewables, Carbon Capture and Sequestration (CCS) and nuclear generation, such as feed-in tariffs. Note that the carbon price effects, mainly working through increases in costs of electricity, support the switch to low-carbon electricity on the supply side, and also reduce the electricity use by the non-electricity sectors. These results include the effects of any carbon leakage from the differentials between carbon prices in different countries and regions distinguished in the model. However they do not include any leakage due to differences in regulatory pressure. This effect is very uncertain, since it could work both ways because the higher standards imposed also protect the domestic market.

The effects build up 2015–2020 and then much more gradually to 2050. The effects on CO_2 of building regulations are the weakest. The effect of industry regulations are more focused in the beginning of the period and those on transport on the end. It is important to note that without these regulations, the overall target reduction in GHGs would not be met. However, their effects are surprisingly small by 2050 in relation to those of the carbon prices. One reason is that regulations are already included in the reference case, being an important component of energy policy in many countries already. Another reason for the small effect of the additional regulations is the rebound effect. We have investigated this effect with E3MG (Barker *et al.*, 2009) and concluded that for a package of energy-efficiency measures covering the transport, residential and services buildings and industrial sectors of the economy considered as non-regret in the IPCC's (Intergovernmental Panel on Climate Change) AR4 WG3 report (IPCC, 2007), the total rebound effect by 2030 would be over 50%. The other reason is that the changes in standards are stronger when they are first introduced and they erode away over time. In contrast, the carbon prices and their effects are increased more rapidly over time. Furthermore the increasing carbon prices induce technological change.

The importance of which sector should decarbonise first can be illustrated by the effects of transport regulations. The electricity sector has the most low-cost opportunities for decarbonisation, partly because it is responsible for most of the emissions of CO_2 from coal, with the highest CO_2 content per unit of useful energy. When carbon prices are introduced, there is an immediate switch to gas and a slower switch to renewable sources in power generation. Without carbon prices or regulation for the power sector, it would continue to burn large amounts of coal as in the reference case. If transport were to be converted to using electric vehicles in this scenario, then the extra electricity required would come from burning coal, and CO_2 emissions could be higher than in the reference case. Such a scenario was compared to the effect of removing transport regulations from the Decarbonisation Scenario. Decarbonising transport on its own in the reference case leads to substantially higher CO_2 emissions over the pathway to 2050. In contrast, when transport is regulated in a scenario with all other sectors decarbonised, the CO_2 emissions are substantially reduced. However there is no comparable difference in the GDP effects, which depend on the investment in low-emission vehicles, which are much the same in all the scenarios with additional transport regulation.

5.5. Reduction of Fossil Fuel Subsidies

Fossil fuels, for example kerosene, are subsidised in some countries. This means that the consumption of fossil fuels is encouraged and supported in one or another form by giving a price advantage to this type of fuels. Figure 5.6 shows the scale of the subsidies, at about 0.5% of world GDP, the fact that they have been rising over the Great Recession, and the large share of petroleum products in the subsidies. IEA (International

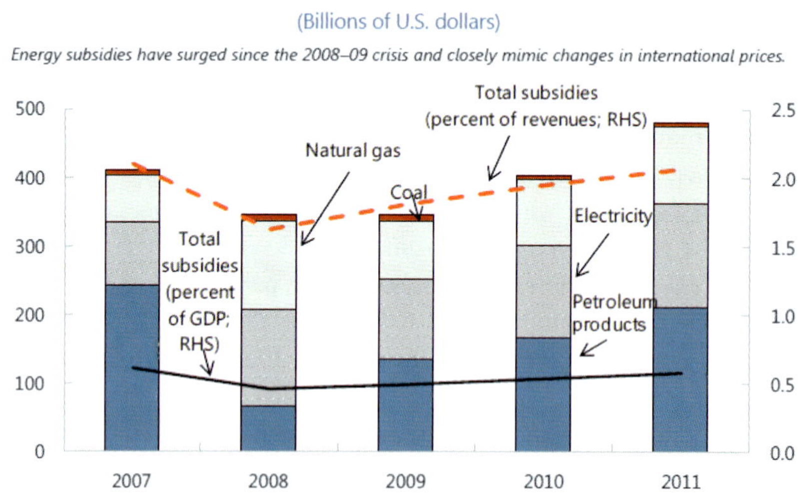

Sources: Staff estimates, Organisation for Economic Co-operation and Development, International Energy Agency, Deutsche Gesellschaft für Internationale Zusammenarbeit, IMF *World Economic Outlook*, and World Bank.

[1]Data are based on most recent year available. Total subsidies in percent of GDP and revenues are calculated as total identified subsidies divided by global GDP and revenues, respectively.

Figure 5.6. Energy subsidies (pre-tax) by energy carrier.

Source: IMF (2013: p. 10).

Energy Agency) *et al.* (2011) give four main reasons for subsidising fossil fuels:

1. Alleviating fuel poverty.
2. Boosting domestic supply.
3. Redistributing national resource wealth.
4. Protecting employment.

Fossil fuel subsidies encourage consumption of fossil fuels and therefore impact directly carbon emissions from fuel burning. Removing fossil fuel subsidies has a similar effect to a price increase or tax. The scale of the impact of higher prices on demand for fossil fuels will depend on the price elasticity of demand. Reduced demand for fossil fuels will result in lower emissions.

The scale of the fossil fuel subsidies in many developing countries is substantial and, in many low-income countries, such subsidies greatly exceed spending on health (World Bank, 2009). Most studies conclude that they lead to waste, over-consumption of fuel and excess pollution, and that although benefits to the poor are often used as a justification, these tend to be small (IEA, 2008; World Bank, 2009). Victor (2009) further argues that whilst their removal would provide substantial win–win outcomes by reducing pollution and improving economic performance, they are often supported by entrenched interest groups (see Shenoy (2010) for a review of the failure of attempts to remove kerosene subsidies in India). The economic case for subsidising alternative, low-carbon products and processes, meanwhile, relies on subsidies leading to the development of new markets and the realisation of economies of scale as demand and production increase so that the new technologies eventually become self-financing. Where governments maintain fossil fuel subsidies alongside new subsidies for low-carbon technologies, however, this creates inherent economic inefficiencies in the climate policy mix.

We have included the further reduction in fossil fuel subsidies as a component in the Decarbonisation Scenario, using the assumptions adopted by the IEA in the World Energy Outlook (WEO) 2010. The reduction has small overall effects on global CO_2 emissions, although larger for the main countries concerned. By 2020 global CO_2 emissions are estimated to be about 0.4% lower than the reference case. This reduction is much less than that calculated by the IEA *et al.* (2011: p.7) of 4.7% by 2020, but not all the subsidies have been removed in the Decarbonisation Scenario and much of the reform has been already included in our Reference Scenario.

5.6. Effects of a Policy Portfolio for Decarbonisation on GDP Growth and Employment

5.6.1. *Introduction and previous modelling for IPCC/EMF*

There is a substantial body of academic literature relating to estimating the economic impacts of implementing decarbonisation scenarios. A wide range of economic models has been used and several comparison exercises have been carried out, for example by the IPCC and the Energy Modelling Forum (EMF) (Clarke *et al.*, 2009; IPCC, 2007). These analyses of model results have found that decarbonisation of the scale described in this chapter will cost anything between 0 and 5% of global GDP by 2050, allowing for biomass energy with carbon capture and sequestration. Although this is quite a wide range, the grouping of results is often used to infer a degree of robustness in the findings of the models.

However, this conclusion fails to take into account that the models that produce the results are almost all computable general equilibrium (CGE) models that have their roots firmly in the theories of neoclassical economics. These models typically all rely on similar assumptions (and datasets) and their results reflect these assumptions. If the assumptions turn out not to be valid, then the impression of robustness is illusory.

As explained in Chapter 4, in the results presented in this chapter we use a model that uses a different set of assumptions and a different economic database. The E3MG model comes from a post-Keynesian background and does not rely on many of the assumptions that are common to CGE models. In general an empirical approach is instead adopted although, as discussed below, it is assumed that finance for capital is available. As we shall see below, this is important in determining (and interpreting) the outcomes from the scenarios.

5.6.2. *The effects of the policy mix on global GDP growth*

In this section we examine the effects of different components of the Decarbonisation Scenario. Figure 5.7 shows the impacts of all the decarbonisation policies on annual GDP, as percentage difference from the reference case. In this scenario, which includes revenue recycling and the effects of regulations including the associated additional investment, annual GDP is higher by 3.3% by 2050. This figure is purely the result of economic interactions and does not include any possible (and uncertain) benefits of a more stable climate or the "co-benefits" related to reduced exposure to local air pollution.

The reasons that we see an increase in GDP are:

- There is a large increase in investment in renewable and energy-efficient technologies, which creates jobs, e.g. in construction (see below)
- The use of revenues from market-based instruments is recycled back to the economy through lower income taxes, boosting disposable incomes and consumption
- The reduction in global oil prices reduces the flow of wealth to oil-producing countries that have high savings rates

This result clearly contrasts with the range of results that has been produced by CGE models. The reasons for this relate to underlying modelling assumptions. In particular, the CGE models usually assume some kind of "optimisation" through efficient price adjustment, which means that there are no unused labour or capital resources in the economy. This means that imposing a constraint (carbon or otherwise) on the economy cannot lead to an increase in GDP. In contrast, E3MG allows for the possibility of unused resources in the global economy.

To assess the effects of the policy mix and to aid the comparison between our results and those produced by CGE models, we examined the contribution of the two groups of policies in the mix, which are generally excluded from standard CGE results in comparative exercises, namely revenue recycling and regulations. The two additional

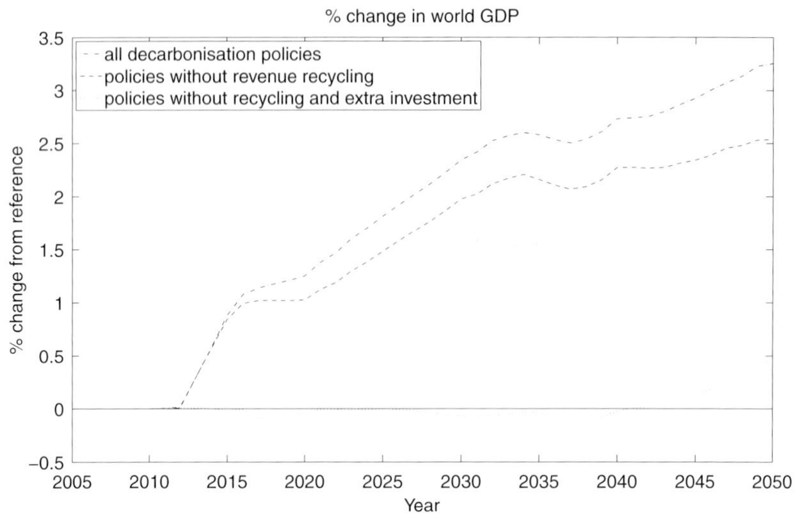

Figure 5.7. The effects of different mitigation policies on GDP as percentage differences from the Reference Scenario.

Source: E3MG.

projections are also shown in Figure 5.7. The dashed line shows the impact on GDP without including the positive effects of the policies of revenue recycling. This approach is quite similar to the "lump sum" approach of revenue recycling that is often (but not necessarily) used in CGE modelling exercises. Our results show that the revenue recycling is responsible for about 0.7% points of the total GDP increase by 2050.

The other variant of the scenario gives an indication of the effect of the assumption of "crowding out" of any extra investment in CGE models. In a CGE modelling approach, a strict version of the savings-investment identity is adopted, in which total global investment in low-carbon technology can only increase if there is either a reduction in investment elsewhere or an increase in savings. The implication is that the available finance in the reference case is allocated optimally for investment purposes. While still respecting the savings-investment identity, the investment constraint is not imposed in E3MG so the assumption is that there is available finance. This could, for example, be finance that is channelled into assets, which has small wealth effects, but little impact on current real economic activity.

The extra investment in our policy mix is associated with regulations, which are also normally excluded from CGE exercises. In this variant we exclude both the recycling of revenues and the group of policies imposing regulations (including their associated extra investment). This represents a worst-case outcome for GDP. The model results show that the positive GDP impact would become almost zero and only positive towards the end of the forecast period. Nevertheless, even under these assumptions with the emissions trading remaining in place, it is notable that there is no substantial cost to society associated with the carbon-price policies.

There is also an important secondary, consequential effect on world demand, which is incorporated into these policy outcomes, but which requires coordinated global action. As the world economy shifts towards low-carbon technologies, the demand for fossil fuel oil will be lower and the world prices of oil will rise more slowly than in the reference case. The lower inflation will raise real incomes and private consumers' expenditures. There are offsets to these increases, the main one being the effects of the increases in costs of electricity to consumers. This consequential increase in GDP is important in contributing to the higher GDP associated with the extra regulations.

Figure 5.8 shows the impacts of the decarbonisation policies on global employment levels. As one would expect, given the GDP impacts, the model results show that total employment could increase by around 40 million jobs (0.9%). For the main part, these are workers that would otherwise be unemployed.

Again, we show the contributions to the results of the two groups of policies, namely revenue recycling and regulations with the extra investment. There is a slightly broader range of results, partly because of the loss of employment in fossil fuel industries. This is the main reason for the employment levels being lower than reference levels in the scenario without the recycling and investment policies. Here carbon prices via emissions trading in the energy-intensive sectors are driving decarbonisation through higher electricity prices. Countries with substantial employment in fossil fuel industries have lower employment compared to the Reference Scenario.

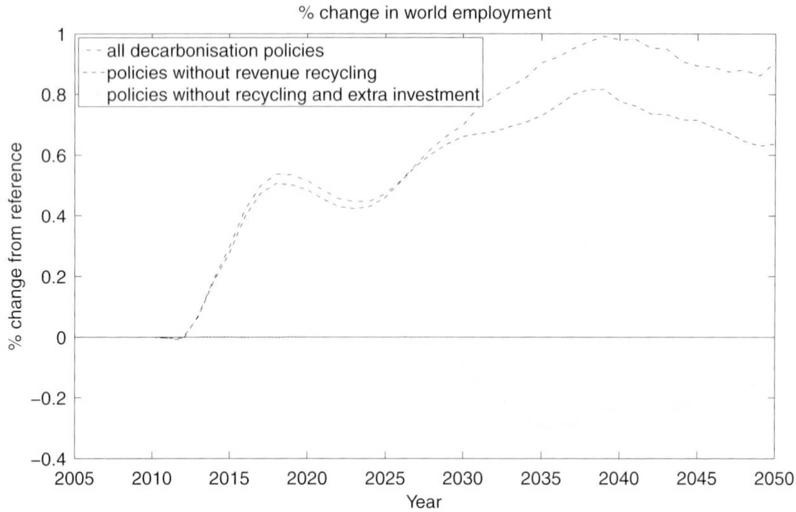

Figure 5.8. The effects of different mitigation policies on employment as % differences from the Reference Scenario.

Source: E3MG.

5.7. Conclusions on the Economic Feasibility of Decarbonisation

The economic feasibility of decarbonisation has been assessed in the chapter by examining the different components of the policy mix chosen to achieve the 2°C target. Most of the discussion has been about the use of carbon prices as the economic main driver of decarbonisation. However we have recognised the political difficulty of applying carbon taxes or ETS to final consumers and in our policy mix we have limited the direct application of carbon prices to the energy-intensive sector and have relied on strengthening regulations and expanding their coverage to reduce energy use and switch to lower-carbon sources of energy.

These policies will have implications for equity and social justice. Climate change is inherently inequitable and unjust; mitigation of climate change helps to reduce this inequity. There are three aspects to the problem. First, extreme weather events have their greatest effect on vulnerable and poor, especially those relying on subsistence agriculture. Second, those people who suffer are, by and large, *not* those who have benefited from the historical and expected future emissions of GHG responsible for the increase in atmospheric concentrations. And third, policies based on conventional cost–benefit analysis are founded on a utilitarian philosophy that averages costs and benefits using average incomes, so the vulnerable and poor are underrepresented.

The policies chosen for GHG mitigation itself also have implications for equity and social justice. Higher prices for electricity will be a burden on those in inadequate housing with obsolete heating and cooling equipment. This inequity can be alleviated by targeting the energy efficiency measures and spending on such buildings. The actual policy mix has been chosen to follow those policies that have actually been implemented by governments, with the additional coverage to other OECD countries and eventually all countries. The policies also include the phased introduction of ETS and increases in the prices of allowances so that the richer countries act first. China is assumed to act at the same time as the richer OECD countries, but with a scheme starting with much lower allowance prices.

The key findings discussed in this chapter can be summarised as follows. First, a carbon price, set through emissions trading primarily in the power sector, is an essential component of the policy portfolio, but it is not enough, even when a carbon tax is imposed on the rest of the economy. Widespread use of regulations both for the power sector and industry sectors and for buildings, equipment and vehicles, are also necessary to support the effects of rising carbon prices on use of fossil fuel energy. These higher prices and regulations need to be supplemented by specific incentives to ensure the take-up of renewable technologies so as to realise economies of scale, reduce costs and increased technological change. The effects of such incentives are already obvious in the reduced unit costs and improved performance associated with the expansion of solar- and wind-generation technologies over the past decade. The need to transform the economy in this way with complementary policies confirms the conclusions by Grubb (2014).

The second key finding is that appropriate policies can lead to "green growth" in the sense that the GHG mitigation generates more output and employment compared to development dominated by fossil fuel energy.

However, the effects on growth and employment are very small compared to the reductions in GHG. They are also uneven across sectors, with the fossil fuel sectors losing output and employment and most other sectors gaining. In view of the uncertainties in the projections and in implementation of the policies, we conclude that with appropriate policies the effect on development and growth of GHG mitigation will be small and difficult to detect.

A third finding, which is discussed in the next chapter, is that decarbonisation involves a wide range of low-carbon technologies. It seems very unlikely that there is a "silver bullet" technology or policy that will bring about effective, efficient and equitable GHG mitigation. A policy mix is required, with progress and innovation on many fronts, if decarbonisation is to be achieved.

References

Barker, T. (2008). The Economics of Avoiding Dangerous Climate Change. An Editorial Essay on the Stern Review, *Climatic Change*, 89, 173–194.

Barker, T., Dagoumas, A. and Rubin, J. (2009). The Macroeconomic Rebound Effect and the World Economy, *Energy Efficiency,* 2, 411–427.

Bosquet, B. (2000). Environmental Tax Reform: Does it Work? A Survey of the Empirical Evidence, *Ecological Economics*, 34, 19– 32.

Clarke, L., Edmonds, J., Krey, V., Richels, R., Rose, S. and Tavoni, M. (2009). International Climate Policy Architectures: Overview of the EMF 22 International Scenarios, *Energy Economics*, 31, S64–S81.

Demailly, D. and Quirion, P. (2006). CO_2 Abatement, Competitiveness and Leakage in the European Cement Industry under the EU ETS: Grandfathering versus Output-based Allocation, *Climate Policy*, 6, 93–113.

Ekins, P. and Barker, T. (2001). Carbon Taxes and Carbon Emissions Trading, *Journal of Economic Surveys (Special Issue on Issues in Environmental Economics, eds. Nick Hanley and Colin Roberts)*, 15, 325–376.

Ekins, P. and Speck, S. (eds) (2011). *Environmental Tax Reform (ETR): A Policy for Green Growth*, Oxford University Press, Oxford.

Fisher, C. (2005). 'Technical Innovation and Design Choices for Emissions Trading and other Climate Policies', in Hansjürgens, B. (ed.). *Emissions Trading for Climate Policy: US and European Perspectives*, Cambridge University Press, Cambridge, UK, pp. 37–53.

Grubb, M. (2014). *Planetary Economics Energy, Climate Change and The Three Domains Of Sustainable Development*, Routledge, Oxford.

Hansjürgens, B. (2005). 'Introduction', in Hansjürgens, B. (ed.) *Emissions Trading for Climate Policy: US and European Perspectives*, Cambridge University Press, Cambridge, UK, pp. 1–14.

IMF (2013). *Energy Subsidy Reform: Lessons and Implications*, International Monetary Fund, Washington, DC.

IEA (2008). *World Energy Outlook 2008*, International Energy Agency, Paris.

IEA, OPEC and World Bank (2011). Joint report by IEA, OPEC, OECD and World Bank on Fossil-fuel and other Energy Subsidies: An Update of the G20 Pittsburgh and Toronto Commitments, Prepared for the G20 meeting, Paris.

IPCC (2007). *Climate Change 2007: Mitigation. Contribution of Working Group III to the Fourth Assessment Report of the Intergovernmental Panel on Climate Change*, Metz, B., Davidson, O. R., Bosch, P. R., Dave, R. and Meyer, L. A. (eds), Cambridge University Press, Cambridge.

Jaffe, A., Newell, R. and Stavins, R. (2005). A Tale of Two Market Failures: Technology and Environmental Policy, *Ecological Economics*, 54, 164–174.

Kelman, S. (1981). Economists and the Environmental Muddle, *The Public Interest*, 64, 106–23.

Keppler, J. H. and Cruciani, M. (2010). Rents in the European Power Sector due to Carbon Trading, *Energy Policy*, 38, 4280–4290.

Knopf, B., Luderer, G. and Edenhofer, O. (2011). Exploring the Feasibility of Low Stabilization Targets, *WIREs Clim Change*, 2, 617–626.

Kyoto Protocol (1997). Kyoto Protocol to the United Nations Framework Convention on Climate Change. Available online at: http://unfccc.int/essential_background/kyoto_protocol/background/items/6034.php. Accessed 24 February 2014.

Neuhoff, K., Martinez, K. and Sato, M. (2006). Allocation, Incentives and Distortions: The Impact of EU ETS Emissions Allowance Allocation to the Electricity Sector, *Climate Policy*, 6, 73–91.

Patuelli, R., Nijkamp, P. and Pels, E. (2005). Environmental Tax Reform and the Double Dividend: A Meta-analytical Performance Assessment, *Ecological Economics*, 55, 564–583.

Shenoy, B. (2010). *Lessons Learned from Attempts to Reform India's Kerosene Subsidy*, International Institute for Sustainable Development, Geneva.

Stavins, R. (2005). 'Implications of the US Experience with Market-based Environment Strategies for Future Climate Policy', in Hansjürgens, B. (ed.). *Emissions Trading for Climate Policy: US and European Perspectives*, Cambridge University Press, Cambridge, UK, pp. 63–77.

Sterner, T. and Hammar, H. (2005). 'Designing Instruments for Climate Policy', in Hansjürgens, B. (ed.). *Emissions Trading for Climate Policy: US and European perspectives*, Cambridge University Press, Cambridge, UK, pp. 17–36

Tietenberg, T. (2002). The Tradable Permits Approach to Protecting the Commons: What Have we Learned? FEEM Working Paper 36.

Victor, D. (2009). *The Politics of Fossil Fuel Subsidies*, International Institute for Sustainable Development, Geneva.

Weitzman, M. (2009). On Modeling and Interpreting the Economics of Catastrophic Climate Change, *Review of Economics and Statistics*, 91, 1–19.

World Bank (2009). *Climate Change and the World Bank Group: Phase I: An Evaluation of World Bank Win–Win Energy Policy Reforms*, World Bank, Washington, DC.

Chapter 6

Feasibility of Decarbonisation from a Technology Perspective[1]

J.-F. Mercure and Pablo Salas

6.1. Introduction

The concept of decarbonisation corresponds to the goal of generating change in the current economic system in order to reduce the emissions of greenhouse gases (GHG) into the atmosphere. Since the largest portion of GHGs are emitted during the transformation of natural energy resources into some form of useable society service (IPCC, 2007b), be it electricity, movement or heat, an important focus of a study of feasibility of decarbonisation must be centred upon the effect of activities of the energy sector of the economy on the environment. Therefore, the subject of interest in this chapter and book is not just the energy sector, but rather, how to generate change in the energy sector such that it can be gradually steered towards sustainability and, most importantly, the stabilisation and reduction of its GHG emissions.

In this chapter, the aim is to describe whether and how the energy sector may be decarbonised. However, in order to explore different possible futures for the energy sector, realistic assumptions are required regarding the engineering feasibility of these possibilities. The issues facing decision-making in the energy sector are all closely correlated to

[1]This chapter is an adaptation of three papers previously published in the journals *Energy* and *Energy Policy*, (Mercure, 2012; Mercure and Salas 2012, 2013), and additional unpublished work.

one another, and must be taken with a global overview in order to ensure the consistency of a particular strategy, rather than focusing on individual sectors. An important example of potentially unsuccessful energy planning oversight would be to incentivise transport electrification without planning the decarbonisation of the power sector, which could lead to increases rather than decreases in GHG emissions due to a possible reduction in efficiency of energy use. This chapter aims to provide simple conceptual tools for analysing the energy sector as a whole and understanding it in the broader context of decarbonisation.

Transforming the energy sector towards sustainability does not simply signify enumerating energy sources with low GHG emissions, but such sources must also be made economically attractive to the instigators of change, the investors and firms who build the capital required for energy extraction, generation or transformation. Therefore, any discussion of energy sources is inseparable from associated costs. Costs, however, are transient numbers, highly influenced by the state and progress of technology. They are constrained by physical limits imposed by nature, through variables such as resource productivity or the technical difficulty of resource extraction. Finally, costs are also influenced by energy policy, with which it is hoped that change can be generated.

This chapter approaches the subject of the technical feasibility of decarbonisation with a study of the dynamics of the energy sector, and identifies the important drivers of change. Using the framework thus introduced, insight is given about how the energy sector might be realistically expected to evolve. The concept of the energy technology ladder is introduced, which is considered here to capture the most important dynamics embedded in such a complex system.

The design of energy roadmaps requires assessments of the constraints on possible future energy landscapes, with their associated uncertainties. An assessment of global energy resources and associated costs is therefore required, and is provided drawing from previous work (Mercure and Salas, 2012, 2013). It is insufficient to provide a simple list of total technical potentials for each energy resource, as this does not involve the costs associated with the exploitation of particular types of energy resources and could lead to the construction of unrealistic energy landscapes. Since details on the particular methods with which energy may be produced can be easily found elsewhere (see for instance GEA, 2012), these are not the primary focus of the present work. Instead, a methodology is presented which allows the determination of global economic energy resource potentials, in the form of cost–supply curves. As the assessment of natural resources inherently involves significant uncertainty, these are given in terms of ranges and confidence levels. Using this methodology, global cost–supply curves are given for all main types of energy resources. Natural resources are divided into four categories: renewable energies, biomass, nuclear energy and fossil fuels. Biomass is a renewable energy source; it is however discussed on its own (see also Chapter 9) as it is a vast subject which deserves broader analysis. From the results of this analysis, given a certain future demand for energy, approximate scenarios for energy use can be constructed. These calculations necessitate assumptions on the technologies involved, and therefore, brief descriptions of some of the current methods of conversion are given.

The study of future energy landscapes can only be performed by considering all types of resources and energy-production systems simultaneously, on a regional level, and by comparing their costs. This is a tremendous task that is too complex to describe properly in this chapter. A simple global analysis can be performed, however, without involving specific regional details, using a global energy demand estimation. From the full set of cost–supply curves, combinations of energy flows from all sources can be derived using assumptions on the future price of energy. This is explored in the discussion of this chapter.

6.2. Energy Sector Dynamics

6.2.1. *Dynamics of change*

This section introduces drivers of change for the energy sector of the global economy. Change is not performed by policy makers, but it is incentivised by them. The real creators of change are the investors, who hope to invest their money into profitable projects, and are constrained by current economic conditions. Therefore, the role for the policy maker is to generate change using various control parameters that influence the choices of investors, which are introduced here, along with their likely effects.

The nature and evolution of the energy sector is governed by a large number of phenomena which stem from economic or physical principles. It is possible using simple assumptions, to group them into three simplifying categories. The energy sector is influenced by the state of technology, the nature and availability of natural resources, and energy policy. Thus, change in the energy sector must stem from changes in these three sets of parameters. Imagining a hypothetical future where no geopolitical events play major roles in influencing the energy sector, issues which are not within the scope of this chapter (see Chapter 8), change is ascribed to three types of effects: technological change, resource use and depletion, and the existence of pricing of environmental damage such as GHG emissions or other energy policies. This section aims at describing what is meant by these terms and how they can be simply understood and modelled.

Technological learning does not concern predictions of new revolutionary ways with which energy might be produced in the future. It concerns the rate at which existing technologies improve such that their costs decrease and that they gain the ability to diffuse more easily. Therefore, no assumptions about the unknown are made, although it is assumed that current trends are maintained. Such extrapolations are extremely helpful in forecasting costs and only require knowledge of progress rates and technology deployment at any particular point in time.

All types of energy are obtained from the transformation of natural resources of some kind. Thus local energy landscapes are closely linked to what sort of raw energy resources can be found in particular localities. While some types of energy sources are readily transported across the planet, other types must be used in specific locations, making each region of the world unique. Resource availability and use can be modelled, as is shown in this chapter, using a framework of cost–supply curves, where existing resources are used up to the point where their marginal costs become approximately equal.

Changes in energy policy decisions may affect the energy sector in different ways. In equilibrium economics, climate change is associated with a market failure which originates from the lack of appropriate pricing of environmental damage such as GHG emissions into the atmosphere (see for instance Stern (2006)). Sustainability and economic efficiency are often thought in this perspective to be possible only through the pricing of these externalities. In a broader context of energy sector planning, other types of policies are encountered, such as feed-in tariffs, subsidies and regulations.

In future scenarios where mitigation intervention takes an important role, the energy sector is expected to evolve through gradual change in energy policies. At each point in time, changes in technology that may occur are those which are economically possible given the current financial conditions, or in other words, those that are made economical by policy instruments. The critical principle upon which mitigation rests is that heavy polluters will likely scale down their activities as they see their operating costs increase steeply, while more expensive but cleaner technologies gradually become competitive. The general resulting effect is one which we have termed the energy technology ladder (Mercure, 2012a, 2012b), where in a system with stringent climate policy, the energy sector sees itself going through several technological transitions, where emitting technologies are gradually replaced by more and more efficient systems. Such a phenomenon allows a cost-effective decarbonisation of the energy sector, but it is likely to be also observed in other sectors of the economy, such as transport.

Three drivers of change are described in the next sections, and the resulting structure of each change is detailed. A description is first given as to how technological learning can be represented simply. A framework is then given which enables taking into account quantitatively natural resource use and depletion, using cost–supply curves. Reserves of energy resources are, however, not independent of technological learning, and thus, for completeness, a discussion is given on how these interact with one another. This effect is directly related with the observed expansion of resources and reserves, as is common, for instance, with fossil fuel resources. A discussion of the price of carbon is then given and how it influences investment decisions. This additional interaction in the energy landscapes of the world triggers desired transformations towards sustainability and low GHG emissions through the energy technology ladder, an effect which is described in the final part of this section.

6.2.2. *Technological learning*

Technological learning is a concept that stems from a generally observed empirical law, which applies to a surprisingly large number of systems in a variety of sectors of the economy. It may be stated very simply: the cost associated with the production of new units of a certain good is a continuously decreasing function of the cumulative investment in the particular technology which enables its production, and that function is a power law with a negative exponent between zero and one (e.g. Arrow, 1962). In other words, when one represents graphically the cost as a function of cumulative investment, or cumulative number of sales, using logarithmic axes, one observes a decreasing linear trend. Its slope is the exponent of the power law, which is related to the progress rate.

This is shown as an example in Figure 6.1 where such data are represented for various energy technologies. Progress rates k are given in brackets, and these are simply related to power law exponents b through $b = \log(1-k)/\log 2$. The effect produced by this law follows common sense: new technologies require very few additional investments or sales in order to improve their production methods, while older, well-established technologies require much larger investments in order to significantly add to the already large pool of accumulated knowledge and production

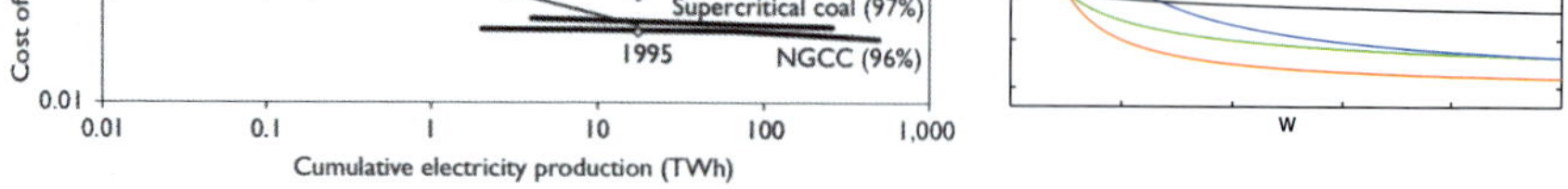

Figure 6.1. Left-hand side: Technological learning for several energy producing technologies. Costs of electricity production are represented as functions of cumulative energy production on logarithmic axes, where linear trends are observed. Progress rates are given in brackets (IEA, 2000). Right-hand side: Technological progress sketched for different values of the exponent b, plotted using linear axes.

knowhow. One can appreciate in Figure 6.1 that this law is observed in some cases over several orders of magnitude. As long as enough data are available to establish with some degree of confidence the progress rate, one may extrapolate into the future how costs may evolve with future numbers of sales or amounts of investment.

This effect is of crucial importance for the evolution of the future energy sector. This is due to the fact that the sector is composed of technologies of all possible levels of development. The newer can have the potential to become less expensive than widely used older technologies, which provides potential for change. These increasing returns give rise to highly non-linear mathematical systems and to the property of path-dependence. Costs arc thus dependent on the full investment history, creating all kinds of possible future cost landscapes. Government support for new technologies may trigger avalanche effects where these technologies ultimately overtake a large share of the market, whereas they may well have disappeared in the absence of sufficient investment. It is indeed often the case that private investment in new technologies does not materialise without help from governments in the form of support policies, since initial costs may be too high and risky for the average investor, a situation termed the "technology valley of death" (Murphy and Edwards, 2003). Therefore, initial pushes may trigger fast improvements in new technologies which would not have occurred otherwise.

6.2.3. *Technical potentials and cost–supply curves*

The extraction potential of all energy sources, whether finite (stock resources) or renewable, is limited by physical and economic constraints. At the end of 2013, it was widely debated when the world might run out of fossil fuels. Similarly, controversy surrounds the issue regarding how much land might be used for energy production in the future, and whether this will compete with food production. Such issues must be considered when studying how and when the world economy might become decarbonised. A simple and general framework is thus presented in this section which describes natural resource use and depletion. This framework has been used in several studies in the past (de Vries *et al.*, 2007; Hoogwijk, 2004; Hoogwijk *et al.*, 2003; Hoogwijk *et al.*, 2005; Hoogwijk *et al.*,

2009; Rogner, 1997). It is extended in this work to include uncertainty analysis and used as a tool that provides the reader with a global overview of world energy resources based on current literature.

The common theme is one of costs increasing with resource use. Occurrences of a particular type of natural resource in the environment may be ranked in order of their difficulty of extraction, which are associated with costs that follow the same order. The assumption is that the cheaper sources are exploited first, which is not necessarily always strictly true, and provides us with a useful simplification.[2] As exploitation progresses to higher and higher levels, more and more expensive sources are exploited, until their cost becomes such that the associated energy is more expensive than some alternative in the market, where the expansion of exploitation stops.

Wind energy is a good example of this progression. The wind resource has been reported larger than several times the expected future world energy consumption (see for instance Lu *et al.*, 2009). However, producing the whole electricity supply using wind energy is not likely to be a cost-effective solution, compared to heterogeneous energy mixes. In order to demonstrate this, land sites suitable for the installation of wind turbines may be ranked in order of the quality of the wind resource on particular available plots of land, expressed by the probability distribution of wind speeds. Such a distribution, when multiplied by the power curve of particular wind turbine generators and summed, yields the proportion of time over which the turbine is expected to generate its rated power output, commonly named the capacity factor. Starting from the windiest sites, as the quality of the wind resource degrades, the associated capacity factor decreases. For a fixed capacity value, associated with a particular turbine model, the expected investment cost per unit of energy rises steeply for poor wind sites. In order to produce all of the world's energy consumption using wind power, it is likely that a large number of poor wind sites are used, requiring a very large number of wind turbines which are likely to produce on average a poor amount of energy per unit. It is more likely that new wind capacity building slows down at the point where good quality wind sites are depleted, a point where the cost of

[2] See the supplementary information in Mercure and Salas (2012) for a full discussion.

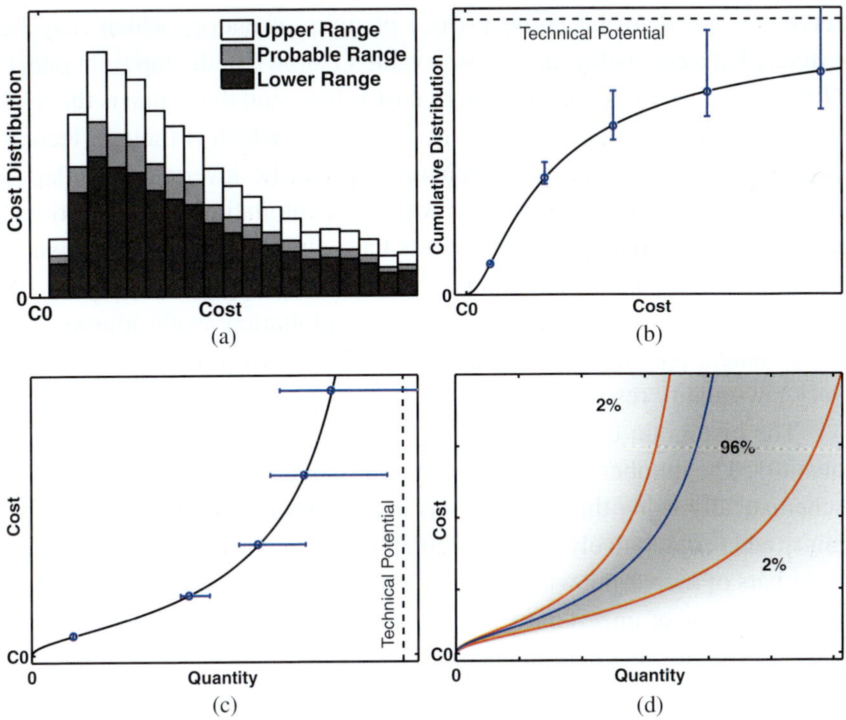

Figure 6.2. Cost–supply framework sketched graphically. Top left: Histogram of the number of units of energy available within cost ranges, with corresponding uncertainty. Top right: Cumulative cost distribution of energy units, which converges towards the technical potential. Bottom left: Cost of exploitation *C(Q)* of an additional energy unit given a number of units *N* which have already been exploited, or cost–supply curve. Bottom right: Probability density of the location of the cost–supply curve in the *C–Q* plane, shown as a grey shading gradient. The maximum of the density corresponds to the most likely location of the cost–supply curve, shown as a blue curve, while the 90% confidence region in the *C–Q* plane is bounded by the red level curves. Reproduced with permission from Mercure and Salas (2012).

additional wind energy generation capacity becomes uneconomical compared to alternatives.

Figure 6.2 describes this concept by using the cost–supply framework. Occurrences of energy sources are first ranked in order of costs, shown as a histogram *n(C)* of the number of energy units available as a function of their cost of exploitation (top-left panel). Such a histogram serves to calculate the number Q(C) of units of energy which may be extracted at a cost below any chosen value C, shown in the top right panel. $Q(C)$ converges to a constant at high cost values, and that value is the total technical potential, the total amount of resource which is thought technically possible to extract. This relationship can be inverted, in order to describe the cost $C(Q)$ associated with the exploitation of an additional unit of this resource given that Q units have already been used, shown in the bottom-left panel. This is the so-called cost–supply curve. It has the important property that the cost values at exploitation levels approaching the technical potential diverge to infinity. This corresponds to the use of very low-quality resources.

The histogram of resources $n(C)$ is, however, not a perfectly known quantity. The number of units in each cost interval is uncertain, expressed schematically using three histograms in three shades of grey. This uncertainty adds cumulatively[3] in the calculation of $Q(C)$ and $C(Q)$, shown with error bars of increasing width. Extending this argument by using an appropriate analysis of uncertainty, regions in the Q–C plane can be associated with varying levels of confidence.

Using the uncertainty ranges to define two additional curves, for the upper and lower 95% confidence levels, a probability density can be defined in cost–quantity space for the location where the real cost–supply curve would most probably lie if it were possible to derive it with certainty. This is shown in the bottom-right panel, where the red curves denote the 95% confidence-level region and the blue curve is the maximum. Such probability densities are normally asymmetric, since the uncertainty over undiscovered resources lies at higher quantity values. Note that the uncertainty is assumed to vanish at the contemporary position in the cost–quantity (C–Q) plane, since current costs and levels of current exploitation are known. This framework is used to present data throughout this chapter.

Both exhaustible and renewable resources may be described by cost–supply curves. It is not the same quantity which is expressed by the variable $Q(C)$, however. For exhaustible resources such as fossil or nuclear

[3]The total uncertainty is the root of the sum of the squares, which increases approximately cumulatively for similar uncertainty values at each cost interval.

fuels, this quantity refers to a number of energy units, expressed in joules or GWh, while for renewables, it corresponds to amounts of energy flows, expressed in GWh/y. Therefore, even if there is no limit to the amount of energy that can be extracted from renewables, there is a limit to the amount which can be obtained in a year. This fundamental difference produces different behaviours for these two classes of resources. Renewable systems can progress forwards or backwards along the cost–supply curve, as units are built or decommissioned. Exhaustible resources, however, can only progress upwards in cost once low-cost resources are combusted, an effect that underlies the one-way gradual increase in the cost of these resources.

All types of energy sources are subject to increasing costs with resource use as described by the cost–supply curve. However these do not come into cost evaluations in the same way for different systems. In some cases, such as for fuel-based technologies, the cost–supply curve occurs in fuel costs, which stems from the stock of resources, for fossil and nuclear fuels, or from land productivity in the case of biomass fuels. In other cases, such as wind, solar or wave energy, the cost–supply curve appears in the form of a capacity factor, which decreases with increasing capacity. For the remaining cases, such as hydroelectricity and geothermal energy, the cost of capital is evaluated on a case-by-case basis, and possible sites are ranked according to their cost estimated by engineering firms.

6.2.4. *Learning changes cost–supply curves*

Technological learning occurs in methods of natural resource exploitation as well as in the building of energy production capital, two effects connected but slightly different. Cost reductions from learning-by-doing in capital building occur through the repetitive production of similar goods and follows power laws. Costs of exploitation, as expressed by the cost–supply curve, may also decrease with the number of units extracted. As shown by experience, these do not necessarily follow power laws, which makes their modelling difficult. They are more susceptible to drastic changes in their trend of reduction originating from isolated discoveries in exploitation methods (IEA, 2000).

Examples of this are plentiful, and include recent discoveries in shale gas extraction methods, or tar sands processing, which were considered uneconomical until specific discoveries or developments were made, such as fracturing or horizontal drilling, which occurred only just several years ago, but have come about through accumulated knowledge and experience. Such effects, mainly observed with stock energy sources such as oil and gas, have important implications. It drives the well-known "expansion" of reserves; the amount of resource which is considered economical to exploit. However, the latter are also expanded as new sources are discovered around the planet. There is thus a continuous flow of energy from discoveries towards resources, and from resources towards reserves (Mckelvey, 1972; Rogner, 1997).

6.2.5. *Emission rates and carbon costs as a source of change*

Low GHG emission energy systems may be supported and promoted in various ways, such as subsidies, feed-in tariffs or green certificates. However, the system which is regarded by neoclassical economists as the most efficient economically is the emissions trading scheme, which attempts to set appropriate pricing of GHG emissions as an externality to all of earth's inhabitants (e.g. Perman *et al.*, 2003; Stern, 2006). According to environmental economics theory, in order to correct such a market failure, externalities should be priced in such a way that polluters, who benefit from such actions, pay enough to compensate those who are affected by the effects of pollution, improving the allocation of resources and the efficiency of the market (Perman *et al.*, 2003).

Such a principle is formalised within an emissions trading scheme, where the emission of metred units of GHGs is accompanied by the purchase of emissions allowances. Assuming that no corruption occurs, for any region of the world over which such a scheme operates, the number of allowances corresponds to the actual amount of emissions of GHGs. By controlling the number of allowances sold, the price of allowances is set free to adjust itself according to demand, but the emissions are set as an exogenous variable, following a chosen emission pathway (such as those in Chapters 4 and 5). Such an emissions path generates a time path for the price of emission allowances, which depends on the amount of possible emission abatement that exists in the economy as a function of abatement cost.

For the individual investor, the purchase of emission allowances corresponds to an additional operation cost for energy production. This cost is proportional to the amount of GHGs particular technologies emit per unit of energy. Where this cost is high enough, it may become economical for utility operators to replace existing high-emission systems with more efficient ones. The price of emission allowances therefore generates a driver for change, whose strength varies depending on the rate of emissions of technologies: highly polluting technologies generate higher carbon costs.

Emission rates are highest for coal-fuelled technologies, but lower for gas turbines. They are nearly zero where combusted fuels are derived from biomass, as biomass derives its embedded carbon through absorption from the atmosphere. Emission rates for fossil fuels are low if carbon capture and storage (CCS) technology is used to sequester CO_2. Therefore, emission rates can also be negative when biomass-derived fuels are combusted while using sequestration of GHGs. As a result, in an emissions trading scheme, such technologies should in principle use a negative number of emission allowances, and therefore have negative costs.[4] When devised in that way, emissions trading schemes strongly favour these types of technologies where the price of allowances is high, as the sequestration of CO_2 becomes a highly profitable activity.

6.2.6. *The levelised cost of electricity*

In the previous sections, three drivers of change were introduced that influence the decisions of investors as to which technology to build to produce energy. In the case of electricity production, these decisions are generally formalised in the industry using the framework called the levelised cost of electricity (LCOE, e.g. in IEA (2010a)). It corresponds to an estimated value above which electricity should be sold in order to at least

[4] In other words, technologies with negative emission rates generate emission allowances, corresponding to the amount of CO_2 that has been removed from the atmosphere, which can be sold onto the carbon market.

break even in terms of profitability. The LCOE is based on the evaluation of two net present values. The first is the sum of the present value of the costs incurred during the lifetime of the electricity generating capital,

$$NPV_1 = \sum_{t=0}^{Lifetime} \frac{I(t)+F(t)+OM(t)+C(t)}{(1+r)^t}, \tag{6.1}$$

where the costs in the numerator correspond, in order of appearance, to investment, fuel, operation and maintenance and emission costs, and r is the discount rate chosen by the industry. The second calculation is the present value of the total earnings which would be obtained by selling the electricity at the break-even value, the LCOE:

$$NPV_2 = \sum_{t=0}^{Lifetime} \frac{LCOE \times E(t)}{(1+r)^t}, \tag{6.2}$$

where $E(t)$ is the amount of energy produced. These two net present values are equal by definition, and the LCOE can be isolated:

$$LCOE = \frac{\sum_{t=0}^{Lifetime} \frac{I(t)+F(t)+OM(t)+C(t)}{(1+r)^t}}{\sum_{t=0}^{Lifetime} \frac{E(t)}{(1+r)^t}}. \tag{6.3}$$

Note that the time dependence in these refers to what investors assume the values will be in the future from their estimations performed in the present. The LCOE is therefore composed of forecasts evaluated at the time of investment decision-making, which may or may not turn out to be correct.

The LCOE is designed to take account of every factor which may influence the cost of producing electricity for an electricity firm, whether cost variations originate from learning, cost–supply curves and depletion or emission allowances costs. Hence, decision-making based on comparisons of the LCOE of all existing energy technologies reflect all relevant issues and react appropriately to changes in all relevant parameters.

6.2.7. *Induced technological change*

In the context of firms competing in an energy market, decisions are taken when planning the building of infrastructure in view of minimising production costs at the firm level. Where some of the cost driving variables change in time, changes in the composition of planned or newly built energy systems should occur as a result of changes of investor strategy. Learning tends to favour new technologies, while cost–supply curves slow the growth of technologies where natural resources are limited, and emission allowance costs favour clean technologies. However, change in energy technology is limited by the lifetime of the capital, especially if it is assumed that no early scrapping of capital occurs. As the components of the LCOE tend to change faster than the rates of capital decommission, the system is never cost-optimal (e.g. Vogt-Schilb and Hallegatte, 2011).

High emission allowance prices should generate large changes in energy systems. As zero- or low-emission energy-producing systems tend to be more expensive than highly polluting ones, it is reasonable to associate a cost to technology switching, the marginal abatement cost. For a given price of emission allowances, one may think that all technology substitutions which lead to energy production cost less than those of current technology should in principle be performed. In a system with infinitely fast construction times and short lifetimes, the marginal abatement cost could be equal to the price of emission allowances. However, in a system with long timescales such as the power sector, not all substitutions targeted by policy will actually take place, but only the projects for which the planning decision-making process happens to take place at that particular moment may be influenced. Other power systems will likely remain in place until the end of their lifetime, a time corresponding to the period over which investors expect to receive an income to pay back their investment. This gives rise to a system which is never optimal and potentially so-called lock-in effects.

Nevertheless, it is in good measure change in energy policy that generates change in the energy system. A fast changing price of allowances or rate of subsidy generates more technology substitutions than a slow one. At each step, of all abatement possibilities, only affordable

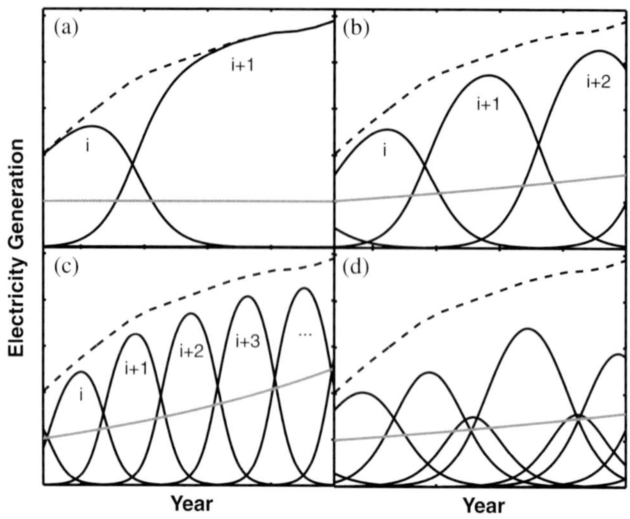

Figure 6.3. Sketch of the concept of the energy technology ladder, where energy technologies are gradually replaced as the price of emission allowances increases. The electricity generation by technology is shown as solid lines, the total electricity demand is the dashed line, while the price of emission allowances is shown as a grey curve. Technology *i* is replaced by *i + 1*, which in turn is replaced by *i + 2* and so on. (a) Without an increase in the price of carbon, isolated technological transitions may occur through learning-by-doing effects. (b) With a modest increase in the price of carbon, several transitions are generated. (c) For a high rate of increase, several more transitions are favoured. (d) A more realistic case: a heterogeneous mix of technologies with scattered values for the rates of uptake and GHG emissions. Reproduced with permission from Mercure (2012a).

measures are generally considered. Since with learning, costs are highly dynamic, the energy system may be expected to undergo a number of transitory states before arriving at a decarbonised state. These transitory states are crucial in easing the transition towards a low-, zero- or negative-emission energy system in a cost-effective way. This is sketched in Figure 6.3, and is termed the energy technology ladder (Mercure, 2012a, 2012b).

Energy generation, in GWh, by different technologies with different emission rates are depicted with solid lines. The sum of these curves yields the total energy generation, which equals the demand, shown as a dashed line in each panel. The price of allowances is shown in arbitrary units as a grey curve. Panel a shows the system with a constant price of carbon. In such a case, learning effects alone may be enough to generate a possible technological transition from technology *i* to technology *i + 1*. However, faster technological change can be driven by both learning and an increasing price of carbon, shown in panel (b), where several successive transitions occur from technology *i* to *i + 1*, and subsequently from *i + 1* to *i + 2*, where each new technology emits lower amounts of GHG per unit of energy and therefore incurs lower carbon costs. The number of successive transitions increases with the rate of change of the price of carbon, as shown in (c), and yields a final system with lower GHG emissions. As the rate of increase of the price of allowances rises, higher abatement solutions become economical to use, going through whole series of temporary solutions. Energy transitions, however, possess intrinsic time constants. In other words, the number of transitions cannot significantly exceed the total time span considered divided by the average lifetime of the capital.

The hypothetical system depicted in the panels (a)–(c) of Figure 6.3 was given equally spaced values for the various emission rates. Real systems are not composed in this way, and therefore the observation is akin to that shown in panel (d), where variations of emission factors and time constants were introduced. It is clear however in which order technological transitions are expected to arise: heavy-emission coal technologies are likely to be, in the short term, replaced with much more efficient combined-cycle gas turbines (CCGT), which have lower emissions. These will, however, probably be replaced in time by low-emission systems, such as CCS systems retrofitted to existing fossil fuel plants. These are likely to be replaced in turn by either renewables such as wind, solar and wave power, by nuclear reactors, or by systems based on biomass sources, in order of costs and pace of learning. However, if negative emissions are to be rewarded by negative costs of allowances, then zero-emission sources will be replaced by combined biomass and CCS systems. Systems with negative emissions could become very important in order to stabilise emissions in time to limit global warming, but their actual future availability should probably not be relied upon (Riahi *et al.*, 2012).

6.2.8. *The economic potential of energy resources*

The economic potential of an energy resource is defined by its opportunity cost. If the opportunity cost is negative, meaning that additional development is less expensive per unit of energy produced than all other alternatives, then it is optimal to build an additional unit of energy resource-consuming capital. If that cost is positive, there is at least one alternative which is more profitable. As argued in previous work (Mercure, 2012a), although it may not be completely realistic to assume perfectly rational expectations for every individual investor; investors in energy generation capacity have the tendency to follow this rule, on average, when taking decisions.[5] In such a framework, if no changes occurred in the global pool of knowledge on natural resource potentials, energy policy and the costs of technology, the level of use of all resources would converge towards a situation where their opportunity costs are nil. However, as argued in the previous section, the energy system is characterised by timescales that are much longer than the typical life span of energy policy decisions and the evolution of knowledge. Therefore, this equilibrium never quite takes place, and the cost of energy production can never really be completely minimised.

It is nevertheless the case that some natural energy resources have higher or lower opportunity costs than others, and that this varies on the global and regional levels. In order to determine which allocations of natural resources generate economically efficient energy systems in particular regions of the world, starting from a particular state of the sector, additional demand for energy flows should be divided between resources according to their opportunity cost. The economic potential of energy resources is therefore defined as the following. Assuming that energy demand increases with time, using a particular projection, or a range of projections, of future energy demand, and allocating the additional demand compared to its current value to natural resources according to their opportunity costs, new levels, or ranges of levels, of natural resource

[5]This relaxed assumption allows for unknown/unknowable sets of motivations influencing individual investor decisions, leading to distributions in behaviour, of which the trend can for instance be taken.

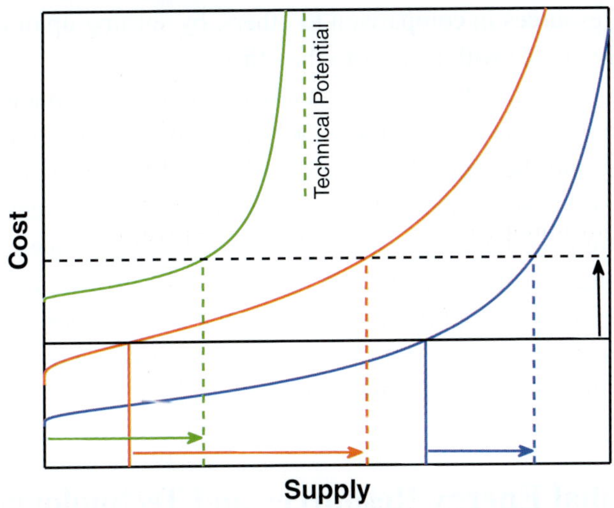

Figure 6.4. Sketch of the economic potential of energy resources as given by their cost–supply curves. Resources are allocated in such a way that the sum of energy flows equals the demand (solid black line). However, for all opportunity costs to be zero, some resources will be utilised to higher levels than others (solid vertical green, red and blue lines). When projecting resource use for a future where energy demand increases, new levels of resource (dashed vertical green, red and blue lines) define their economic potential (colour arrows), as opposed to their technical potential, the level of use where a cost–supply curve diverges. The new levels of resource use define a new overall marginal cost (dashed black line).

use can be derived, as well as a new overall marginal cost, since additional energy units are likely to be more expensive at higher levels of use. The difference between their future individual levels of use and their current values defines their economic potential given the future date for which the projected demand was taken (see Figure 6.4). Opportunity costs are defined by comparisons of the cost–supply curves associated to each resource at the current level of resource use,[6] and the economic potentials depend on one another. Therefore, economic potentials can be increased or decreased using various forms of government support such as taxes, subsidies or an emissions trading scheme, which generate preferences for

[6] Rigorously speaking, every joule of energy should be allocated to energy resources following their opportunity costs, all of which change every time a joule of energy is allocated to one or another of the available resources.

particular resources in comparison to others, by shifting up or down their cost–supply curves with respect to the others.

This is represented formally in a model for energy technology substitution based on cost considerations, which calculates rates of technological change, and therefore the change in the use of the associated natural resources. Many such models based on investor behaviour seeking lowest cost opportunities have been defined and used, for instance the TIMER model using a logit substitution function (de Vries *et al.*, 2001), MESSAGE using an optimisation of the total cost of energy production (Messner and Strubegger, 1995), and FTT:Power using a dynamic family of logistic differential equations (Mercure, 2012a).

6.3. Global Energy Resources and Technologies

6.3.1. *Renewable energy resources*

All types of energy come from a primary natural resource, for instance wind, sunlight, fissile elements, oil or sugarcane. While some of these resources originate from perpetually flowing sources such as the sun, others stem from exhaustible repositories such as oil wells or coal mines. Six main types of renewable energy sources currently exist: hydro, solar, wind, geothermal, ocean (wave and tidal) and biomass energy.

Despite the variety of renewable energy systems and resources, the energy sector is dominated by fossil fuels. The costs of energy production using the former remain in most cases above those associated with the latter. However, as described earlier, investment costs decrease with cumulative capital investment through learning-by-doing, according to learning rates, and for most renewable sources (excepting hydropower), such costs can be expected to decrease much faster than those of mature fossil fuel alternatives (see Figure 6.1). Thus, in order to assess the engineering feasibility of decarbonisation, in addition to the technical potential of each energy source, consideration must be given to the associated cost of energy production at each level of use. The framework of cost–supply curves is used for this purpose in order to convey a global overview of the potential scale of exploitation of each type of resource discussed, as a function of all alternatives, using the approach of economic potentials described earlier. All types of energy resources not discussed in this chapter have either technical potentials too low to make significant contributions to the global energy supply (less than 10 EJ/y) or have such costs that puts them outside any rational expectation of becoming competitive with the alternatives discussed.[7]

6.3.1.1. *Wind power*

Wind originates from thermal gradients in the earth's atmosphere, as well as the Coriolis force, the combination of which drives air currents in complex manners. Wind speeds depend strongly on altitude as well as geographic features, the climate and the roughness of the ground. In general, wind speeds increase logarithmically with elevation at low altitude (Sørensen, 2011), and, for a specific elevation and geographic location, occur statistically following a well-defined Weibull probability distribution which decreases both towards zero and large wind speeds (Grubb and Meyer, 1993). Average wind speeds on sites useable for energy production, for instance in the UK, range between 5.1 and 9 m/s at 10 m elevation,[8] the lower boundary determined by technology and the upper limit by the decreasing supply of locations with high wind speeds (Sørensen, 2011). The power density offered by land areas must be calculated using technical characteristics of particular turbines. For a particular site, one integrates over all wind speeds the product of the site wind speed probability distribution and the turbine power curve. However, a correlation exists between the yearly average wind speed and the number of full load hours (Abed and El-Mallah, 1997). The minimum possible distance between turbines in a wind farm are determined by losses produced by the wake of neighbouring turbines which results in lower wind speeds and increased turbulence, limiting the density of energy which can be extracted per unit of land area (see for instance Mackay,

[7] With learning-by-doing, any new expensive energy technology can, in principle, become competitive with contemporary systems. However, without cumulative investment, they never do. Systems too expensive to attract any type of support are unlikely to enter the market.

[8] 5.1 m/s corresponds to the lower boundary of the class 3 wind site category. This limit is used for instance by Archer and Jacobson (2005) to define wind sites of commercial value.

2008). M Hoogwijk *et al.* (2004) assumed a maximum density of energy production of 4 MW/km^2.

Various research groups and institutions specialise in wind power assessments (see for instance the work of Archer and Jacobson (2005), Fellows and Gow (2000) and Lu *et al.* (2009)), and several have calculated the global distribution of wind power (Archer and Jacobson, 2005; Fellows and Gow, 2000; Grubb and Meyer, 1993; Hoogwijk *et al.*, 2004; Lu *et al.*, 2009; WEC, 1994), resulting in a range of values between 70 and 2,509 EJ/y for onshore wind power, and about 57 EJ/y for offshore wind power (Krewitt *et al.*, 2009). A global onshore value of 346 EJ/y has been derived by Hoogwijk *et al.* (2004), in whose work, used for the present analysis, estimations of wind speed distributions were applied to points on a global onshore grid. Energy potentials obtained from yearly average wind speeds determined on every point of the grid were subsequently aggregated into various cost–supply curves for specific regions of the world. Wind farm sites are less interchangeable than solar energy sites, as the geographical distribution with average wind speeds suitable for energy production are scattered, and the majority of areas in any region of the world possess mediocre average wind speeds (Troen and Petersen, 1989). Therefore, the distribution of wind energy resources in terms of cost exhibits a very different structure. As opposed to solar energy some specific geographical areas are strongly sought after by wind farm developers, while some others are hardly noticed by the industry. The cost–supply curves derived from the distribution of wind resources possess a strong curvature, which represents a clear hierarchical ordering of the probability of finding good wind farm sites within regions, for instance at grid points in the work of Hoogwijk (2004). The profitability of a wind farm venture depends strongly on the quality of the wind resource, determined through capacity factor and average turbulence values. Geographical locations suitable for potential wind farms are thus not substitutable for one another as they are for solar farms, described below, or biomass energy crops, analysed in the next section. The higher slope of the cost–supply curve at every cost value for wind produces a level of penetration for the sector which depends more critically on the marginal cost of alternatives, or equivalently the price of electricity, than, for instance, solar and biomass energy.

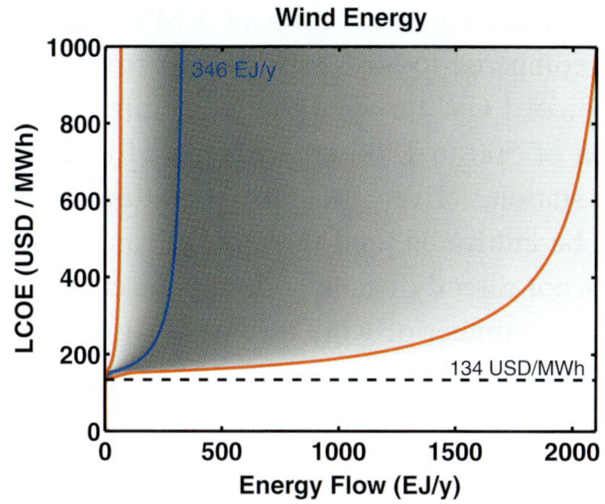

Figure 6.5. Global cost–supply curve for wind energy. Reproduced with permission from Mercure and Salas (2012).

Figure 6.5 presents the global economic potential for wind power. It gives an aggregate cost–supply curve, using for the median curve the data of Hoogwijk *et al.* (2004), with a technical potential of 346 EJ/y. Lu *et al.* (2009), on the other hand, estimated a technical potential of 2,470 EJ/y by calculating wind potentials over the global onshore area and excluding low wind areas by restricting capacity factors to values above 20%. Archer and Jacobson (2005) calculated a potential of 2,257 EJ/y in an assessment where the land included was restricted to class 3 wind energy sites but did not include alternate uses of the land, a value used in this section to define the upper 95% confidence level. Meanwhile, WEC (1994) estimated 70 EJ/y based on an evaluation of the number of sites with average wind speeds above 5.1 m/s, but with a chosen value of only 4% of that land available for wind turbine installation, to account for alternative land uses, a value that was taken for the lower 95% confidence level. All other existing studies result in values within this range (Fellows and Gow, 2000; Grubb and Meyer, 1993).[9]

It is to be noted that using up a large fraction of that potential results in large areas becoming covered by wind farms. Since typical individual

[9]Fellows and Gow (2000) conclude with an estimate for 2020 of 148 EJ/y, while Grubb and Meyer (1993) calculated a global potential of 191 EJ/y.

wind turbines currently have capacities of 3 MW but capacity factors of about 25–35%, compared to 80% for coal power stations, replacing one coal power station of 1 GW for wind energy requires 700 to 1100 turbines, covering an area of 500 to 1500 km^2, compared to about 1 km^2 for the original power station.[10] Even though agricultural land used by wind farms may still be cultivated, and therefore a competition for land with agriculture does not directly occur, a decarbonised world likely implies that given the large number of wind turbines that may be required, these will invade traditional rural landscapes.

In the case of offshore wind power, although distributions of wind speeds at offshore locations tend to have a higher median, the air flow possesses less turbulence and wind speeds vary less in time, all of which contribute to producing a higher power density in terms of land area. The total area where such turbines can be installed is small compared to onshore areas, unless floating turbines become widely available (Weinzettel *et al.*, 2009). The potential of offshore wind energy is of about 57 EJ/y (Krewitt *et al.*, 2009), approximately six times lower than the median potential of onshore energy given in Figure 6.5, while its cost per unit energy is significantly higher. Its contribution falls well within the range of uncertainty given above and does not affect significantly the global cost–supply curve estimate.

6.3.1.2. *Solar energy*

Solar radiation over the earth's surface is of about 1.2×10^5 TW, or 3.6×10^6 EJ/y (Crabtree and Lewis, 2008). The fraction of that energy that can be harvested with existing systems has been estimated by several studies (de Vries *et al.*, 2007, Hofman *et al.*, 2002; Hoogwijk, 2004), and ranges between about 1340 to 14800 EJ/y. Even though the total generation of energy from solar technologies has been increasing steadily during the last two decades, they still represent very low percentages in their respective categories; solar heating systems account for 7% of the total energy used for heating in 2010, and solar electricity generation

[10]The variation originates from both assumed ranges in capacity factors of 25–35% and turbine densities of 2.2–4 MW/km^2 (Hoogwijk *et al.*, 2004; Mackay, 2008).

represented only 0.15% of the total electricity generation during the same year (IEA, 2012).

Solar energy can be harvested using either of two existing technologies, photovoltaic (PV) devices (Avrutin *et al.*, 2011) or concentrated solar power (CSP) (ETSAP, 2010a). Single-crystal silicon PV diodes currently have light conversion efficiencies of up to 25%, while III–V semiconductor cells such as GaAs systems have efficiencies of up to 28%, and solar cells using concentrated sunlight can convert light as efficiently as 43.5% (Green *et al.*, 2011). The resulting electricity generation energy density ranges between 5 and 100 MW/km^2, depending on the type of devices used and the geographical location. On the other hand, CSP technology uses a traditional steam turbine where water was heated using sunlight concentrated with various arrangements of mirrors, and have efficiencies of around 13–24% (ETSAP, 2010a) and energy densities near 25 MW/km^2 (IPCC, 2011b). These systems are, however, restricted to high irradiance areas, and therefore have a lower global technical potential.

Areas with good energy production potential using PV solar technology are, in general, highly substitutable for one another. This stems from the nature of the technology, which can be installed equally easily in diverse locations, and depends mostly on average local irradiance, which varies very little over large areas of land. The total technical potential within such areas depends primarily on the amount of land available, and the cost–supply curve is relatively flat until the available low-cost land is completely covered by solar energy units, at which point the cost increases steeply. This property generates a situation where, were the technology to become competitive in local energy markets, large energy potentials could be unlocked in particular regions of the world at similar costs. The opportunity cost of the land is the limiting factor to the technical potential.

Figure 6.6 shows the global economic potential for solar energy, using PV as technology. It is an aggregation of curves determined in various world regions, based on the work of Hoogwijk (2004) but rescaled to match the values of de Vries *et al.* (2007). The global technical potential is of 3384 EJ/y. Regional cost–supply curves were drawn from an analysis performed using the land-use model IMAGE 2.2 to

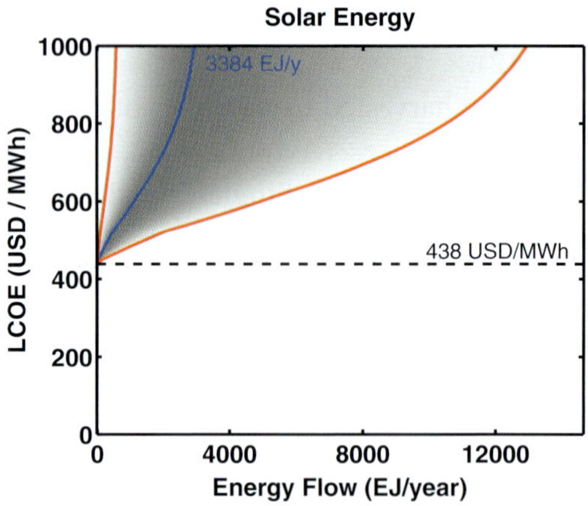

Figure 6.6. Global cost–supply curve for solar energy using photovoltaic technology. Reproduced with permission from Mercure and Salas (2012).

determine land-suitability factors and the opportunity cost of the land at every point of a global grid, while regional estimates of solar irradiation were used to determine the energy potential, which have been taken here for the median cost–supply curve. The lower 95% confidence-level curve, with a technical potential of 1340 EJ/y, is the one evaluated by Hoogwijk (2004), while the upper 95% confidence level, with a technical potential of 14778 EJ/y, stems mostly from increases in land availability following future changes in land requirements for alternative purposes, amongst which are improvements in agricultural efficiency (de Vries *et al.*, 2007).

The use of a large fraction of the technical potential for solar energy using PV systems signifies covering up large amounts of land with solar panels. Mackay (2008) calculates an average productivity value of 10 MW/km^2 for the UK, while Hoogwijk (2004) used values between 6 and 25 MW/km^2, depending on the geographical location, with capacity factors of up to 50%. When compared to coal power plants of capacity of 1 GW and capacity factor of 80%, this implies that the replacement of one such power plant by solar panels requires an area between 50 and 500 km^2.

6.3.1.3. *Hydroelectricity*

Hydropower stems from water pressure gradients that are produced by the run-off of rainfall through landscape topographies, using dams to restrict water flow and accumulate water at elevation levels higher than those given by the landscape, producing a potential for electricity production using turbines. Hydroelectricity is the most deployed renewable electricity technology, with a global installed capacity reaching an estimated 1010 GW, which produced around 2% of the total primary energy supply in 2010 (IEA, 2012; REN21, 2011). As a fraction of its total technical potential, it is also the most developed of all renewable resources, to the extent that around 23% of the global hydroelectric technical potential is currently in use. However, its exploitation around the world is not even: 25% of the European technical potential has already been developed, while Africa uses only 7% of its hydroelectric resource (IJHD, 2011).

Costs of hydroelectric systems are highly site specific and were found to have varied between around 400–4500 US$/kW in 2002 dollars in an extensive global analysis made by Lako *et al.* (2003). These values are influenced by many different factors, which include material and labour costs, but also critically the opportunity cost of the land. The latter refers to the consequences of flooding large areas of land, and the resulting displacement of communities and agricultural activities, and thus varies strongly from region to region. For this reason, the deployment of hydropower is often decided on political rather than financial grounds. Hydroelectric resources are not substitutable between geographic areas, as potential river basins possess widely different properties, generating a strong hierarchical ordering of resource units and a strong slope in the associated cost–supply curve.

Figure 6.7 shows the global cost–supply curve for hydroelectricity, where the dotted line indicates current deployment of 12 EJ/y, 23% of the median technical potential of 66 EJ/y, calculated in this work from data gathered by IJHD (2011). The high deployment-to-potential ratio is an indication that the remaining number of suitable sites for building dams is relatively limited. Using this curve, the cumulative capacity yields a cost of production of about US$68/MWh. However, the

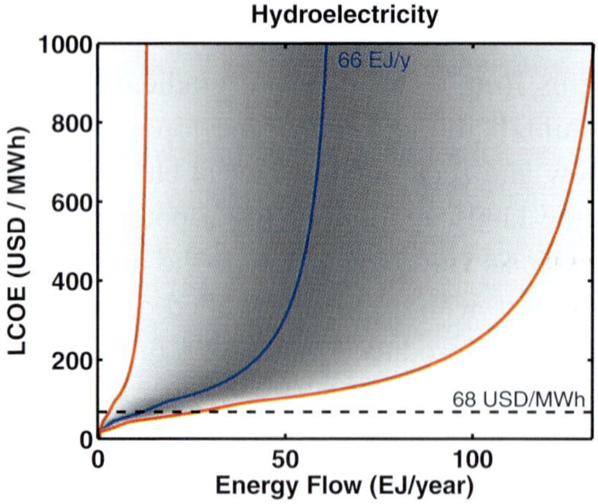

Figure 6.7. Global cost–supply curve for hydroelectricity. Reproduced with permission from Mercure and Salas (2012).

development of hydropower projects hardly follows an order based on cost, but follows instead that dictated by political considerations, which are out of the scope of this work. Therefore, this value is only indicative, and projects with LCOE values between US$23 and US$460/MWh have been recently developed (IEA, 2010a). It is not unreasonable, however, to assume that future developments will follow approximately a cost order, since the development of the limited number of remaining available sites involves either increasing opportunity costs in inhabited areas due to increasing local populations, or large transmission costs associated with increasing distances to uninhabited areas. The cost–supply curve was derived using the theoretical, technical and economic local potential values from IJHD (2011). A distribution of recent cost values was obtained from Lako *et al.* (2003), the range of which was interpreted (in 2008 dollars) as what is currently assumed to be economic. The remaining technical potential (above the economic potential) was thus assumed to involve higher costs. The upper 95% confidence-level curve was derived by considering the global theoretical potential of 144 EJ/y, while the lower 95% confidence level was derived by assuming no additional construction of hydroelectric dams, limiting hydroelectric generation to 12 EJ/y.

6.3.1.4. *Geothermal energy*

Renewable geothermal resources, stored beneath the earth's surface in the form of heat, are constantly replenished by the radioactive decay of isotopes of uranium, potassium and thorium (Macdonald, 1959; Wasserburg *et al.*, 1964). Although geothermal heat has been used for millenia (Cataldi, 1993), and its utilisation for electricity generation commenced at the beginning of the last century (Lund, 2005), its current deployment is small in comparison with other sources of energy. It represents only 0.3% and 1.1% of the total electricity generation and heating production respectively, figures that are not expected to change drastically in future years, according to the projections given by the IEA (Cataldi, 1993; IEA, 2012; Lund, 2005).

Geothermal resources are classified into four categories: hydrothermal (liquid and vapour dominated), hot dry rock (where fluids are not produced spontaneously), magma (molten rock in regions of recent volcanic activity) and geopressured (hot high-pressure brines containing dissolved methane) (Mock *et al.*, 1997). The most commonly used type is hydrothermal, although high expectations exist regarding the development of enhanced geothermal systems (EGS), oriented towards the hot dry rock type through hydraulic stimulation (Tester and Anderson, 2006). According to estimations made by Aldrich *et al.* (1981) based on the report of EPRI (1978), the estimated geothermal heat accumulated under the continental masses to a depth of 5 km depth is of approximately 1.46×10^8 EJ, most of it assumed to be stored in rocks and water, with a proportion of 6:1 in favour of the former. Even though this is a vast amount of heat, only a small part of it is recoverable for productive purposes.[11]

Stefansson (2005) found a high correlation between the number of active volcanoes in a region and the estimate of the size of hydrothermal resources for electricity generation. Therefore, using the total number of

---

[11] Note that the average replenishment of the geothermal heat underground is several orders of magnitude inferior to the stock of heat currently available: around 65 mW/m^2 at the continental level, producing an average thermal energy recharge rate of about 315 EJ/yr (Pollack *et al.*, 1993). This value can be considered as the theoretical potential of geothermal energy if viewed in terms of sustainable extraction of geothermal resources over an extended period. However, the amount of time over which geothermal resources could be used at higher rates than this is likely to be more than 1,000 years.

volcanoes active in the world, and subtracting those located on the sea floor or in arctic regions, he estimated the global potential for hydrothermal energy to approximately 200 GW (producing 6.0 EJ/y of electricity with 95% capacity factor). Using this information, along with the statistical analysis between wet and dry systems developed by Goldstein *et al.* (2009), Bertani (2010, 2012) estimated the total geothermal technical potential to be 1200 GW (36 EJ with 95% capacity factor), including hydrothermal as well as EGS technologies.

While geothermal resources are available all over the world, their accessibility differs from site to site according to various technical characteristics including the geological structure of the ground and the depth and type of heat reservoirs. In the vicinity of tectonic plate boundaries, narrow zones characterised by significant volcanic activity (so-called volcanic belts), geothermal gradients are particularly high, between 40 and 80°C per km of depth, enabling the extraction of high-temperature geothermal resources. In contrast, areas with low volcanic activity are characterised by low and uniform geothermal gradients: around 25°C per km of depth (Aldrich *et al.*, 1981; EPRI, 1978). The technical difficulty for the extraction of geothermal resources in active areas is highly site specific (Pasqualetti, 1983), and thus such sites are not substitutable for one another. On the other hand, the geothermal gradients in the rest of continental masses have smaller variations, and are therefore almost perfectly substitutable for one another. Cost–supply curves were produced for both types of land and both hydrothermal and EGS technologies, generating four curves which were subsequently aggregated in each E3MG region. The combination of both properties for both technologies gives a distinct curvature to the aggregated cost–supply curve.

The global aggregation of curves yields the cost–supply curve presented in Figure 6.8. The associated global technical potential is of 35 EJ/yr, assuming a 95% capacity factor. The lower 95% confidence-level curve assumes a technical potential of 4 EJ/y taking into consideration two elements. In the case of hydrothermal technology, a conservative potential estimate of 70 GW (2 EJ/y) was derived based on well-known sites that have been already characterised by direct intervention or informed calculations (Bertani, 2010, 2012). In the case of EGS technology, given its very early stage of development (Tester and Anderson, 2006), Bertani estimated a technical potential of 70 GW

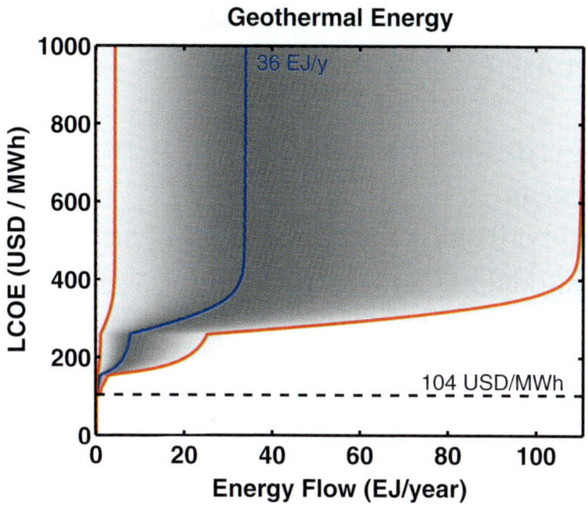

Figure 6.8. Global cost–supply curve for geothermal energy. Reproduced with permission from Mercure and Salas (2012).

(2 EJ/y). The upper 95% confidence level also considers two elements: in the case of hydrothermal technology, it is assumed that hidden resources are five to ten times higher than identified resources (Stefansson, 2005), so the hydrothermal potential may rise up to 1000–2000 GW, which corresponds to a maximum of 60 EJ/yr. In the case of EGS, the technical potential is calculated by an extrapolation of resources in the US to the global level using the proportion between the heat stored at depths of less than 10 km in the US with the known EGS primary energy potential in the same region, estimated by Tester and Anderson (2006) as 1×10^6 EJ of heat stored per 2.61 EJ/yr of EGS primary energy potential. Using the estimation of the heat stored at depths less than 10 km on the global scale produced by Rowley (1982) of 403×10^6 EJ, this estimation results in a global technical potential of 57 EJ/yr, yielding a total of 117 EJ/y.

6.3.1.5. *Ocean energy*

The term "ocean energy" denotes renewable energy produced using seawater as a resource, where unlike for wind energy or hydroelectricity, not only the kinetic energy of seawater can be used to produce electricity, but also temperature gradients in the ocean and salinity differences near river mouths. Using ocean energy as a general classification type, it can be divided into four main sources of energy (IPCC, 2011c; ETSAP, 2010b):

- **Wave energy**, driven by transfers of energy from the wind to the surface of the ocean
- **Tidal energy**, driven by the rise and fall of sea levels due to gravitational forces (tidal range) and the resulting water currents
- **Ocean thermal energy**, driven by temperature gradients between upper and lower ocean layers
- **Salinity gradient energy**, derived from salinity gradients between ocean and fresh water at the mouths of rivers

Section S.3.6 of the supplementary material in Mercure and Salas (2012) provides a review of theoretical potentials for these sources, resulting in a total that could be as high as 609 EJ/y. Technical potentials, however, are much lower and uncertain, since the current development status for ocean energy technologies excluding tidal is preliminary, and cost data are in some cases unavailable. Specific geographical and configurational requirements for tidal and salinity gradient technologies involve, as is the case for hydroelectricity, calculating the technical potential by summing the potential values from a large number of individual studies. Such studies have not been performed exhaustively on a wide scale yet. Meanwhile, wave and ocean thermal potentials are based onto global extrapolations carried out using physical measurements. Global energy potentials calculated in various studies are given in Table 6.1, and additional details can be found in the Section S.3.6 of the supplementary material in Mercure and Salas (2012). Energy potentials for ocean thermal and salinity gradient energy are theoretical and highly uncertain, and no reliable cost estimates were found. These types of resources were therefore not included in the present calculations for the most probable cost–supply curve, due to the risk of generating misleadingly optimistic potentials given the lack of reliable information. Wave and tidal systems are better established. Therefore, using cost values obtained from the IEA and the ETSAP, a cost–supply curve for ocean energy based on an aggregation of separate cost–supply curves for wave and tidal energy was produced, and is given in Figure 6.9, with a small technical potential of 22.5 EJ/y (ETSAP,

Table 6.1. Technical potentials for different types of ocean energy used to define the cost–supply curve. The uncertainty ranges are defined using the Min and Max values, while Mode represents the most probable value.

Technology	Min EJ/y	Mode EJ/y	Max EJ/y	Study
Wave	6.3			(IPCC, 2007d)
		18.9		(WEC, 1994)
			65	(UNDP, 2000)
Tidal	1.8			(Hammons, 1993)
		3.6		(Hammons, 1993)
			7.2	(WEC, 1994)
Total	**8.1**	**22.5**	**72.2**	
Ocean	3.2			(Charlier and Justus, 1993)
Thermal		32		(Charlier and Justus, 1993)
			85	(Nihous, 2007)
Salinity	5.8			(Skramesto *et al.*, 2009)
Gradient		7.2		(Krewitt *et al.*, 2009)
			83	(Cavanagh *et al.*, 1993)
Total	**17.1**	**61.7**	**240.2**	

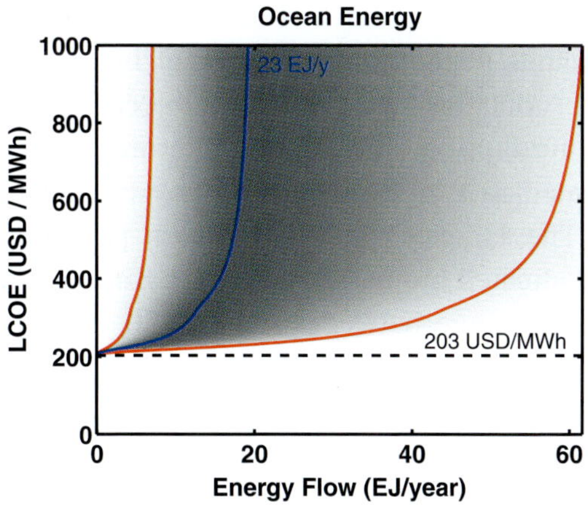

Figure 6.9. Global cost–supply curve for ocean energy. Reproduced with permission from Mercure and Salas (2012).

2010b; IEA, 2010a). The lower and upper boundaries of the uncertainty range were obtained from the extremal values of 8.1 and 72.2 EJ/yr respectively given in Table 6.1.

6.3.1.6. *Bioenergy*

Bioenergy, energy derived from plant biomass, is currently the most widely exploited renewable energy resource, with 53 EJ/y, 10% of global annual primary energy use (IEA, 2012). While biomass combustion using integrated gasification combined-cycle technology (BIGCC) is expected to become the most efficient biomass-based electricity production method (Rhodes and Keith, 2005), the combination of biomass and CCS technology has been shown to produce negative CO_2 emissions (Gough and Upham, 2011). Additionally, the emission factors of some power plants using conventional coal technologies have been reduced by co-firing coal and biomass fuels. Meanwhile, liquid biofuels derived from biomass, such as ethanol and biodiesel, have the potential to replace oil-derived transport fuels with minimal changes in vehicle internal combustion engine technology and jet engines (IPCC, 2011a).

Biomass currently used for electricity and biofuel production largely originates from forestry and agricultural residues, and other forms of commercial or household mixed solid waste. Volumes of waste available could amount up to 100 EJ/y but they are highly uncertain. Waste amounts depend on efficiency of biomass use (such as in food or timber production) and therefore subject to significant changes depending on future policy. These are therefore difficult to model, as are their associated costs of production (Hoogwijk *et al.*, 2003; Smeets *et al.*, 2007). The largest share of biomass potential lies with the production of dedicated biomass crops. Global technical potentials for primary bioenergy range between 0 and 1550 EJ/y (de Vries *et al.*, 2007; Hoogwijk *et al.*, 2005; Smeets *et al.*, 2007; van Vuuren *et al.*, 2009; Wolf *et al.*, 2003). Such crops include perennial woody short-rotation coppiced trees, such as willow, poplar or eucalyptus, perennial grasses such as miscanthus, elephant grass and switchgrass, starch-rich crops such as wheat, corn, sugar beet and sugarcane, and oil-rich crops such as rapeseed and palm. Depending on their nature, they can be transformed into energy carriers by using, among many processes, combustion, gasification or anaerobic decay for electricity production, fermentation or the Fischer–Tropsch process for transport fuel production (IEA Bioenergy, 2009).

6.3.1.7. *Potential biomass resources and cost–supply curves*

Biomass production for energy purposes makes use of agricultural land and thus may have a high opportunity cost. The technical potential that lies in agricultural land is large, but energy production from biomass is in direct competition for land with food production, a situation which has the potential to drive significant increases in world food prices (Dornburg *et al.*, 2010). Following the methodology of Hoogwijk *et al.* (2005, 2009), Smeets *et al.* (2007) and Wolf *et al.* (2003), the explicit assumption of specific intervention for the protection of food production is taken and therefore leads to the use for energy purposes exclusively of the agricultural land which remains after the global demand for food has been met. Additionally to agricultural land, however, other types of geographical areas with low productivity exist which can be used for less demanding bioenergy crops. Hoogwijk *et al.* labelled this type of land as "rest land", while the agricultural land was labelled as "abandoned land". The nature and use of the land were evaluated in this extensive work yearly up to year 2100 using the land-use model IMAGE 2.2. Land use depends strongly on assumptions regarding world population, diet habits, global urbanisation and trade of agricultural products. The four main IPCC (2000) scenarios, A1, A2, B1 and B2, were taken as assumptions for all exogenous variables in these calculations, and results are presented for each. Large differences arise between scenarios, with technical potentials ranging between around 305 EJ/y for the A2 scenario to 679 EJ/y for the A1 scenario, which results in large uncertainties for values of the global biomass technical potential. However, other ranges have been estimated, with more pessimistic projections of 0 to 648 EJ/y by Wolf *et al.* (2003) and optimistic values of 367 to 1548 EJ/y by Smeets *et al.* (2007). The low end of the range given by Wolf *et al.* stems from high projected food demand and low agricultural productivity, while the high end is due to mostly vegetarian diets and high productivity. Meanwhile, the high end of the range of Smeets *et al.* originates from "super high" agricultural productivity, high availability of the land and landless animal production systems.

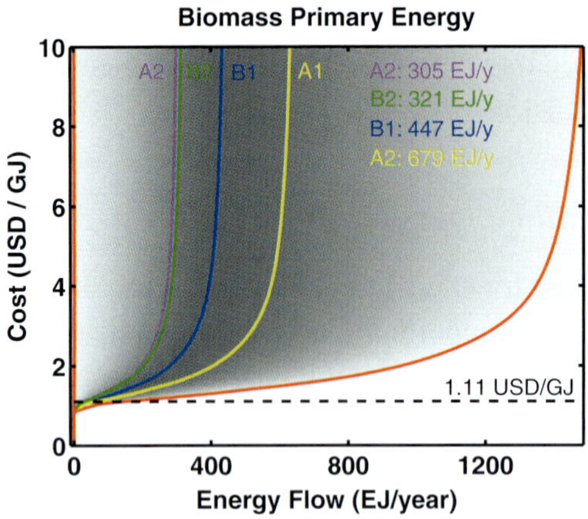

Figure 6.10. Global cost–supply curve for biomass energy. Reproduced with permission from Mercure and Salas (2012).

Figure 6.10 presents the global economic potential of bioenergy in terms of primary energy before conversion to electricity or liquid biofuels, derived from the data of Hoogwijk *et al.* (2005, 2009), Smeets *et al.* (2007) and Wolf *et al.* (2003), using both abandoned and rest land, but includes also a small component from baggage of 3 EJ/y. Four cost–supply curves are given, calculated by Hoogwijk *et al.* (2009) for the A1, A2, B1 and B2 SRES scenarios, shown as solid curves. The lower 95% confidence level was taken as the pessimistic value of zero from Wolf *et al.* (2003), while the upper 95% confidence level was taken as the optimistic value of 1548 EJ/y from Smeets *et al.* (2007). For a decarbonisation scenario, the cost–supply curve derived using B1 was considered the most appropriate, and thus taken as the median in this plot, but for any other purposes, the choice of curve must be consistent with the particular working assumptions.

6.3.1.8. *Bioelectricity and biofuels*

The use of bioenergy necessitates a transformation from the primary biomass feedstock into a form of fuel useable in particular applications. In such a transformation, efficiency is critical and there exist an inevitable loss. For example, in biofuels for transportation, either plant starch or

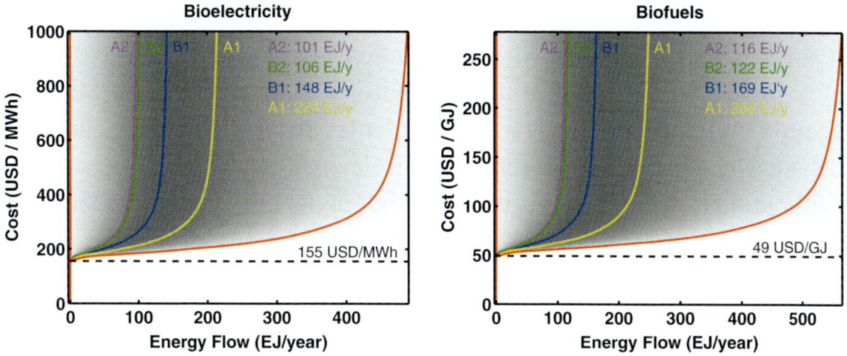

Figure 6.11. Potential of biomass energy as a secondary energy carrier, electricity. The technical potentials given refer to situations where the whole biomass primary resource was used for electricity production in the four SRES scenarios (left-hand side) or for biofuels (right-hand side).

sugar is converted to ethanol, or oil extracted from a plant is transformed into biodiesel. Both these processes may reject a significant part of the plant material used. Meanwhile, for electricity either plant material is combusted directly or it is gasified and then combusted, both operations leading to efficiency losses. Such efficiencies are technology dependent (Dornburg *et al.*, 2010; Hoogwijk *et al.*, 2005, 2009).

An analysis of bioenergy technologies is beyond the scope of the present chapter. However, in order to compare the bioenergy potential to other sources of energy, the primary biomass resource of Figure 6.10 must be transformed into a usable secondary carrier of energy, given in Figure 6.11. In particular, if one is to compare the potential bioelectricity output to other renewables, the primary resource must be transformed from energy in the form of plant material to energy in the form of electricity.

6.3.2. *Nuclear energy*

6.3.2.1. *Nuclear technology*

Nuclear energy remains an important component of the power sector. It is often seen as neither sustainable nor friendly to the environment.

Through several accidents, rising costs and its output of dangerous waste, it has gradually diminished its reputation since its early development after the Second World War. Many environmentalists campaign strongly against building new nuclear power plants. Several countries currently debate whether or not to renew their fleet of nuclear power stations. Without taking a marked position in this debate, this section explores the contribution that nuclear energy can make as part of the effort for decarbonisation.

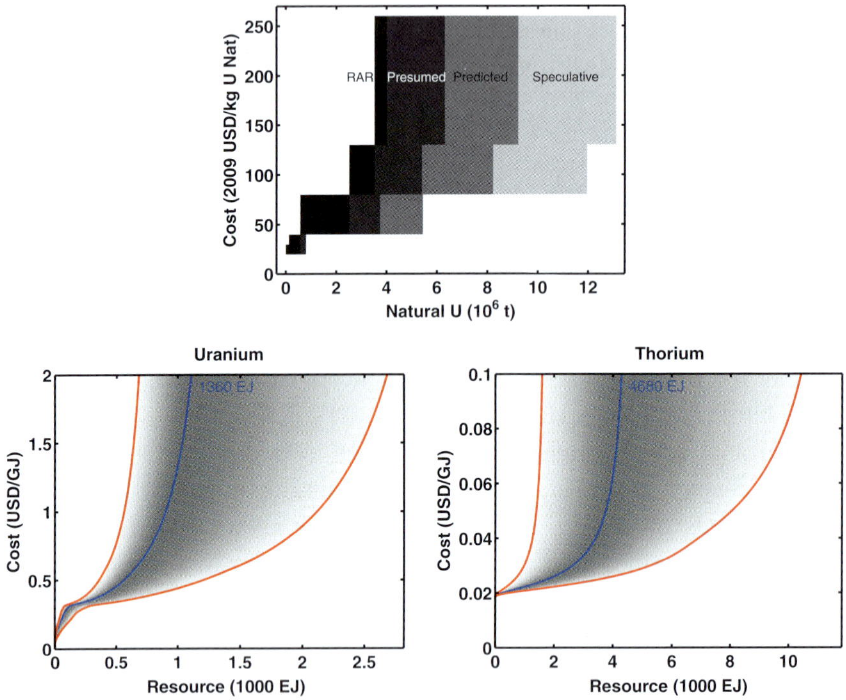

Figure 6.12. Cost–supply curves for nuclear fuels. Top panel: Cost–supply curve for uranium and fuels of military origin in mass units represented as boxes corresponding to cost and uncertainty ranges given by the IAEA (2009). RAR indicates reasonably assured reserves and various shades of grey indicate decreasing confidence levels. Left-hand panel: Cost–supply curve for uranium derived from the same data, using the methodology described in Mercure and Salas (2012). Resulting uncertainty ranges are highly asymmetric. Right-hand panel: Cost–supply curve for the resources of thorium. Reproduced with permission from Mercure and Salas (2012).

A significant but not enormous potential for low-carbon energy lies in deposits of fissile materials around the planet. The use of such resources results in the production of radioactive waste, and is associated with safety and proliferation risks. However, as opposed to fossil fuels, it does not produce as an externality the emission of GHGs associated with climate change (other than during construction). Thus the dangers associated with nuclear power stations, proliferation and nuclear waste may be measured against the danger of emitting an amount of CO_2 corresponding to the same quantity of energy produced. In particular, where decisions may be taken to close down nuclear power stations, for instance in the view of building a more sustainable energy sector, it is not unlikely that they will be replaced in the medium or long term instead by fossil fuel power stations, including through the importation of electricity from neighbouring countries with different energy policies.

Nuclear energy is not a sustainable source of energy, and an analysis of the size of the finite resource of nuclear energy is required. Fuel burn-up efficiencies are highly dependent on particular technologies used, which are thus briefly described. Two fuel cycles for energy production have been developed in the very early days of the nuclear age; the uranium and the thorium cycles. While only the uranium cycle has become successful commercially, thorium could become important in a future where uranium resources became scarce, as reserves of thorium larger than those of uranium are known to exist. Thus, following the framework of cost–supply curves, an attempt to provide an analysis of known reserves of fissile materials is given. Nuclear technologies as they are currently used only extract a very small fraction of the calorific content of nuclear fuels. Fast nuclear reactors have the ability to burn fuels up to 50 times more efficiently (Bonche, 2002), and therefore change this picture dramatically.

The cost–supply curve for uranium is provided using the current average efficiency of fuel use with thermal reactors. It is, however, stressed that, were uranium resources to become scarce, both options of fast reactors and the thorium fuel cycles exist to extend reserves of fissile materials probably by two orders of magnitude.

6.3.2.2. *Global resources of fissile materials*

Five sources of fissile materials for nuclear reactors are known to exist. These are enumerated in order of cost. The first comes from stocks of highly concentrated ^{235}U or ^{239}Pu originating from decommissioned nuclear arsenals diluted with ^{238}U. The second source is lightly enriched $^{235}U/^{238}U$ produced from mined natural deposits. The third originates from U and Pu recovered from spent fuel (using the PUREX process (Bonche, 2002)). The fourth source is thorium using the $^{232}Th/^{233}U$ fuel cycle. The fifth source is U which occurs in very low concentrations in seawater. Producing a cost–supply curve involves creating a scenario for the nuclear sector, and requires careful consideration of uncertainty. In particular, if innovative use of fast reactors is invoked, higher fuel efficiencies can be obtained.

In order to construct a cost–supply curve for U and Pu, the nuclear industry was assumed to continue to use current methods and thermal reactors, and therefore, only consider fuel originating from naturally mined U and from nuclear arsenals. Many authors stress that deposits of thorium worldwide are three times larger than those of uranium (Abu-Khader, 2009; Bonche, 2002; Sinha and Kakodkar, 2006; Suess and Urey, 1956). However, less prospection efforts have been carried out and, as a consequence, the current reasonably assured reserves of Th, in tonnes of natural Th, are lower than those of U, a situation which is likely to change as interest in Th grows (IAEA, 2009). The nuclear fuel cycle for Th is more efficient than that of thermal reactors based on U, hence it leads to larger amounts of energy per tonne of natural Th and thus lower fuel costs per unit of energy.

The uncertainty ranges are highly asymmetric due to the tendency of the size of speculative resources to increase with the level of uncertainty. It is observed that in terms of energy, reserves of Th are larger than those of U, and that these are also less expensive per unit of energy, due to the higher burn-up rate of the Th system. It must be emphasised again that U resources could, in principle, be used with much higher burn-up rates, were fast reactors to be deployed globally. The resources of U do not include seawater U, as data on these are scarce and highly speculative. Finally, it is to be noted that the fuel component of the LCOE for nuclear reactors is very small compared to the investment costs, which results in very small influence of the fuel costs onto the decision-making, unless nuclear resources are depleted.

6.3.3. *Fossil fuels*

6.3.3.1. *Introduction*

The process of decarbonisation of the world economy involves, in large parts, gradually phasing out the use of fossil fuels. Fossil fuels are in essence accumulations of hydrocarbons that originate from biomass material that has lived during various eras of geological history. Their combustion results in the addition of large amounts of carbon into the existing natural carbon cycle. This is known to cause a gradually increasing perturbation to the equilibrium in the energy balance of the planet, through additional radiative forcing (IPCC, 2007a). In 2010, 81% of the world's energy use in all forms, be it heat, movement or electricity, was produced from the three types of fossil fuel: oil, gas and coal (IEA, 2012). The goal of decarbonisation is thus a very ambitious one, since it requires changing the largest part of the world's energy production infrastructure. The methods of energy production based on fossil fuels are simple and deeply ingrained. The problem of decarbonisation could have been simpler had the world been quickly running out of fossil fuels. In such a scenario, the various economies of the world would have been faced with the necessity to change energy sources. This section demonstrates that this is not the case and, moreover, waiting for fossil fuel depletion to decarbonise the global economy leads to GHG concentrations which are thought to lead to extreme climate change (IPCC, 2007b). In particular, reserves of coal are likely to be plentiful beyond a horizon of 100 years in most scenarios of energy demand and use, and the amount of carbon content in coal reserves alone are enough to produce extreme climate change. Known gas resources are also large, in particular when newly explored shale gas and methane hydrates are included. Therefore, the amount of low-cost fossil fuel resources such as these currently available constitutes a significant obstacle to decarbonisation. As the global economy is gradually steered away from fossil fuel use and demand for hydrocarbons decreases, it is furthermore likely that their prices fall, generating a rebound effect. In this section, current fossil fuel reserves and resources are reviewed.

Cost–supply curves are derived by combining all types of resources associated with either oil, gas or coal. These are based on a database involving mixed sources and a particular methodology, both described in the supplementary information of Mercure and Salas (2012).

6.3.3.2. *Global fossil resources and cost–supply curves*

As occurs with all types of exhaustible natural resource, fossil fuel resources and reserves are known to continuously expand, even though they are gradually consumed (see, for instance, Rogner (1997) and McKelvey (1972)). This is due to periodic resource discoveries and improvements in the methods of extraction. Therefore, what is considered economical to extract changes over time. Reserves are distinct from resources, the former referring to the resources which are known to exist with almost complete certainty and to be economical to extract, while the latter refers to those which are thought to exist with various degrees of confidence, and those currently thought too expensive to extract. As technological improvements and additional knowledge affect the economics of different methods of extraction, there is a flow from resources towards reserves, and therefore reserves expand. Meanwhile, discoveries continuously add to resources. As prospection for hydrocarbon sources remains very active, this makes the production of cost–supply curves more difficult than for renewables. In the case of renewables, for instance, if the distribution of wind speeds distributed around a certain geographical area is known, a well-defined cost–supply curve can be derived, which is not likely to change very much except through a rescaling of costs due to changes in wage rates or inflation, etc. For fossil fuels, however, the cost–supply curve is at best a snapshot in time of what is known to exist and be recoverable with current technology.[12]

In order to assess the feasibility of decarbonisation, it is necessary to explore cost–supply curves for fossil fuels, even if they are derived from

[12] Rogner (1997) demonstrates very effectively the effect of discoveries and technological improvements.

current knowledge, and therefore expected to change in the future. It is unlikely that fossil fuel resources turn out smaller than what is currently known. On the contrary, it is probable that they turn out larger as methods of extraction are devised for types of occurrences which were until recently not thought possible to use, such as gas hydrates or oil shales. Therefore, the cost–supply curve uncertainty ranges are highly asymmetric. The associated extraction costs, which increase as low-cost conventional sources are depleted, nevertheless decrease due to technological improvements, and it is therefore not immediately obvious whether costs are likely to go up or down in the future. Since the extraction of fossil fuels, particularly for oil and gas, involves new and different challenges for every type of resource, technological learning does not quite follow the power law behaviour associated with building over and associated with repeated building of the same industrial product over time (IEA, 2005), and thus does not necessarily improve following a power law of the number of, for instance, barrels of oil extracted.

Four cost–supply curves given in Figure 6.13, where the top-left panel provides that for liquid hydrocarbons, the top-right panel gaseous hydrocarbons and the bottom two solid hydrocarbons, hard (bottom-left panel) and soft (bottom-right panel) coal. These enable the appreciation of the relative scale between conventional and unconventional sources of oil and gas. Resources of oil and gas are large only when uncertain unconventional sources are considered, and these include oil shales, oil sands and extra heavy oil for liquid hydrocarbons, and shale gas, tight sands gas, coal bed methane and methane hydrates for gaseous hydrocarbons. In the case of oil, these sources have the common property that their extraction currently generates significant environmental problems, and moreover use significant amounts of energy, which increases their effective GHG emission factors. Meanwhile, the exploitation of shale and tight gas sources requires the fracturing of the underlying rock bed, leading to a greater possibility of environmental concerns including hydrocarbon contamination of aquifers and drinking water. As opposed to oil and gas, coal cost–supply curves are extremely flat and involve only one type of resource, extracted with well-established mining methods. The resources of coal are found to be very large in comparison to all other exhaustible sources analysed in this chapter. As all exhaustible resource cost–supply

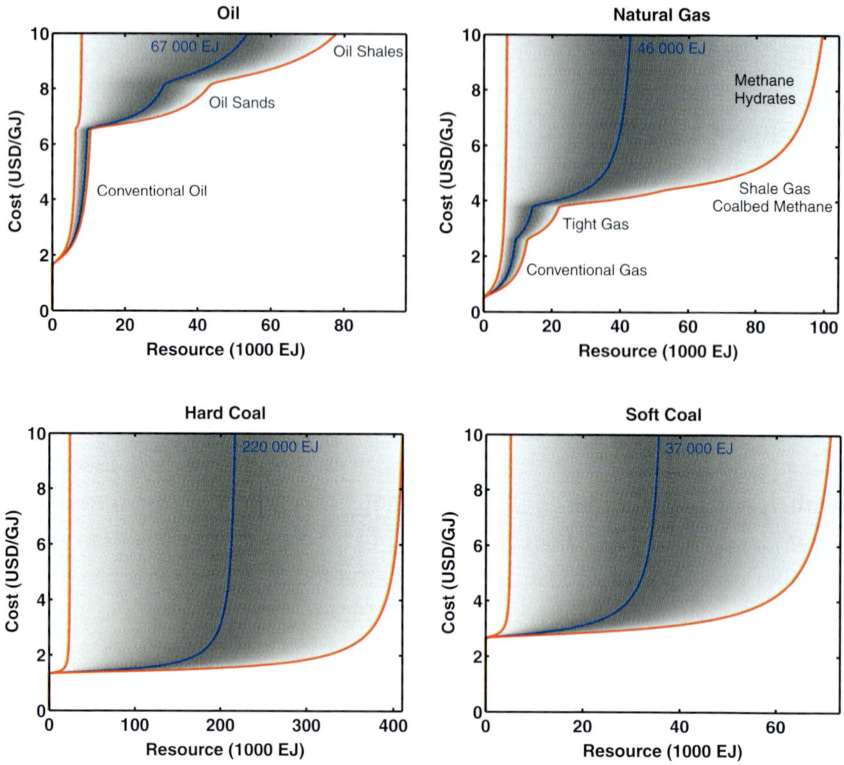

Figure 6.13. Cost–supply curves for liquid hydrocarbons (top left), gaseous hydrocarbons (top right), and solid hydrocarbons (bottom left and right), the last of which have been divided into two categories according to their calorific content, denoted hard (bottom left) and soft (bottom right) coal. Reproduced with permission from Mercure and Salas (2012).

curves are expressed in units of energy, a framework is given below to convert them into energy flows using assumed rates of extraction, in order to compare their size with those of renewable energy sources.

6.3.3.3. *Fossil fuel resources, GHG emissions and climate change*

It is well established that the complete combustion of all fossil fuel resources in any future scenario leads to dangerous climate change. This

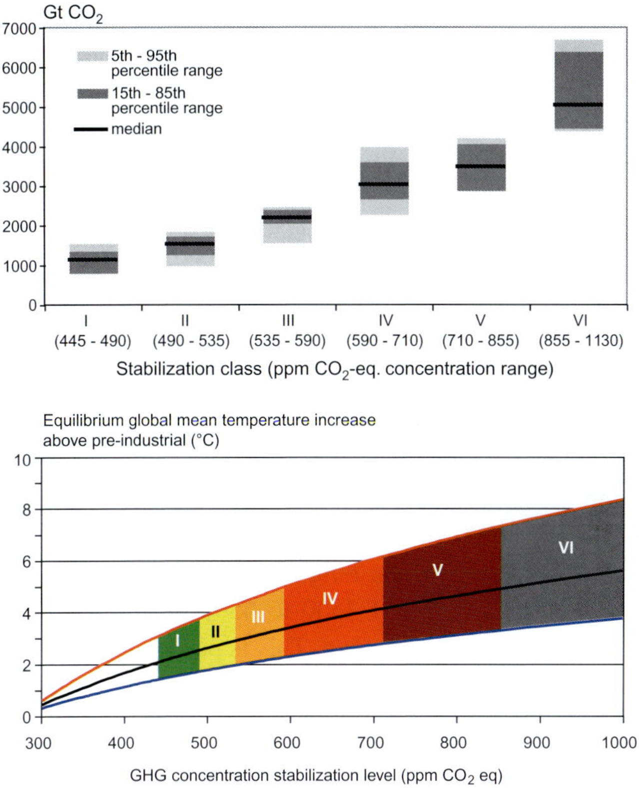

Figure 6.14. Approximate conversion of cumulative emissions up to 2100 to ranges of global warming. Left-hand panel: Approximate conversion from cumulative emissions to concentration ranges. Right-hand panel: Approximate relationship between concentration ranges and global warming ranges. Reproduced from IPCC (2007c).

result is depicted here very simply by using current knowledge regarding anthropogenic climate change originating from industrial CO_2 emissions, which was obtained from performing statistics over a large number of emission scenarios and their resulting global warming (IPCC, 2007c). Figure 6.14 reproduces two graphs from IPCC (2007c), which express an approximate conversion between ranges of cumulative emissions from now to 2100 and ranges of CO_2 concentration in the atmosphere (left-hand panel), and in the second (right-hand panel) a conversion between these ranges of concentration of atmospheric CO_2 to ranges of global warming.

These relationships are approximate since global warming depends on the time profile of the emissions rather than simply their cumulative amount in 2100. These relationships include also all uncertainties which arise from these models. Although very approximate, these relationships allow a clear demonstration of the impact of the energy sector on climate change.

Stocks of fossil fuels, if completely combusted, release specific amounts of CO_2 depending on their nature. Such amounts may be expressed in terms of tonnes of CO_2 per tonne of fuel, as calculated for instance by Keeling (1973). Assuming only positive global CO_2 emissions up to 2100, it is straightforward to calculate the cumulative emissions that would occur if the global economy depleted all the resources of, independently, hard coal, soft coal, oil and gas. When including all reserves, resources and speculative resources, this results with, respectively, about 32000 Gt, 5100 Gt, 5600 Gt and 5500 Gt, for a total of about 52000 Gt of CO_2. These values lead to a warming that is near or beyond the range of 4–8°C predicted for a range of 4500 to 6500 Gt of CO_2, which corresponds to dangerous climate change. This is particularly true if combusting all reserves of coal, which contain much more carbon than necessary in order to seriously damage the climate system.

6.4. Carbon Capture and Storage and its Constraints

6.4.1. *Global CO_2 storage capacity*

The capture of CO_2 and its storage is a viable activity given that storage space is available at a cost which does not render particular power-production activities unprofitable. An analysis of the global capacity for storing CO_2 as a function of cost is thus provided. The geographical mismatch between the capture and storage of CO_2 is an important consideration of local nature which cannot be analysed here; and is therefore omitted. It is stressed, therefore, that costs for transport of the gas to areas where storage is available may be prohibitive for certain power plants, in which case the decarbonisation of these electricity production activities may be achieved at lower cost by replacing these power plants by other low-emission

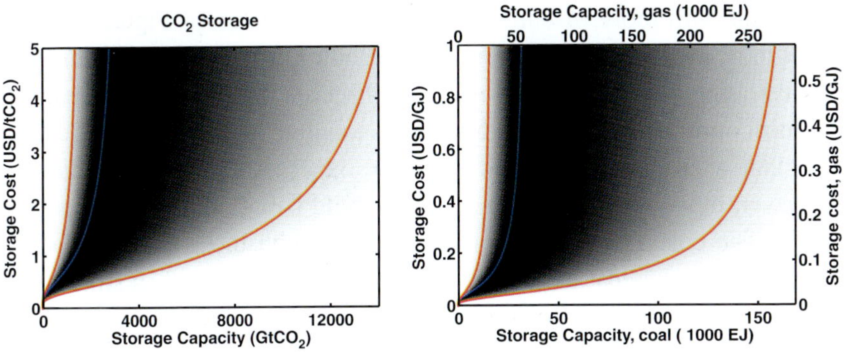

Figure 6.15. Cost–supply curves for the sequestration of CO_2. No transport or capture costs are included here. Left-hand side: Cost–supply curve in units of mass of sequestrated CO_2. Right-hand side: Same cost–supply curve expressed in units of energy content of coal, bottom and left axes, and gas, top and right axes.

systems rather than adding CCS capability. Assuming therefore small transport costs where it is used, a cost–supply curve is derived for the storage of CO_2, using the global technical potential for storage along with cost values expressed in terms of an additional cost per unit of energy.

Figure 6.15 presents a global cost–supply curve for the sequestration costs and global storage space available, where storage potential values were obtained from (IEA, 2011a) and cost values from (IPCC, 2005). The left-hand panel expresses the cost–supply curve in units of mass of sequestrated carbon dioxide gas. One tonne of CO_2 sequestrated is associated with a certain amount of energy which was obtained from the fuel that was combusted, but that amount depends on which type of fuel was used. More energy is obtained per tonne of CO_2 produced when gas is combusted than with coal, since gas has a much higher calorific content than coal. Therefore, the cost–supply curve may be expressed in terms of energy of combustion, and for filling a specific amount of storage space for CO_2, more or less energy may have been obtained depending on which fuel was combusted. Therefore, the right-hand panel of Figure 6.15 shows the same cost–supply curve relationship expressed in units of energy of combustion of two fuels, coal and gas, where the left and bottom axes refer to the combustion of coal, while the top and right axes refer to the combustion of gas. The values observed in this graph can thus be directly compared with those of natural gas and coal reserves of Figure 6.13, in order to determine how much of these resources can be combusted while sequestrating their CO_2 emissions. It is observed that the storage space in the most optimistic scenario is not large enough to accommodate all of the CO_2 that would be produced if the entire reserves of coal were burned. However, most of the methane reserves could be combusted and sequestrated. The cost axes may also be compared with those of Figure 6.13, revealing that the cost of sequestration per unit of energy is a small fraction of the mining or extraction costs of coal and gas.

6.4.2. *CCS in combination with biomass and negative emissions*

Carbon capture and storage technology applied to fossil-fuelled power plants is a good mitigation option which can contribute significantly to the effort of decarbonisation. This technology also has the potential to produce much larger reductions in CO_2 emissions if it is applied to biomass-fuelled power plants such as BIGCC. Effectively, when grown, biomass draws its mass of carbon from the atmosphere, and thus acts as a carbon sink. As it is combusted, it releases again its carbon content to the atmosphere; however, if its carbon content is sequestrated after combustion, net emissions become negative. The same three technologies, pre-, post- and oxyfuel-combustion, may be applied to biomass systems. For instance, post-combustion capture does not depend on the fuel, while pre-combustion capture can be applied to BIGCC plants.

Systems with negative emissions possess an enormous mitigation as well as financial potential. Within an emissions trading scheme involving emission allowances, systems that have net negative emissions *create* allowances which can be sold onto the carbon market. In an emission pathway that leads to gradual decarbonisation, as the number of allowances made available decrease proportionally to the chosen emission path, the price of carbon is very likely to increase unless massive abatement measures are taken. While fossil-fuelled power plants gradually become less and less profitable, biomass plants using CCS become *more and more* profitable. These systems have therefore the potential to eventually become more attractive financially than fossil-fuelled plants in later stages of decarbonisation (see, for instance, Mercure (2012a)). Many calculated scenarios of decarbonisation involve systems with negative emission factors (IPCC, 2007c).

6.5. The Feasibility of Decarbonisation in Terms of Natural Resources

6.5.1. *Meeting demand and supply for power*

Power demand cannot be met by just any combination of supply. The composition of the electricity sector must follow certain rules which are dictated by technical constraints. Electricity demand follows a more or less predictable pattern resulting from the daily habits of both households and the industry. The ratio of maximum to minimum demand may reach values such as 1.5, and the profile of the demand must be met exactly at every instant within a fraction of a second in order to maintain grid stability, voltage and its frequency of 50 or 60 Hz. Many electricity-generation technologies can only operate with a constant output, and using only such systems would require the grid to dissipate about 16% or more of its output.[13] Moreover, some systems, such as wind and solar power, have an output which is not controlled but fluctuates according to the supply of the natural resource. The mismatch between these must be provided by electricity technologies which are able to change their output from zero to their maximum capacity within seconds or minutes. This is depicted in Figure 6.16.

The electricity system in each country is composed of power stations possessing a range of possible power ramping rates, from strictly zero, as with nuclear reactors, to slow systems such as coal thermal power stations, to very fast systems such as hydroelectric dams, gas turbines and diesel generators. The marginal costs of producing electricity for each type also tend to follow the same order, with nuclear reactors in the lower range and diesel generators being the most expensive per MWh. In free electricity

[13] This value is approximate and can be obtained by assuming a sinusoidal daily demand variation and integrating the difference to a constant power supply high enough to cover peak time.

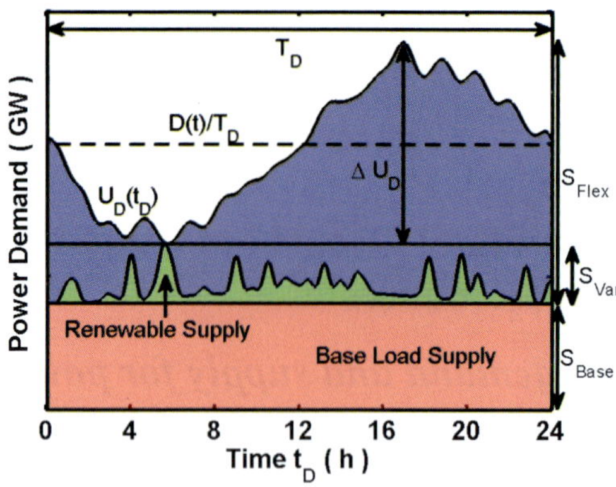

Figure 6.16. Sketch of the typical structure of the daily demand and supply of electricity along. The demand U_D fluctuates along the time of day t_D. The supply is composed of a base load generation from base load-only technologies, such as nuclear or coal power stations (red). Variable renewables such as wind and solar produce the amount shown in green. The remaining difference between supply and demand must be met by flexible technologies, which are able to cope with changes from shutting down entirely to producing a very large output, shown in blue, and this is usually met by gas turbines or hydroelectric dams. Flexible technologies can, if needs be, replace base load-only technologies, shown as a dark blue band. Reproduced with permission from Mercure (2012a).

markets, the price of electricity also varies during the day, bid by companies for every unit of expected electricity demand. For units of energy which all types of power stations are able to provide, the so-called base load supply, the price is lowest, and it increases during peak time as less and less power stations are available to cover the last few units requested. Diesel generators are generally kept as a last resort. Therefore, peak time power stations produce energy during only a fraction of the time of day, but are rewarded by higher electricity prices when they do run.

6.5.2. *Scenarios for the exploitation of exhaustible resources*

The comparison of exhaustible and renewable energy resources can only be done through the choice of a path in time up to 2100 for the rate of exploitation of exhaustible resources, which determines how much of the resources are used today and how much are kept for the future, or never used altogether. This is performed in order to transform the stocks previously calculated (Figure 6.12 and Figure 6.13), measured in EJ, into energy flows, measured in EJ/y, which can be compared against renewable energy production. The rate of extraction is however not likely to be constant, in particular in a scenario of decarbonisation. It depends on global energy demand and on rates of technological substitution between all types of energy production technologies.

The price of energy carriers is the key that unlocks different types of natural resources, as it determines which sources are economical to exploit, and which are not. As an example, while the price of the barrel of oil remained low, the Canadian oil sands were only considered until recently as a potential resource for the future. However, as soon as the price of oil reached the threshold of around 70 US$/barrel, their exploitation became profitable and quickly expanded to a wide-scale development (Bruno *et al.*, 2010; IEA, 2011b). Given a certain demand for energy, long-term average prices of natural resources increase in such a way as to unlock enough resource production to meet the demand. Therefore, the choice of a particular price path for the future yields a specific production profile of energy. The production increases at each price increment which unlocks a new type of resource, going up the cost–supply curve, but decreases as less expensive types of resources are gradually depleted. This concept is at the core of one of our previous studies (Mercure and Salas, 2013).

The choice of a price path for natural resources transforms resource stocks into resource flows. This is the procedure which is carried out in the following discussion. The use of a constant rate of increase in the price of resources, a linear price, is convenient as it produces an equivalence between price and time. As the price increases (and time moves forwards), the amount of resources available for production increases gradually, until a maximum is reached, where production begins to decrease as depletion begins. The rate of increase of the price determines how much new resources become available each year. Therefore, the curves produced using the cost–supply curves for stocks of exhaustible resources through this process have the property that they resemble ordinary resource flow cost–supply curves, as in the case of renewables, but decrease to zero flow values at high price values where depletion is complete. If the price increases more slowly, their use could extend further into the future and *vice versa*. Therefore in reality, these curves should depend on a dynamic price determined by the context, which is outside of the scope of this work. Thus only specific curves are given here, the full methodology and data having been made available in Mercure and Salas (2012, 2013).

6.5.3. *Calculation of cost–flow curves*

Figure 6.17 depicts schematically the calculation of cost–flow curves. In panel (a), a hypothetical initial (or current) resource distribution is given (in black) along with a probability of resource extraction $f\ (P - C)$ dependent on the price P (in red). The gradual depletion of the resource is shown in panel (b), where curves for the remaining distribution are given after various time spans (in black), each occurring at particular values of the price $P(t)$. As the price P increases, the mid-point of the probability function (in red) moves to the right, and the amount of resources remaining decreases. By summing the flow of resources extracted at all cost values C, a time (or cost[14]) dependent flow $F(t)$ is obtained (panel c)), which decreases in the long term where depletion occurs. This flow may be interpreted as the amount of resource that is *unlocked* by the carrier price. Therefore, panel (c) may be expressed with either time or price as the horizontal axis. In order to determine the marginal increase in price that occurred in the process of increasing production by a certain amount, the axes are inverted, shown in panel (d). Given that large uncertainties exist in the resource distribution (as described previously), the flow is expressed in terms of a confidence-level range. For that purpose, two boundary curves are given which delimit the 95% confidence-level region in the cost(time)–flow plane (in red), along with a median curve (in blue).

This cost–flow curve possesses properties similar to cost–supply curves for renewable resources. It provides an energy flow given a

[14] Given that the long-term average price (excluding fluctuations) of the carrier strictly increases in time.

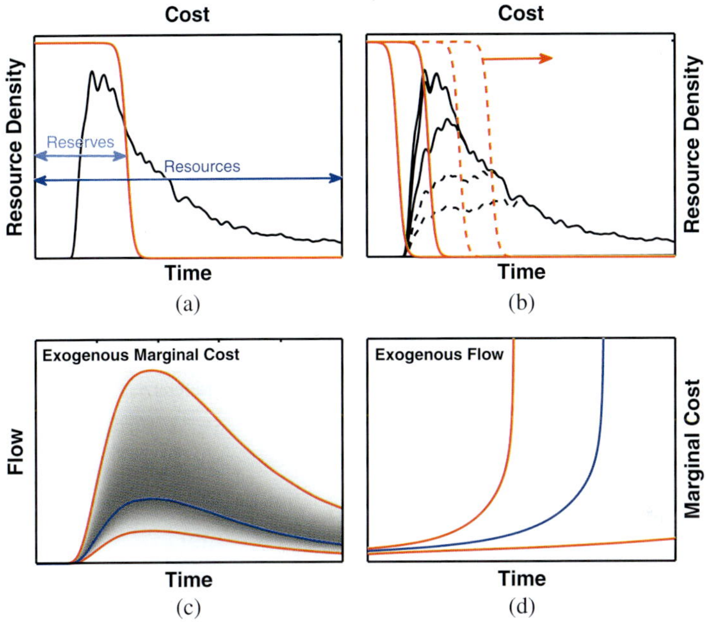

Figure 6.17. Process of stock resource extraction. (a) Original (current) resource distribution as a function of extraction cost (black curve) and probability function for resource extraction given the price of the resource commodity that delimits reserves out of resources (red curve), its maximum being equal to 1. (b) Dynamic process of resource depletion as the price of the resource commodity increases. The black curves corresponds to the distribution of resources left after increasing amounts of time have passed and increasing amounts of resources have been consumed, associated with increasing prices for the energy commodity. (c) Flow curve and uncertainty range for the amount of resource unlocked by an increasing exogenous price of the commodity as a function of time. (d) Marginal cost of production of the commodity for an exogenous demand (as a function of time). The area between the red curves indicates the 96% confidence-level region, while the blue curve represents the median curve. Reproduced with permission from Mercure and Salas (2013).

carrier price. However, as opposed to these, stock resource flow curves converge to zero at high cost values (or long times) due to depletion. Thus, the dynamic nature of this model stemming from the time path of prices must be stressed, and produces the timescale at which depletion occurs.

6.5.4. *Cost–flow curves for exhaustible resources*

Figure 6.18 presents flow curves for four exhaustible resources, oil, natural gas, coal and uranium derived using linear price increases. Coal resources include both hard and soft coal, while thorium resources were not included with uranium since no commercial thorium reactors currently operate. Increases in production are observed at low price values in

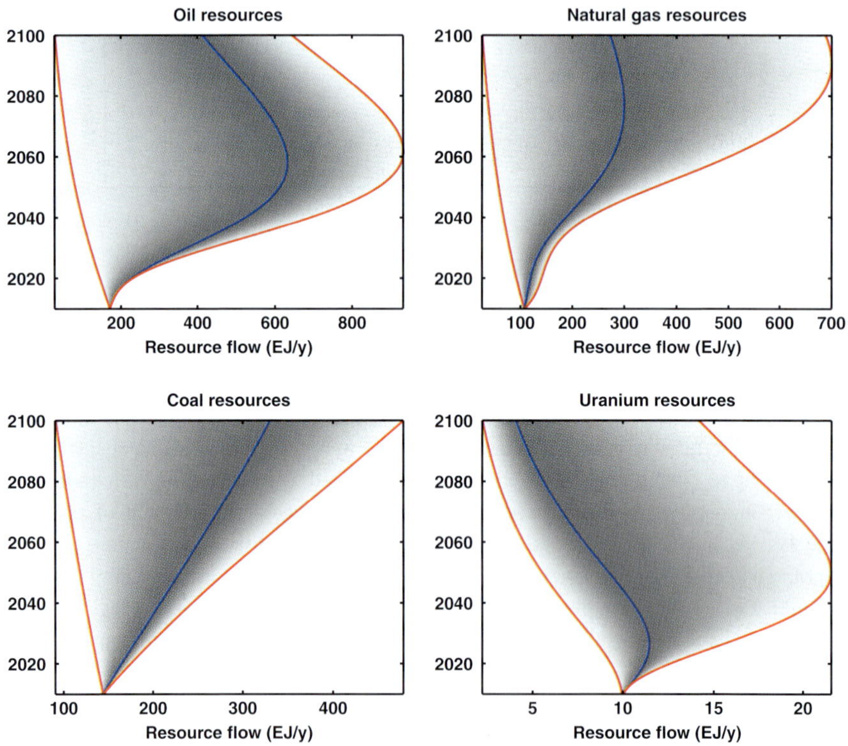

Figure 6.18. Resource flow curves as calculated from the cost–supply curves for fossil fuels and uranium given in Figure 6.13 and Figure 6.12. The curves for natural gas, coal and uranium are expressed in terms of a cost of electricity production, the LCOE, for the purpose of comparison with renewables. Reproduced with permission from Mercure and Salas (2013).

all cases, until a peak is reached, where depletion begins.[15] This can be thought of similarly as in peak oil theory (Hubbert, 1962).[16]

The potential for fossil fuels, if used within this century, is large enough to cover more than twice current energy demand over the next few centuries. As described previously, combusting all the available fossil resources within the next century will most likely result in dangerous climate change. It is therefore clear that decarbonisation is not likely to be eased by scarcity of any of the three fossil fuels, and that dangerous climate change is very likely to occur before depletion begins.

Note that the flow curves of Figure 6.18 are simplified scenarios of future energy use, based on simple linear prices, in order to discover what amounts are unlocked by price and where depletion occurs, but are not meant to represent forecasts. They are merely indications of the scale of resource use and possible depletion following current rates of exploitation. The price should in principle depend on demand and the rate of resource extraction and transformation, and move in such a way that the supply matches the demand; a much more complicated process. Therefore, energy flows could be higher in earlier years with depletion setting in earlier, or the opposite, depending on the composition of future energy landscapes.

6.5.5. *Possible range of future total primary energy demand*

Projections of the total primary energy demand (TPED) up to 2100 are intrinsically highly uncertain. Such calculations require many assumptions regarding global economic development, population growth and energy policy, among many other factors. Additionally, such results

[15] Note that in all cost–flow curves, the cumulative sum of the energy produced up to times very far in the future, where the price becomes very high, yields back the total technical potentials of Figure 6.12 and Figure 6.13.

[16] However, as opposed to peak oil theory, cumulative production does not follow a logistic curve. This stems from the different assumptions used in Hubbert peak theory, where the effect of price on resource use is not specified.

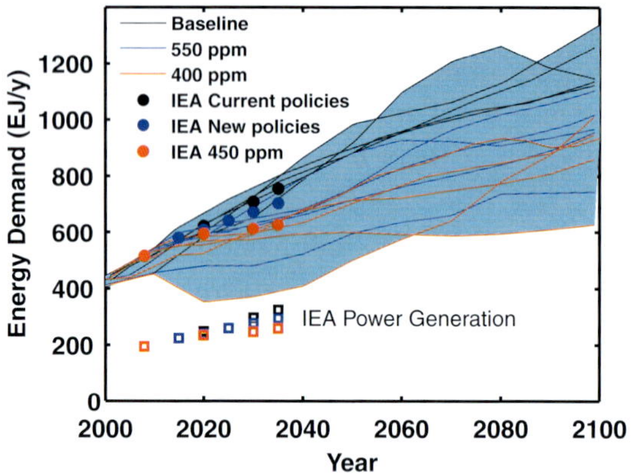

Figure 6.19. Range of total primary energy demand up to 2100 derived from projections by five models as part of the ADAM project, MERGE, E3MG, POLES, TIMER and REMIND (adapted from Edenhofer *et al.* (2010)), and projections of total primary energy demand and the fraction used for electricity production from the World Energy Model of the IEA (2010b).

depend on the type of model used for their calculation. In order not to adopt the assumptions underlying any particular model, it is possible, and desirable, to use an aggregation of various recent studies. Projections of energy use up to 2100 made by five European models (MERGE, E3MG, POLES, TIMER and REMIND) have been collected in the ADAM model comparison exercise, under assumptions for three types of scenarios according to their associated final atmospheric CO_2 concentrations of 400 and 550 ppm as well as for a baseline scenario (Edenhofer *et al.*, 2010). The resulting TPED projections are given in Figure 6.19. Additionally, projections of the TPED up to 2035 by the IEA (IEA, 2010b) are given using circles for three different scenarios (which are not the same as those adopted in the ADAM project), the 2008 value being of 514 EJ/y. The fraction of TPED used for producing power projected by the IEA up to 2035 is given using squares, the 2008 value being of 193 EJ/y which results with 73 EJ/y of electricity. Note that in mitigation scenarios, the TPED is expected to decrease partly due to accounting effects: since the contribution of renewables to the TPED is counted in terms of electricity, the transfer of supply from fossil fuels to renewable or nuclear sources reduces the accounted TPED by the amount lost in the transformation of fossil fuels into electricity.

6.5.6. *Cost–supply and cost–flow curves as a tool for energy planning*

The cost–supply curves of renewable energy resources (Figure 6.5 to Figure 6.11), and the cost–flow curves for exhaustible resources (Figure 6.18) are to be used in the following way. If energy technologies are to compete in a market, their profitability depends on the price of energy carriers and on their cost–supply curve. In particular, in the power sector, the price of electricity plus individual subsidies determines the economic potential of energy sources. The sum of these economic potentials must correspond to global energy demand at any point in time, and therefore, given chosen subsidies, the price of carriers must be such that this is the case. This allows the construction of approximate future energy landscapes and associated energy policies and carrier prices.

World power systems all have different prices of electricity, and therefore this methodology is only very approximate. Moreover, the methodology presented here suggests performing a cost optimisation of the energy system, or in other words, finding the lowest cost solution for future energy use at every point in time. In view of technological learning and timescales associated with energy technology substitution as described previously, this is not strictly an appropriate treatment. The energy sector does, however, gradually converge towards the lowest-cost solution given energy policy decisions and its inherent set of timescales. Therefore, as stated earlier, this methodology yields approximate but reliable results, especially when taking into account uncertainties associated with global technical potentials. This is the primary motivation for providing such detailed uncertainty analysis.

Hydroelectricity possesses low costs, albeit exhibiting large regional variations, but a fairly low technical potential. This inevitably leads to a limited number of new hydroelectric dams being added globally during the next century. Solar energy possesses the largest technical potential of all renewable sources. Large uncertainties exist, however, which depend on future technological developments and improvements in energy density per unit of land area used, especially when considering CSP technologies. Moreover, the cost of solar energy is high, and its penetration into energy markets depends highly on energy subsidies. Wind energy, possessing a better established and tested technology, has a less uncertain technical potential. Biomass energy possesses a similar technical potential, however it depends strongly on global land-use scenarios and food demand associated with scenarios of population growth, and therefore also features large uncertainties. Geothermal energy possesses a significant but not major technical potential, but at affordable costs.

The scale of these curves may be compared with the scale of energy flows available from fossil fuels shown in Figure 6.18, given assumptions regarding future time profiles for the prices of these resources. It can be observed that in the price scenarios given in this section (6.5), the energy flows associated with oil, coal and gas are of the same order as those of renewables. It can be thus concluded that decarbonisation appears, in principle, to be technically possible, since energy demand is not expected to increase by orders of magnitude by the end of the century, as shown in Figure 6.19. However, to draw such a conclusion requires a closer inspection into conversion efficiency factors, provided in the next section.

6.5.7. *The replacement of fossil fuels by renewable energy*

The set of cost–supply curves for renewables, along with the set of cost–flow curves for stock resources, were derived to be used in energy planning. By looking at the scale of resource use, and the increases in cost that additional exploitation entails, one can devise various ways with which transfers between energy sources could occur in the future. In particular, one can derive the effect of potential supply of governmental support or taxation for various energy sources, shifting up or down any of these curves. As an example, a large potential of solar energy could be unleashed, but it would require a very important amount of support. Additionally, the cost–flow curves for stock resources indicate that increases in carrier prices are to be expected in scenarios where their expansion is projected, in order to unlock additional more expensive resource types. These increases may produce incentives to switch to renewables.

As one particular type of energy policy, aimed towards the reduction of GHG emissions, one may be interested to explore the replacement of certain amounts, possibly the major part, of fossil fuels by renewable energy sources. This transfer must be calculated using efficiency conversion factors, summarised in Table 6.2. Such a substitution occurs in the form of electricity as a secondary energy carrier, and thus involves larger absolute amounts in units of energy when measured in hydrocarbon form. When replacing electricity produced by coal, gas and oil by renewables,

Table 6.2. Conversion factors for electricity quantities equivalent to fossil fuels for various types of energy use, using 2008 as base year. *Fraction* indicates the fraction of fuel used in various sectors, while *Conversion* indicates the associated conversion factor for the equivalent amount of electricity. In a process of global decarbonisation, where most of these sectors become electric, approximate equivalent amounts of electricity can be calculated using these values to convert amounts of fossil fuels. This is given for a range of projected energy demand.

	Coal		**Gas**		**Oil**	
	Fraction	**Conversion**	**Fraction**	**Conversion**	**Fraction**	**Conversion**
Electricity	65%	42%	39%	57%	7%	45%
Industry	19%	100%	18%	100%	8%	100%
Buildings	4%	33%	23%	33%	8%	33%
Land Transport	0%	—	0%	0%	54%	25%
Aviation	0%	—	0%	0%	7%	—
Shipping	0%	—	0%	0%	13%	—
Other	12%	100%	20%	20%	10%	100%
Weighted Conversion	60%		68%		37%	
Current use	139 EJ		107 EJ		170 EJ	
Electricity eq.	83.4 EJ		72.8 EJ		62.9 EJ	
Remaining	—		—		34 EJ	
Total 2008	219 EJ					
Total 2100	241–504 EJ					

the conversion factors are on average of about 42%, 57% and 45%, respectively (IEA, 2010b). However, for uses of fossil fuels where the heat is directly used (for instance the metallurgical industries, boilers, furnaces, but with the exception of building heating), their electrification, where possible, may not necessarily produce large changes in efficiency, and thus would likely require amounts of electricity corresponding to their calorific content. In the case of building heating, efficiency improvements of 200–500% can be obtained by using heat pumps, 300% being the most realistic value to expect (Mackay, 2008).

Meanwhile, if the fraction of the supply of oil used for road transport (about 54%, see IEA (2011b)), were to be replaced by electricity in a scenario where the global road transport fleet becomes electrified, the amount of electricity required would likely be lower than the amount of oil used, in units of energy, due to the low efficiency of the internal combustion engine. It is however unlikely that the aviation and shipping fleets can be electrified (IEA, 2009), since distances travelled without refuelling are large and the associated amounts of energy required to be kept onboard are better provided for using biofuels rather than electricity. The improvement in energy consumption of plug-in hybrid cars, when powered by the electricity grid, compared to reference diesel/petrol cars was found to be of a factor around 4 (van Vliet *et al.*, 2011), with 0.37–0.43 MJ/km for hybrids compared to 1.8–1.9 MJ/km for internal combustion engines (note, however, that the world average fuel consumption of the passenger road transport fleet is of about 35 MJ/km (IEA, 2009)). Their CO_2 emissions were found to be lower than the references by a factor of about 2 when the electricity is produced by coal power plants, or higher for lower-emission power systems. In the case, however, where oil is replaced by biofuels, similar engines remain in use (internal combustion or jet engines), and if efficiencies do not improve, the replacement of oil by biofuels occurs by equal amounts.

6.5.8. *The decarbonisation of the global economy before 2100*

According to the IEA, TPED in 2008 is of 514 EJ (IEA 2010b). It can be seen from IEA data (IEA, 2013), that about 416 EJ is supplied in the form of fossil fuels, with 139 EJ of coal, 170 EJ of oil and 107 EJ of gas (see Table 6.2). Out of the remaining 98 EJ, 51 EJ are supplied by biomass sources, 30 EJ by nuclear reactors and 11 EJ by hydroelectricity. The potential process of decarbonisation of the global economy requires a transfer of these 416 EJ of fossil fuels towards renewables or nuclear energy, within their resource limits. Additionally, one may observe from Figure 6.18, panels (a), (b) and (c), in comparison to the cost–supply curves of renewable energies, that very large amounts of energy flows from fossil fuels (>700 EJ/y) are available at relatively low costs, especially from coal. Thus, decarbonisation is not as straightforward a process as it may appear, since energy demand is expected to grow significantly in all scenarios up to 2100.

In the process of decarbonisation, the use of coal and gas is likely to be phased-out in most energy intensive activities through their electrification, assuming that the power sector is also undergoing decarbonisation.[17] The use of oil, however, can only be partially electrified since a significant amount is used in aviation and shipping. Using the conversion factors of Table 6.2, which enable the calculation of the amounts of electricity required to replace the use of fossil fuels in all main activities, weighed average conversion factors can be calculated that determine the total amount of electricity required in order to replace each type of hydrocarbons. Given the respective current use of fossil fuels, the total amount of electricity required in order to keep all economic activities in place would be 219 EJ/y. Thus, for complete electrification of the global economy, renewable and nuclear sources must be able to produce at least an additional 219 EJ of electricity, compared to the current production of 73 EJ/y. This additionally leaves 34 EJ of oil to be replaced by biofuels for aviation and shipping, which converts to approximately equal amounts.

These values are however only lower bounds since they correspond to current energy use. As expressed in Figure 6.19, the TPED in 2100 is likely to be within the range of 600–1200 EJ/y. This scales up all values by a factor of between 1.1 and 2.3. Therefore, the additional load onto renewable and nuclear energy sources from phasing out fossil fuels is likely to lie between 241 and 504 EJ/y, in the case where most of the

[17] Otherwise large efficiency losses could occur.

transport is electrified. Since the current amount of oil used in all forms of transport, about 126 EJ, is larger than the technical potential of biofuels in at least two SRES scenarios, at least a certain fraction of the global land transport fleet must become electric.

As seen in the cost–supply curves of renewable energies presented in the previous sections, these values are large in comparison to the technical potentials of all low-carbon energy sources with the notable exception of solar energy. In particular, in the case where biofuels only are used in transport, the load of energy demand from transport onto the bioenergy sector is likely to be large, thus restricting the amount available for electricity. Without subsidies for solar energy, the amounts of energy available from renewables at LCOEs below US$200/MWh are of about 170 EJ/y for wind, 60–150 EJ/y for electricity from biomass (less the amount used for transport), 6 EJ/y for geothermal, 40 EJ/y for hydroelectricity and none for ocean energy.

6.5.9. *The decarbonisation of the global economy before 2100 (biomass)*

Since powering global land transport is much more efficient when it is electrified, it is assumed here that it becomes so before 2050. Therefore, some of the biomass resource is available for electricity production. It is then possible to evaluate from Figures 6.5–6.11 reasonable estimates for energy flows from each renewable resource. By simply adding them up, it can be observed to first order whether the energy flows required to achieve decarbonisation, according to the last section based on the projected energy demand, can be produced by the global renewable resources.

Following the lowest estimate of the biomass technical potential in the A2 scenario, a flow of around 50 EJ/y can probably be realistically expected in all scenarios, to be used for either electricity production or transport. Due to cost considerations, flows of no more than 40 EJ/y are likely to be obtained from hydropower. Geothermal energy may generate a flow of about 15 EJ/y, and wind power can be expected to produce a large amount, possibly around 100 EJ/y. Marine energy could possibly provide about 15 EJ/y at relatively high costs. Excluding solar energy, this results in a flow of renewable energy of 240 EJ/y. Given the likely requirement of between around 240 and 500 EJ/y, the possible remaining energy demand must be provided for by costly solar energy, by a value between 20 and 260 EJ/y, possibly using government support to help it bridge the "valley of death" through learning-by-doing (Murphy and Edwards, 2003) and become competitive.

6.5.10. *Decarbonisation, land use and the natural environment*

These approximate large numbers, however, signify important underlying changes to the local environment where these energy flows are produced. As stated in Sections 6.3.1.1, 6.3.1.2 and 6.3.1.7, values of the energy productivity of the land for wind, solar and biomass energy are low, therefore producing large energy flows from these sources involves large areas of land devoted to energy-production activities. The replacement of a 1 GW coal power station by wind energy is likely to require between 700 and 1,100 turbines covering an area of between 500 and 1500 km^2. Meanwhile, in the case of solar energy, this would require an area of between 50 and 500 km^2 covered in solar panels, and in the case of biomass this becomes of the order of 5000 km^2. In a decarbonisation scenario without the use of CCS technology, a global capacity of between 3600 and 7600 GW of coal, gas and oil power plants (scaling up data from IEA (2013) by between 1.1 and 2.3) are likely to be required to be replaced by renewable energy systems.

Assuming a future where 100 EJ/y are produced by wind power, 0–260 EJ/y by solar energy and 50 EJ/y by biomass, the respective land requirements would be of about 2.3–5.9 Mkm^2, 0.2–4.1 Mkm^2 and 8.3 Mkm^2, for a total between 10.8 and 18.3 Mkm^2 (see Table 6.3). Since the total onshore land area covers approximately 149 Mkm^2, this represents 7 to 12% of global land area to be dedicated purely to energy production. According to the Global Land Cover (2000), 15.7% of the global land area is currently devoted to agriculture. Although energy-production activities need not be restricted to such areas but may be done in uninhabited regions such as deserts, transmission cost considerations will often generate the constraint to use areas in the vicinity to urban centres, located predominantly within agricultural areas. Thus agriculture would not be in

Table 6.3. Ranges of global land energy demand values. The global land area is of about 149 Mkm^2. CF is the capacity factor, or percentage of total potential output from a renewable source that is obtained in practice.

	Density MW/km^2	CF%	Density TJ/km^2	Use EJ/y	Area Mkm^2	Global%
Wind	2.2–4	25–35	17–44	100	2.3–5.9	1.5–4.0
Solar	6–4	< 50	63–95	20–260	0.2–4.1	0.1–2.8
Biomass			6	50	8.3	5.6
Total					10.8–18.3	7.2–12.3

competition uniquely with biomass energy crop production, but also with wind and solar power.

Thus, without additional reductions in energy demand before 2100, the use of the land for energy purposes is likely to increase to such an extent that a significant fraction of the global inhabited land area would become dedicated to such activities. Thus, reducing energy consumption below the current supply of 500 EJ/y with an increasing global population is a challenging but necessary task, especially given that in developing regions with high population growth, energy intensity tends to be lower than in developed economies. CCS technology may also be considered as a useful tool to restrict land use from renewables. The extent of land use for energy production may threaten the ability to preserve natural areas in order to protect ecosystems and biodiversity. Thus, significant additional modelling work should be done in order to address the feasibility of maintaining both land-intensive energy-production activities from renewables without significantly affecting agriculture and food production or ecosystems and biodiversity, in terms of global land use. This could be achieved with a land-use model where the allocation of the land responds to the global energy demand as well as the food demand.

6.6. Conclusions

This chapter addressed the feasibility of the decarbonisation of the world economy from a technology perspective. In previous chapters of this book, emission reductions are studied from an economic point of view where the amounts of fuel used is explored in detail using a data-driven approach. The conversion from fuel use to emissions is a simple one since the carbon content of fuels is known to a very high accuracy; however the amount of fuel used for achieving a particular purpose depends on technology, in other words, the method one uses for that purpose, and therefore technology is also an important consideration beyond economics.

GHG emissions originate in a large part from energy-related activities. Achieving emission reductions can mostly be done through generating change in the energy sector, and this is predominantly done by incentivising the key actors that are responsible for the operation. For this purpose, changes in energy policy are used, and these can stimulate transformations in the sector that entail technology diffusion and learning-by-doing.

Low-carbon sources of energy exist that can replace the current use of carbon-rich energy sources. However, using economic models one could construct scenarios of the future where the use of resources lies beyond what is available on the planet, or that are unlikely in terms of costs. With this chapter we thus attempted to provide simple tools to assess whether particular scenarios are feasible given the planet's energy-resource endowment. Renewable and non-renewable resources must be treated differently when exploring energy flows. Given the tools provided, one can intuitively explore various ways that the world could re-allocate its current energy use across available low-carbon sources.

Using the data provided, we gave a picture of how these resources could be allocated in a case of complete future fossil fuel phase-out in an exercise designed to explore whether this possible at all. Through this we obtained numbers concerning the amount of land that would be required to be dedicated to energy generation globally, which amounts to a significant proportion of the global land cover. This indicates that reductions in global energy consumption are most probably a key aspect of future energy policy in a context where emissions are very low. Going beyond this, however, requires detailed modelling of the global energy sector and land use, a subject beyond the scope of the present chapter.

References

Abed, K. and El-Mallah, A. (1997). Capacity Factor of Wind Turbines, *Energy*, 22, 487–491.

Abu-Khader, M. (2009). Recent Advances in Nuclear Power: A Review, *Progress in Nuclear Energy*, 51, 225–235.

Aldrich, M., Laughlin, A. and Gambill, D. (1981). *Geothermal Resource Base of the World: A Revision of the Electrical Power Research Institute's Estimate*, Los Alamos Scientific Laboratory Report, LA-8801-MS, University of California, Los Alamos, New Mexico.

Archer, C. and Jacobson, M. (2005). Evaluation of Global Wind Power, *Journal of Geophysical Research-Atmospheres*, 110, 12110, 20 pp.

Arrow, K. (1962). The Economic-Implications of Learning by Doing, *Review of Economic Studies*, 29, 155–173.

Avrutin, V., Izyumskaya, N. and Morkoc, H. (2011). Semiconductor Solar Cells: Recent Progress in Terrestrial Applications, *Superlattices and Microstructures*, 49, 337–364.

Bertani, R. (2010). 'Geothermal Power Generation in the World. 2005–2010 Update Report', in *Proceedings of the World Geothermal Congress 2010,* Bali, Indonesia, 25–29 April 2010, pp. 1–41.

Bertani, R. (2012). Geothermal Power Generation in the World. 2005–2010 Update Report, *Geothermics*, 41, 1–29.

Bonche, P. (2002). *Le Nucléaire Expliqué par des Physiciens*, EDP Sciences, Paris.

Bruno, K., Baizel, B., Casy-Lefkowitz, S., Shope, E. and Colarulli, K. (2010). *Tar Sands Invasion*, Corporate Ethics International, Earthworks, Natural Resources Defense Council and Sierra Club.

Cataldi, R. (1993). Review of Historiographic Aspects of Geothermal Energy in the Mediterranean and Mesoamerican Areas Prior to the Modern Age, *Geo-Heat Centre Quarterly Bulletin*, 18, 13–16.

Cavanagh, J., Clarke, J. and Price, T. (1993). 'Ocean Energy Systems', in Johansson, T., Kelly, H., Reddy, A. and Williams, R. (eds). *Renewable Energy. Sources for Fuels and Electricity*, Island Press, Washington, DC.

Charlier, R. and Justus, J. (1993). *Ocean Energies: Environmental, Economic and Technological Aspects of Alternative Power Sources*, Elsevier Oceanography Series, Elsevier, Amsterdam.

Crabtree, G. and Lewis, N. (2008). 'Solar Energy Conversion', in Hafemeister, D., Levi, B. G., Levine, M. D. and Schwartz, P. (eds). *Physics of Sustainable Energy: Using Energy Efficiently and Producing it Renewably*, American Institute of Physics, Melville, NY, pp. 309–321.

de Vries, B., van Vuuren, D., den Elzen, M. and Janssen, M. (2001). The Targets IMage Energy Regional (TIMER) Model. RIVM report 461502024, RIVM, Bilthoven.

de Vries, B., van Vuuren, D. and Hoogwijk, M. (2007). Renewable Energy Sources: Their Global Potential for the First-half of the 21st Century at a Global Level: An Integrated Approach, *Energy Policy*, 35, 2590–2610.

Dornburg, V., van Vuuren, D., van de Ven, G., Langeveld, H., Meeusen, M., Banse, M., van Oorschot, M. M. P., Ros, J., van den Born, G. J., Aiking, H., Londo, H., Mozaffarian, H., Verweij, P., Lysen, E. and Faaij, A. 2010. Bioenergy Revisited: Key Factors in Global Potentials of Bioenergy, *Energy and Environmental Science*, 3, 258–267.

Edenhofer, O., Knopf, B., Barker, T., Baumstark, L., Bellevrat, E., Chateau, B., Criqui, P., Isaac, M., Kitous, A., Kypreos, S., Leimbach, M., Lessmann, K., Magné, B., Scrieciu, S., Turton, H., van Vuuren, D. P. (2010). The Economics of Low Stabilization: Model Comparison of Mitigation Strategies and Costs, *The Energy Journal*, 31, Special Issue 1, 11–48.

EPRI (1978). *Geothermal Energy Prospects for the Next 50 Years*, Electric Power Research Institute, Palo Alto, CA.

ETSAP (2010a). Concentrating Solar Power. Energy Technology Systems Analysis Programme. IEA ETSAP and IRENA Technology Brief E10, Paris.

ETSAP (2010b). Marine Energy. Energy Technology Systems Analysis Programme. IEA ETSAP and IRENA Technology Brief E08, Paris.

Fellows, A. and Gow, G. (2000). The Potential of Wind Energy to Reduce Carbon Dioxide Emissions. European Wind Energy Conference, Copenhagen, 2–7 July.

GEA (2012). *Global Energy Assessment — Toward a Sustainable Future*, Cambridge University Press, Cambridge.

Global Land Cover (2000). Global Land Cover 2000. Available online at: http://bioval.jrc.ec.europa.eu/products/glc2000/glc2000.php. Accessed 20 October 2013.

Goldstein, B., Hill, A., Long, A., Budd, A., Ayling, B. and Malavazos, M. (2009). Hot Rocks Downunder: Evolution of a New Energy Industry, *Geothermal Resources Council Transactions*, 33, 185–198.

Gough, C. and Upham, P. (2011). Biomass Energy with Carbon Capture and Storage (BECCS or Bio-CCS), *GHG Science and Technology*, 1, 324–334.

Green, M., Emery, K., Hishikawa, Y., Warta, W. and Dunlop, E. (2011). Solar Cell Efficiency Tables (Version 38), *Progress in Photovoltaics*, 19, 565–572.

Grubb, M. and Meyer, N. (1993). 'Wind Energy: Resources, Systems and Regional Strategies', in Johansson, T., Kelly, H., Reddy, A. and Williams, R.

(eds). *Renewable Energy: Sources for Fuels and Electricity*, Island Press, Washington, DC, pp. 157–212.

Hammons, T. (1993). Tidal Power, *Proceedings of the IEEE*, 81, 419–433.

Hofman, Y., de Jager, D., Molenbroek, E., Schilig, F. and Voogt, M., (2002). The Potential of Solar Electricity to Reduce CO_2 Emissions. Ecofys Report No PH4/14, Utrecht.

Hoogwijk, M. (2004). *On the Global and Regional Potential of Renewable Energy Sources*, PhD dissertation, Universiteit Utrecht.

Hoogwijk, M., de Vries, B. and Turkenburg, W. (2004). Assessment of the Global and Regional Geographical, Technical and Economic Potential of Onshore Wind Energy, *Energy Economics*, 26, 889–919.

Hoogwijk, M., Faaij, A., van den Broek, R., Berndes, G., Gielen, D. and Turkenburg, W. (2003). Exploration of the Ranges of the Global Potential of Biomass for Energy, *Biomass & Bioenergy*, 25, 119–133.

Hoogwijk, M., Faaij, A., Eickhout, B., de Vries, B., Turkenburg, W. (2005). Potential of Biomass Energy out to 2100, for Four IPCCSRES Land-use Scenarios, *Biomass & Bioenergy*, 29, 225–257.

Hoogwijk, M., Faaij, A., de Vries, B. and Turkenburg, W. (2009). Exploration of Regional and Global Cost-Supply Curves of Biomass Energy from Short-rotation Crops at Abandoned Cropland and Rest Land under Four IPCC SRES Land-use Scenarios, *Biomass & Bioenergy*, 33, 26–43.

Hubbert, M. K. (1962). *Energy Resources. A Report to the Committee on Natural Resources*, National Academy of Science, Government Printing Office, Washington, DC.

IAEA (2009). *Uranium 2009: Resources, Production and Demand*, IAEA/OECD/NEA, Vienna.

IEA (2000). *Experience Curves for Energy Technology Policy*, IEA/OECD, Paris.

IEA (2005). *Resources to Reserves: Oil and Gas Technologies for the Energy Markets of the Future*, IEA/OECD, Paris.

IEA (2009). *Transport, Energy and* CO_2, IEA/OECD, Paris.

IEA (2010a). *Projected Costs of Generating Electricity 2010*, IEA/OECD, Paris.

IEA (2010b). *World Energy Outlook 2010*, IEA/OECD, Paris.

IEA (2011a). *Technology Roadmap. Carbon Capture and Storage in Industrial Applications*, IEA/OECD, Paris.

IEA (2011b). *Oil Information 2011*, IEA/OECD, Paris.

IEA (2012). *World Energy Outlook 2012*, IEA/OECD, Paris.

IEA (2013). *World Energy Balances dataset 2013*, IEA/OECD, Paris.

IEA Bioenergy (2009). *Bioenergy — A sustainable and reliable energy source, Main report*, IEA/OECD, Paris.

IJHD (2011). *2011 World Atlas and Industry gGuide*, IJHD, Wallington.

IPCC (2000). *Emission Scenarios*, Cambridge University Press, Cambridge.

IPCC (2005). *Carbon Dioxide Capture and Storage*, Cambridge University Press, Cambridge.

IPCC (2007a). *Climate Change 2007: Mitigation. Contribution of Working Group III to the Fourth Assessment Report of the Intergovernmental Panel on Climate Change*, Metz, B., Davidson, O. R., Bosch, P. R., Dave, R. and Meyer, L. A. (eds), Cambridge University Press, Cambridge.

IPCC (2007b). 'Technical Summary', *Climate Change 2007: Mitigation. Contribution of Working Group III to the Fourth Assessment Report of the Intergovernmental Panel on Climate Change*, Metz, B., Davidson, O. R., Bosch, P. R., Dave, R. and Meyer, L. A. (eds), Cambridge University Press, Cambridge, pp. 25–93.

IPCC (2007c). 'Chapter 3: Issues Related to Mitigation in the Long-term Context', *Climate Change 2007: Mitigation. Contribution of Working Group III to the Fourth Assessment Report of the Intergovernmental Panel on Climate Change*, Metz, B., Davidson, O. R., Bosch, P. R., Dave, R. and Meyer, L. A. (eds), Cambridge University Press, Cambridge, pp. 169–250.

IPCC (2007d). 'Chapter 4: Energy Supply', *Climate Change 2007: Mitigation. Contribution of Working Group III to the Fourth Assessment Report of the Intergovernmental Panel on Climate Change*, Metz, B., Davidson, O. R., Bosch, P. R., Dave, R. and Meyer, L. A. (eds), Cambridge University Press, Cambridge, pp. 251–322.

IPCC (2011a). 'Bioenergy', *Special Report on Renewable Energy Sources and Climate Change Mitigation*, Edenhofer, O., Pichs-Madruga, R., Sokona, Y., Seyboth, K., Matschoss, P., Kadner, S., Zwickel, T., Eickemeier, P., Hansen, G., Schlömer, S. and von Stechow, C. (eds), Cambridge University Press, Cambridge, pp. 209–331.

IPCC (2011b). 'Direct Solar Energy', *Special Report on Renewable Energy Sources and Climate Change Mitigation*, Edenhofer, O., Pichs-Madruga, R., Sokona, Y., Seyboth, K., Matschoss, P., Kadner, S., Zwickel, T., Eickemeier, P., Hansen, G., Schlömer, S. and von Stechow, C. (eds), Cambridge University Press, Cambridge, pp. 333–400.

IPCC (2011c). 'Ocean Energy', *Special Report on Renewable Energy Sources and Climate Change Mitigation*, Edenhofer, O., Pichs-Madruga, R., Sokona, Y., Seyboth, K., Matschoss, P., Kadner, S., Zwickel, T., Eickemeier, P., Hansen, G., Schlömer, S. and von Stechow, C. (eds), Cambridge University Press, Cambridge, pp. 497–534.

Keeling, C. (1973). Industrial Production of Carbon Dioxide from Fossil Fuel and Limestone, *Tellus*, 25, 174–198.

Krewitt, W., Nienhaus, K., Klessmann, C., Capone, C., Stricker, E., Graus, W., Hoogwijk, M., Supersberger, N., van Winterfeld, U. and Samadi, S. (2009). Role and Potential of Renewable Energy and Energy Efficiency for Global Energy Supply. Report No UBA-FB 001323/E. Ecofys, Utrecht.

Lako, P., Eder, H., de Noord, M. and Reisinger, H. (2003). Hydropower Development with a Focus on Asia and Western Europe, Overview in the Framework of VLEEM 2, ECN Policy Studies. Report ECN-03-027, Petten.

Lu, X., McElroy, M. and Kiviluoma, J. (2009). Global Potential for Wind-generated Electricity, *Proceedings of the National Academy of Sciences of the United States of America*, 106, 10933–10938.

Lund, J. (2005). 100 Years of Renewable Electricity: Geothermal Power Production, *Renewable Energy World*, 8, 252–259.

Macdonald, G. (1959). Calculations on the Thermal History of the Earth, *Journal of Geophysical Research*, 64, 1967–2000.

Mackay, D. (2008). *Sustainable Energy without the Hot Air*, UIT Press, Cambridge.

Mckelvey, V. (1972). Mineral Resource Estimates and Public Policy, *American Scientist*, 60, 32–40.

Mercure, J.-F. (2012a). FTT: Power A Global Model of the Power Sector with Induced Technological Change and Natural Resource Depletion, *Energy Policy*, 48, 799–781.

Mercure, J.-F. (2012b). On the Changeover Timescales of Technology Transitions and Induced Efficiency Changes: An Overarching Theory, Working Paper available on ArXiv at: http://arxiv.org/abs/1209.0424.

Mercure, J.-F. and Salas, P. (2012). An Assessment of Global Energy Resource economic potentials, *Energy*, 46, 322–336.

Mercure, J.-F. and Salas, P. (2013). On the Global Economic Potentials and Marginal Costs of Non-renewable Resources and the Price of Energy Commodities, *Energy Policy*, 63, 469–483.

Messner, S. and Strubegger, M. (1995). User's Guide for {MESSAGE III}. Available online at: http://www.iiasa.ac.at/Admin/PUB/Documents/WP-95-069.pdf. Accessed 2 November 2013.

Mock, J., Tester, J. and Wright, P. (1997). Geothermal Energy from the Earth: Its Potential Impact as an Environmentally Sustainable Resource, *Annual review of Energy and the Environment*, 22, 305–356.

Murphy, L. and Edwards, P. (2003). Bridging the Valley of Death: Transitioning from Public to Private Sector Financing. National Renewable Energy

Laboratory. Available online at: http://www.nrel.gov/docs/gen/fy03/34036.pdf. Accessed 2 November 2013.

Nihous, G. (2007). A Preliminary Assessment of Ocean Thermal Energy Conversion Resources, *Journal of Energy Resources Technology-Transactions of the ASME*, 129, 10–17.

Pasqualetti, M. (1983). The Site Specific Nature of Geothermal-Energy — The Primary Role of Land-use Planning in Non-electric Development, *Natural Resources Journal*, 23, 795–814.

Perman, R., Ma, Y., Common, M., Maddison, D. and Mcgilvray, J. (2003). *Natural Resource and Environmental Economics*, 3rd Edition, Prentice Hall, Upper Saddle River, NJ.

Pollack, H., Hurter, S. and Johnson, J. (1993). Heat-flow from the Earth's Interior: Analysis of the Global Data Set, *Reviews of Geophysics*, 31, 267–280.

REN21 (2011). *Renewables 2011. Global Status Report*, REN21 Secretariat, Paris.

Rhodes, J. and Keith, D. (2005). Engineering Economic Analysis of Biomass IGCC with Carbon Capture and Storage, *Biomass and Bioenergy*, 29, 440–450.

Riahi, K., Dentener, F., Gielen, D., Grubler, A., Jewell, J., Klimont, Z., Krey, V., McCollum, D., Pachauri, S., Rao, S., van Ruijven, B., van Vuuren, D. and Wilson, C. (2012). 'Chapter 17: Energy Pathways for Sustainable Development', in *Global Energy Assessment: Toward a Sustainable Future*, Cambridge University Press, Cambridge, and the International Institute for Applied Systems Analysis, Laxenburg, Austria, pp. 1203–1306.

Rogner, H. (1997). An Assessment of World Hydrocarbon Resources, *Annual Review of Energy and the Environment*, 22, 217–262.

Rowley, J. (1982). 'Worldwide Geothermal Resources', in Edwards L. and Chilingar, G. (eds). *Handbook of Geothermal Energy*, Gulf Publishing, Houston, TX, pp. 44–174.

Sinha, R. and Kakodkar, A. (2006). Design and Development of the AHWR — the Indian Thorium Fuelled Innovative Nuclear Reactor, *Nuclear Engineering and Design*, 236, 683–700.

Skramesto, O., Skilhagen, S. and Nielsen, W. (2009). Power Production Based on Osmotic Pressure, *Waterpower*, 16, 1–10.

Smeets, E., Faaij, A., Lewandowski, I. and Turkenburg, W. (2007). A Bottom-up Assessment and Review of Global Bio-energy Potentials to 2050, *Progress in Energy and Combustion Science*, 33, 56–106.

Sørensen, B. (2011). *Renewable Energy: Physics, Engineering, Environmental Impacts, Economics and Planning*, Academic Press. Waltham, MA.

Stefansson, V. (2005). 'World Geothermal Assessment', in *Proceedings of the World Geothermal Congress*, Antalya, 24–29 April 2005, World Geothermal Congress, International Geothermal Association, Bochum, Germany.

Stern, N. (2006). *The Economics of Climate Change*, Cambridge University Press, Cambridge.

Suess, H. and Urey, H. (1956). Abundances of the Elements, *Reviews of Modern Physics*, 28, 53–74.

Tester, J. and Anderson, B. (2006). 'Impact of Enhanced Geothermal Systems (EGS) on the United States in the 21st Century', in *The Future of Geothermal Energy*, Massachusetts Institute of Technology report INL/ EXT-06-13438-6 prepared for U.S. DOE/ID Contract DE- AC07-05ID14517, Massachusetts Institute of Technology, MA, pp. 11–21.

Troen, B. and Petersen, E. (1989). *European Wind Atlas*, Risø National Laboratory, Roskilde.

UNDP (2000). *World Energy Assessment*, UNDP, New York.

van Vliet, B., Brouwer, A., Kuramochi, T., van den Broek, M. and Faaij, A. (2011). Energy use, Cost and CO_2 Emissions of Electric Cars, *Journal of Power Sources*, 196, 2298–2310.

van Vuuren, D., van Vliet, J. and Stehfest, E. (2009). Future Bio-Energy Potential Under Various Natural Constraints, *Energy Policy*, 37, 4220–4230.

Vogt-Schilb, A. and Hallegatte, S. (2011). *When Starting with the Most Expensive Option Makes Sense: Use and Misuse of Marginal Abatement Cost Curves*, World Bank.,Washington, DC.

Wasserburg, G., MacDonald, G., Hoyle, F. and Fowler, W. (1964). Relative Contributions of Uranium, Thorium + Potassium to Heat Production in Earth, *Science*, 143, 465–467.

WEC (1994). *New Renewable Energy Resources: A Guide to the Future*, Kogan-Page Ltd, London.

Weinzettel, J., Reenaas, M., Solli, C. and Hertwich, E. (2009). Life Cycle Assessment of a Floating Offshore Wind Turbine, *Renewable Energy*, 34, 742–747.

Wolf, J., Bindraban, P., Luijten, J. and Vleeshouwers, L. (2003). Exploratory Study on the Land Area Required for Global Food Supply and the Potential Global Production of Bioenergy, *Agricultural Systems*, 76, 841–861.

Chapter 7

Feasibility of Reducing Emissions by End-use Sector

Scott Kelly, Andrew Skelton and Aleix Altimiras-Martin

7.1. Introduction

So far in this book, we have considered how broad policy instruments such as carbon taxes, emissions trading and investment in the evolution of technologies for power can place the global economy onto a path towards decarbonisation. Such approaches to policy are cases of a top–down analysis in which policies influence supply and demand characteristics of energy in sectors of the economy rather than in specific firms, the supply chains of products or specific parts of the built environment and infrastructure. These results showed that research, innovation and investment in new and promising low-carbon technologies throughout the economy will be critical in reducing emissions over the long term if supported by both economic and regulatory strategies.

While such broad policy instruments can be effective if implemented successfully across the economy, there remain four issues crucial to this success:

- Regulatory controls often have a legal grounding in laws targeting a specific environmental compartment (such as air) or a specific sector of the built environment (such as homes).
- Relationships between energy supply and demand, and the technologies best suited to these relationships, can vary between sectors; this includes very different ratios of peak- to base-load demand in a sector

or in the kinds of energy suited to an industrial practice, which in turn influences the kind of energy supply required (quite apart from economic considerations such as the price of fuel).

- The mix of energy production capacity that is feasible in supplying both base-load and peak-load depends on the magnitude of energy demand, which in turn depends on material use in industrial systems. If both material and energy demand were to decrease through changes in product design and end-use, low-carbon technologies become more feasible for supplying a large percentage of the remaining energy demand. Hence material and energy demand reduction, in turn related to the recycling of carbon-intensive materials, must accompany reduction in the carbon intensity of energy production.
- Manufactured goods have complex supply chains, and design or technological choices in one part of that chain may or may not be consistent with the needs of other parts of the chain; hence the details of a supply chain must be understood if policies are to target industrial production and if large multinational firms are to be provided incentives to reduce emissions throughout their supply chains located around the world.

Policy instruments may therefore need to target specific end-use (including intermediate production) sectors in the economy and built environment, and reflect the possibility that rates of diffusion even under a single instrument may be quite different in these sectors due to differences in the suitability of technological innovations and design changes. Such an analysis requires a bottom–up approach more typical of engineering analysis than economic/policy analysis.

In addition, technological innovation will not be restricted to energy systems. Modern industrial practices that are part of the complex systems of production of goods and services rely as much on materials as they do energy. Decarbonisation will not be driven purely by decarbonising the energy source, but also by improving energy and material efficiency in industrial production, thereby reducing carbon intensity throughout supply chains and production layers.

All of this points to a challenge not yet explored in the book: when one looks at carbon emissions from the perspective of the end-user of energy, considerations emerge that go beyond — or at least frame — the economics of low-carbon energy production, or the application of broad policy instruments that provide general signals throughout the economy. This chapter examines that challenge by placing the focus on the end-user of the material or energy and asks how that focus might improve the development of targeted economic, technology and emission policies as a complement to the broader measures assessed in earlier chapters. End-users of materials and energy ultimately decide whether to adopt new low-carbon technology or change behaviour, and the end-users in different sectors of the economy have different technological needs and social/business responsibilities that policies must reflect to be effective. Analysis of the links of energy and material use between sectors can also help identify cross-sectoral strategies that bring synergies to the movement towards a low-carbon global economy.

By adopting a sectoral approach for end-use emissions, we can identify the sectors of the global economy where decarbonisation will have the largest impact, if a choice must be made between the sectors on which to focus limited economic and regulatory resources. In this chapter, we consider how different sectors within a country can be targeted using different strategies and why it might be necessary to have different strategies across different countries as well. For example, in wealthy countries, reducing the rate of emissions from the buildings sector will require a monumental effort in retrofit programmes to the existing building stock, with a similarly large issue of how these retrofits are to be incentivised and financed. In poorer countries, on the other hand, reducing carbon emissions from buildings will require the implementation of strict sustainability guidelines for new buildings — as this is where the majority of future emissions will come from in these countries as they develop economically. This is more an issue of design and construction standards than of economic instruments, although the latter can help drive the former.

This chapter is separated into four distinct sections representing each major end-use sector of the economy (see Figure 7.1 for a comparison of emissions from these sectors). Each of these contributes in a different way to the carbon emissions of a country or the global economy, and each has a unique set of design, technological and business requirements that shape the solutions to lower material and energy use. We start with a discussion

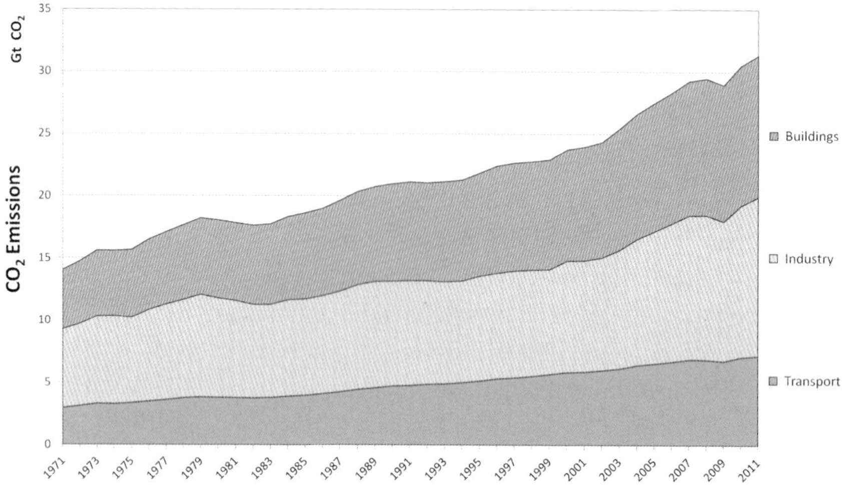

Figure 7.1. CO_2 emissions by end-use sector globally.

Source: IEA (2013).

on the residential sector, the sector with the largest potential to reduce end-use energy consumption. We show that significant technological and behavioural opportunities exist for decarbonisation in terms of the quantity of carbon dioxide that can be abated, and at relatively low-cost, when compared to other CO_2-emitting sectors of the economy (and in comparison to investments in decarbonising the energy system itself). Our attention then turns to transport, the sector with the second-largest carbon footprint. A particular challenge in the transport sector is that the transport of goods around the world underpins the global economic system, which then raises the issue of embodied carbon in traded goods and of carbon leakage. Following on from the section on transport, we look at the role of the industrial sector and how material and energy flows that occur within the economy can be altered to reduce energy and/or carbon intensity through converting industrial systems from cradle-to-grave into cradle-to-cradle systems. The final section looks at the role of multinational enterprises (MNEs) to bring about a transition to a low-carbon world. Particular attention is given to the growing importance of globalisation and the idea of using MNEs as agents of change by targeting their operations through regulatory efforts that drive carbon reduction through the supply chains.

7.2. Buildings

At a global level, buildings consume 35% of the world's resources in construction, are responsible for around 40% of global energy use and are accountable for 30% of total greenhouse gas (GHG) emissions (WGBC, 2010). According to the Intergovernmental Panel on Climate Change (IPCC), global emissions from buildings are likely to double by 2030 without considered and appropriate action to reduce emissions (Levine and Urge-Vorsatz, 2008). Energy and emissions consumed and produced by buildings fall into two distinct categories. The first is the embodied energy and emissions contained within the materials that are used for the construction of the building and the second is the energy consumed over the lifetime of the building providing occupants with electricity, heating and cooling. There is also potential to reduce emissions through the reduction of other non-CO_2 GHGs, particularly through chlorofluorocarbons (CFCs) and hydrocarbons, which account for 23% of GHGs from the buildings sector (Levine and Urge-Vorsatz, 2008). In the IPCC Fourth Assessment Report (AR4), residential and commercial buildings are shown to have the greatest potential impact for reducing GHG emissions and for having some of the lowest costs of abatement when compared against other sectors. In many cases, investment in energy efficiency measures in buildings is shown to be cost-effective and even profitable over the lifetime of the measure. However, international progress remains slow. In this chapter we suggest that increasing this rate of progress in part requires us to stop thinking about the delivery of energy in the form of electricity and gas and instead to think about the types of services that energy provides, such as cooling, warmth and light. It is only when we think about the services that energy provides rather than solely the energy units, that we can start to think about the different opportunities available for delivering a low-carbon future.

7.2.1. *The problem*

In 2009, for the first time in history, the number of people living in cities surpassed those living in rural areas. Each year cities swell by some 70 million people, a trend that is set to continue well into the future. This is important because cities produce over 70% of global CO_2 emissions with the bulk share of all future increases attributable to urbanisation (IEA, 2008a). Between 2011 and 2050 there will be an extra 2.3 billion people on the planet, each requiring additional shelter, warmth and light. The rate of growth in construction — and hence in the construction sector of the economy — to meet this new demand is unparalleled in history. At the same time, individuals all over the world are aspiring to unprecedented levels of economic prosperity and wealth, which will lead to even greater levels of consumption of materials and additional demand for energy services. This trend has already started to manifest with non-OECD (Organisation for Economic Co-operation and Development) countries now consuming more electricity than OECD countries (see Figure 7.2). Buildings are now the largest consumer of electricity, and this is likely to continue, as growing affluence demands new levels of comfort, particularly in warmer climates increasing demands for air conditioning. This effect is only exacerbated as the planet begins to warm due to climate change.

Emission reductions from buildings offer some of the largest benefits at least cost when compared with technology change in other sectors. A peculiar feature of the residential sector is that the majority of emissions attributable to the residential sector come from "indirect emissions"

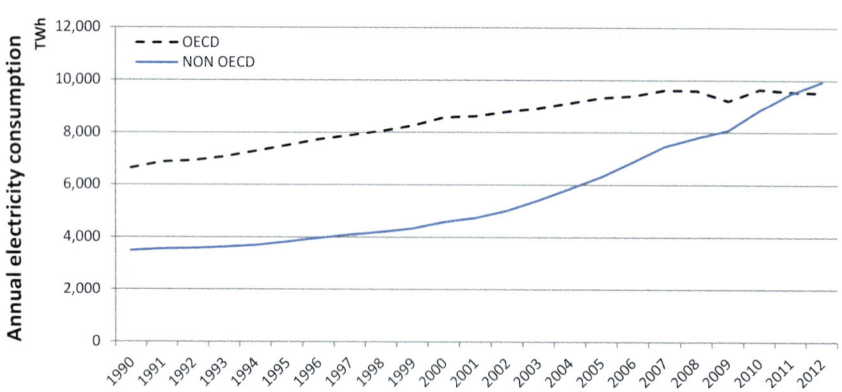

Figure 7.2. Global energy consumption in OECD and non-OECD nations.

Source: Enerdata Statistical Yearbook 2013 (see http://yearbook.enerdata.net).

through electricity production. For example, in 2004, total global emissions from buildings was 8.6 Gt/year, but indirect emissions — that is emissions predominantly coming from the electricity sector — accounted for 5.6 Gt/year (de la Rue du Can and Price, 2008). Therefore, reducing energy demand has the added benefit of avoiding costly investment in energy infrastructure such as new generation facilities and additional transmission lines.

A key finding of the IPCC AR4 was the ability to substantially reduce emissions using mature technologies and energy efficiency measures that exist today and have been successfully used in the past. According to the IEA, the residential sector alone has the most potential for mitigation accounting for 54% of cumulative emission reductions from a business-as-usual (BAU) scenario by 2030 (IEA, 2008b). Furthermore, as shown by Houser (2009), achieving large reductions in global CO_2 emissions from buildings is possible without compromising present living standards.

Efficiency savings within the built environment are often cost-effective over the lifetime of the measure. In the IPCC AR4 it is calculated that at least 31% of global GHG emissions from the built environment can be abated cost-effectively by 2030 (Levine and Urge-Vorsatz, 2008). Cost-effectiveness calculations of carbon-saving technologies are frequently shown to be sensitive to the anticipated future price of fossil fuel (and CO_2). Such calculations, including the analyses in earlier chapters of this book, assume minimal to moderate increases in the international price of fossil fuel (and CO_2), where even modest increases to the counterfactual fuel or CO_2 price will lead to improved cost-effectiveness projections for low-carbon technologies. Indeed, many governments around the world are now prioritising the decarbonisation of this sector for meeting climate change targets (e.g. Germany, UK, US etc.).

Decarbonising the global building stock is a complex, non-trivial problem. The pattern of energy consumption in buildings is strongly related to building type and climatic zone. All over the world, buildings have evolved differently, with the choice of best designs being affected by climate, the availability of construction materials, cultural norms and building function. Large differences exist between countries and regions in construction methods, building codes and planning laws. As a result, buildings across the world are dissimilar and heterogeneous. Such differences confound the problem of decarbonisation and act to highlight the need for bespoke solutions for different climatic regions, cultures and available resources, and hence differences in the degree to which price signals will be able to affect design, construction and operation of buildings. For example, decarbonising an end-of-terrace home in Scotland, where the majority of CO_2 is produced from the combustion of natural gas through a central heating system will require a very different solution — with different marginal abatement costs — to decarbonising a high-rise apartment in Abu Dhabi where the majority of CO_2 emissions arise from air conditioning. To be effective, policies must reflect this relationship between a building, its environment and its energy source.

The diversity of buildings and their different uses and lifetimes pose a distinct challenge for energy and conservation. For example, in high-income countries most of the energy consumed in a building occurs during the building's operational phase, consuming between 70–80% of the building's overall energy consumption (Iddon and Firth, 2013). Similarly, plug-load and electricity consumption from lights and appliances are the fastest-growing source of emissions in these countries. For instance, electricity and natural gas are the main energy commodities in OECD countries providing at least 72% of total household energy requirements (IEA, 2008c). Therefore, in high-income economies, the largest CO_2 benefit will come from the renovation of existing buildings to new sustainability criteria and the design of high-efficiency plug-load devices. The most cost-effective improvements will come from better insulation, improving building leakiness and the improved efficiency of heating systems resulting in lower heating and cooling energy demand requirements (IEA, 2008b).

While the majority of primary energy consumed in high- and middle-income countries is from fossil fuels, biomass remains the dominant source of energy in low-income countries (United Nations Environment Programme, 2008). In low-income rural areas of Africa, India and China the main energy source for more than 70% of the population is traditional biomass including wood, animal dung and crop waste. As many of these countries industrialise the trend will be towards increasing fossil fuel use. According to the IEA (2002), by 2030 biomass will still represent over half of all residential energy consumption in developing countries. According to UNEP (United Nations Environment Programme, 2008) the use of wood and biomass are still major contributors of deforestation and desertification processes, which are known to exacerbate the effects of climate change. Detrimental health effects from the combustion of biomass in unventilated homes are also a growing concern for occupants from these countries. Furthermore, the projected growth in construction of new buildings in developing countries offers important opportunities to reduce future emissions. Improved codes of practice, enforceable energy efficiency standards, sharing of best practice and technology transfer all provide low-income countries with opportunities for low-carbon development that are supported by, but not synonymous with, the broader economic and regulatory policies considered in earlier chapters. Not only will the construction of greener buildings reduce primary energy demand and save consumers money, but such measures will drastically reduce the requirement to build expensive new energy infrastructure unnecessarily.

The effects of climate change on the built environment are often overlooked in the building industry where there is an implicit assumption of current or stationary climates over time (Wang *et al.*, 2011). However, a changing climate will inevitably lead to changing patterns of heating and cooling demand that depend on location. Most predictions allowing for the effects of climate change show significant increases in energy demand, particularly for air conditioning (Wang *et al.*, 2011), which will require investment in power infrastructure capable of delivering baseload electricity. The built environment will increasingly be relied upon not just to provide additional comfort from new extremities in high and low temperatures, but to protect individuals from record high temperatures in summer and freezing conditions in winter. A changing climate will also lead to an increase in the occurrence of extreme weather events such as flooding, cyclones and droughts, requiring changes in building design to withstand these stresses. Indeed, major insurance companies around the world are actively adjusting risk premiums due to recent increases in the occurrence of extreme weather events (Association of British Insurers, 2009). Recent extreme weather events such as Hurricane Katrina and the Australian Floods have destroyed entire regions leaving thousands of people homeless. Many affected areas are now uninsurable (McCorkell, 2011). Thus, buildings will not only be an important contributor for climate change mitigation but will also play an active role in adaptation, with solutions potentially increasing the embodied energy and carbon of buildings as they are reinforced to withstand extreme weather events.

7.2.2. *The solution*

In order to decarbonise the global built environment four distinct strategies are evident:

1. Reduce energy consumption.
2. Switch to low-carbon fuels and renewable energy.
3. Reduce embodied emissions of building materials.
4. Control (limit) non-CO_2-related emissions.

If climate change targets are to be met, a combination of these strategies is required working in concert with the broad economic and regulatory policies of earlier chapters. Of these four strategies, the option with potential to provide the most immediate and largest GHG reductions at the lowest marginal abatement cost is to reduce wasteful energy consumption from buildings. Whatever technological solution is offered the best performance occurs when building occupants are committed to using energy-efficient systems in an appropriate way. Many factors influence an individual's energy consumption behaviour such as age, gender, economic status and socio-demographic conditions. Therefore, education and awareness campaigns are an important catalyst for reducing residential energy consumption as a complement to the economic and regulatory strategies of the earlier chapters.

The building sector involves a wide range of actors (architects, developers, owners, managers and tenants) and an extended life-cycle of products and user preferences, thus making it one of the most complex environmental policy target groups (OECD, 2003). However, there are opportunities to achieve large-scale improvements with the possibility for each strategy to be tailored to local needs and wider government objectives (United Nations Environment Programme, 2008). Such strategies may constitute new legislation, research and development, economic incentives, technical access and educational programmes to change behaviour.

In Scandinavian countries, for example, legislative instruments have been used to control the allowable energy consumption level from dwellings and enforce national building codes to regulate the physical, thermal and electrical requirements of a building (McCormick and Neij, 2009). In China the government has established a tax and fees rebate scheme (known as "feebates") for new buildings that meet the new low energy consumption targets presently set at 65% less than average dwelling consumption (United Nations Environment Programme, 2008). In India, the government has instigated 30% reduction in energy for all government buildings. In Europe, the Energy Performance of Buildings Directive requires all EU Member States to have in place an agreed methodology for calculating the energy performance of buildings. In the UK this has resulted in Energy Performance Certificates (EPC), which grade the efficiency of homes on a scale from A to G. Each time a building is leased or sold this certificate must be provided (Kelly *et al.*, 2012). These all are signals into the property market that are not captured entirely by broader economic and regulatory policies, although they are influenced by them.

Most solutions require the active involvement of government to create a suitable framework to encourage energy-efficient buildings. Leaving the private sector to address energy efficiency without external economic or legislative signals in most cases is not feasible due to the relatively low price of energy and minimal price on carbon (Hope, 2008). It is also important that governments develop strategies in consultation with the building industry (e.g. investors, architects, property developers, construction companies, landlords and tenants) to develop effective instruments that will find acceptance in the market. Such strategies include:

- The implementation of effective government policy to influence the behaviour of not just the building sector but financiers, innovators and planners.

- Benchmarking strategies such as national standards and certification processes, triple bottom-line reporting and international comparisons between buildings and technologies so the market understands what is feasible.
- Regulations for the construction and renovation of buildings and for realistic and measureable energy efficiency standards to drive rates of uptake of low-carbon solutions.
- Economic tools such as taxes, fees, rebates, preferential lending, tax breaks and even cost-neutral policies like the "Green Deal" in the UK and "feebates" in China to bring low-carbon innovations over the investment "valley of death".
- Education through national school programmes, awareness campaigns, labelling products and conspicuous rating of buildings to raise the consciousness of efficiency to building occupiers.
- Deeper understanding into human behaviour with respect to how people interact with technology and buildings.
- Leading-by-doing through the application of stringent building requirements to public buildings aiming to positively build economic capacity and influence the building sector to move in the right direction.
- Support for the transfer of appropriate building technologies to low-income countries to encourage new low-carbon development pathways.

The global contribution of GHG emissions from buildings is high and growing. We have shown the implications of energy consumption from buildings are significant and can no longer be ignored. The penetration of energy efficiency and low-carbon technologies in housing is low at present — often despite significant economic policies that make them feasible — and much of the related business potential is still untapped. By introducing the concept of "energy service provision" in which firms are not simply selling energy but selling the services needed to reduce energy use, it is possible to provide heating, cooling and lighting more cost-effectively while producing less CO_2. Stimulating this shift from a market based solely on energy provision (and decarbonising that energy provision) to one based on consideration of both supply and demand for energy is a necessary component of policies aimed at decarbonising the global economy in the building sector.

7.3. Decarbonising Transport

The transport sector is responsible for around 25% of global CO_2 emissions, with three-quarters of all transport emissions attributable to road transport alone. All remaining transport-related emissions come from international shipping, aviation and rail (IEA, 2010). Transport relies predominantly on petroleum where 95% of total annual petroleum resources are consumed. Over the past three decades emissions from the transport sector rose faster than any other end-use sector, rising on average by 3% per year and effectively doubling transport-related emissions during this period (Krzyżanowski *et al.*, 2005). Between 2008 and 2030 emissions from transport are estimated to rise by 80%, where aviation alone is set to double in less than 20 years as is freight traffic by sea and car ownership and use (Ribeiro and Kobayashi, 2008: p. 325). In the EU the demand for transport fuels is growing faster than any other final energy demand. Figure 7.3 shows the number of vehicles owned per 1,000 people in the population increases substantially with increasing wealth.

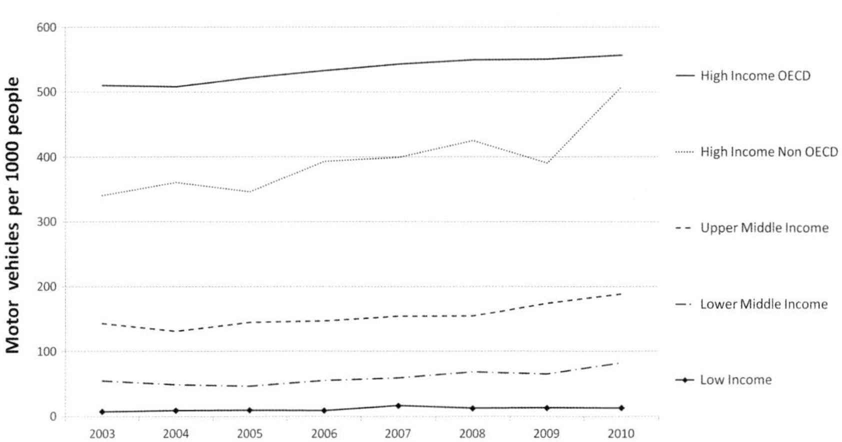

Figure 7.3. Per capita vehicle ownership by income grouping.

Source: World Bank Data Repository (http://data.worldbank.org).

7.3.1. *The problem*

Although the negative externalities arising from transport are often emphasised (and are considered in Chapter 9 of this book), the positive externalities of a well-integrated transport system are also significant. For example, personal mobility is closely linked to economic activity and therefore transport infrastructure is a stimulus to economic growth and poverty alleviation. Mobility is also one of the more universally acknowledged prerequisites for achieving improved standards of living and other human development goals. Increased personal mobility improves access to essential services like health care and education and has the added benefit of making life more enjoyable (WBCSD, 2009). The transport sector and related industries contribute to economic growth through the production of vehicles, the sale of fossil fuels, the development of transport infrastructure and the sale of end-use transport services.

In the EU, transport-related expenditure is estimated as being responsible for at least 14% of total household expenditure (WBCSD, 2009). From a macroeconomic perspective transport infrastructure provides for and maintains a functioning market-based economy. The transport of goods and services between producers and consumers allows for the interaction between economic agents that has the effect of opening up markets, creating efficiencies and lowering transaction costs. Without safe, secure and reliable transport, people could only plan to utilise resources within the geographic area that surrounds them, severely limiting the range of potential opportunities available to them. Policies aimed at reducing transport emissions will be successful only if they recognise these benefits and identify means to meet the same end-use benefits at lower carbon intensity.

Although the benefits of transport can be significant, there are also substantial costs that are borne by society and the environment. Enhanced mobility is often associated with increased air pollution, risk of death or serious injury, congestion, noise, disruption to communities, disturbance of ecosystems and importantly the emission of GHGs (WBCSD, 2004). Negative externalities that are a direct result of transport and its related industries therefore require careful scrutiny, both in regard to climate change and other measures of community sustainability. Research in recent decades has identified the adverse effects of transport-related emissions on human health (Krzyżanowski *et al.*, 2005), with between 12,000 and 24,000 people dying prematurely every year in the UK as a direct result of short-term exposure to air pollution. Those particularly at risk are young children, pregnant women, the elderly and people suffering from pre-existing heart and lung infections.

In 2000 an estimated 1.2 million people died and a further 7.8 million people were seriously injured around the world as a result of road traffic accidents, thus making transport the ninth most important factor contributing to the worldwide burden of disease and injury (WBCSD, 2004). Congestion caused by increased road traffic has a negative effect on economic growth whilst simultaneously increasing the rate of pollution and emissions caused by traffic. Chronic congestion increases the outlay required for fuel and increases travel times, therefore increasing the cost of goods and services sold in the economy. Noise pollution or unwanted sound caused by traffic is becoming an increasing concern, particularly in urban centres. The transport sector is also a major consumer of raw materials for the construction of roads, bridges and railways, consumer of land resources for roads, railways and airports and non-renewable energy sources such as petroleum and oil. In 1996 transport infrastructure was estimated to account for 1.2% of total land area in the EU (WBCSD, 2004). The environmental externalities of transport in Europe are estimated to consume 5.5% of GDP and little progress is being made to internalise these costs into transport prices (EEA, 2001).

The problems caused by transport are particularly acute for developing countries where rapid growth, urbanisation, poverty and shortages of resources are a challenge for enabling people to move about freely. The majority of the world's population still does not have access to personal vehicles. However, this is rapidly changing. In China alone the number of cars in the national fleet increases by an average of 20% per year (Ribeiro and Kobayashi, 2008). The dominant mode of transport in developing countries is by bus or motorised two-wheelers (i.e. scooters). While economists agree that mass automobile ownership increases dramatically at incomes of around US$5,000 per capita, the emergence of low-cost automobiles such as the Tata Nano in India will accelerate the rate of uptake of automobiles in emerging economies (WBCSD, 2009). As incomes grow through global economic development, there will be increasing demand for safer and more reliable vehicles. This is exacerbated by large volumes of second-hand vehicles shipped from more industrialised countries to the developing countries — vehicles that tend to be significantly lower in fuel efficiency and emission controls than market-leading technologies.

Surveys show that the most attractive form of transport is motorised personal vehicle, which is seen as a status symbol as well as being faster, more flexible and more comfortable than other forms of transport. High levels of inequality and access to cheap motorised transport mean that affordable motorisation can take off more quickly than national transport infrastructure in most developing countries. The result is growing congestion, increased air pollution, increased cardio-respiratory disease (see Chapter 9 on co-benefits) and a rise in traffic-related fatalities (Ribeiro and Kobayashi, 2008). Furthermore, growth in low-density development on the outskirts of large cities means there are greater distances to access jobs resulting in a marked decrease in walking and cycling as a share of total travel in developing countries (WBCSD, 2009).

Industrialised economies account for the vast majority of accumulated transport-related emissions. In developing countries automobile travel accounts for 15–30% of all trips, in Europe this is 50%, and in the US it is 90%. Within industrialised economies, a major component of the growth in passenger traffic is due to increasing travel distances to destinations such as work, shops, schools and leisure activities, a problem tackled as much by spatial planning as by broader economic signals such as carbon prices. In wealthy countries, long-distance travel is growing sharply (see Figure 7.4), particularly for air travel as high incomes act to boost international tourism. The effects of globalisation and the rise of multinationals add to the growing number of business-related trips taken across the world.

Industrialisation and globalisation have stimulated freight transport. High pressure to improve freight delivery times has seen a sacrifice in energy efficiency in favour of just-in-time shipments and a shift to faster

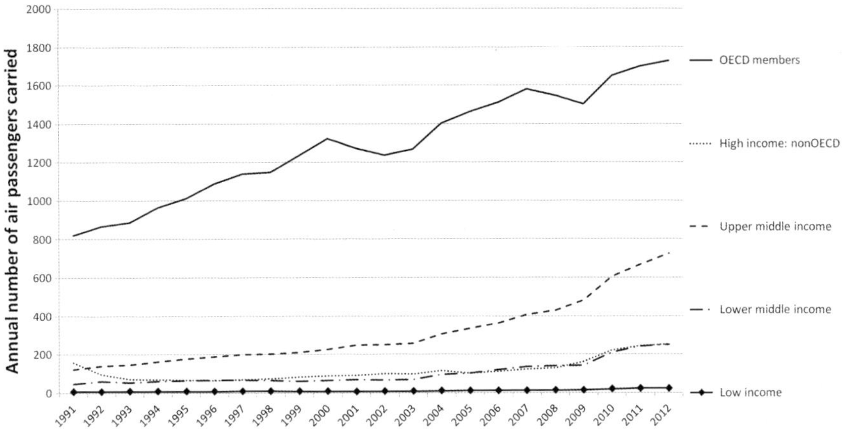

Figure 7.4. Air transport by income grouping.

Source: World Bank Data Repository (http://data.worldbank.org).

and more energy-intensive modes of transport (Ribeiro and Kobayashi, 2008). Internationally there is considerable variation in freight transport depending on geography, availability of infrastructure and level of economic development. Freight transport in the US has the highest amount of freight traffic in the world spread across all modes of transport (shipping, rail and aviation). In contrast, Russia is dominated by rail and pipelines whereas Europe's freight system is dominated by trucking — one of the most carbon-intensive forms of transport — with a market share of 72% (in tonne-km of goods transported, where 1 tonne-km is shipping of 1 tonne of goods for 1 km). China's freight system uses rail with large contributions also coming from trucks and shipping. Shipping alone accounts for 35% of transport emissions globally (Ribeiro and Kobayashi, 2008) and must therefore play a central part in any strategy to decarbonise.

Projections show that, to date, incremental improvements in efficiency are far outweighed by growth in transport activity (IEA, 2010). Under these circumstances transport-related demand will increase by 2% per year where at least 90% of this increase in demand is projected to be in petroleum-based fuels, meaning that CO_2 emissions will grow along with fuel consumption. During this same period there will be a significant regional shift in the structure of transport-related demand. While transport-related emissions are still predicted to grow in mature markets, by far the majority of future growth will come from emerging economies where improved access to transport is part of a larger strategy of economic development.

7.3.2. *The solution: A new approach to thinking about travel*

As with the residential sector, energy demand in the transport sector is made up of many small and highly dispersed consumers, with behaviours that are influenced only partially by broad economic and regulatory signals. For this reason and others, transport is particularly problematic to decarbonise.

For road transport, the most common short-term solutions include incremental improvements to vehicle technology and behaviour change, but we also need to start thinking about more long-term systemic solutions. Reducing road transport emissions can be achieved in five distinct ways:

1. Reduce weight, friction and air resistance acting on the vehicle.
2. Increase the efficiency of converting energy into work and recapturing the losses to use as mechanical energy.
3. Change to less carbon-intensive fuels.
4. Reduce emissions from non-CO_2-based vehicle exhaust and car air-conditioning systems.
5. Reduce the desire and need for people to travel long distances.

Studies by Allwood and Cullen (2012) show the energy required by a standard car can be reduced by up to 91% by reducing the mass of the car (by 75%), reducing the friction coefficient (by 93%), reducing the drag coefficient (by 75%) and shortening the frontal area of the car (by 25%). Each of these measures reduces energy demand for transport even if the source of that energy remains unchanged. They are measures affected by vehicle design and the engineering of industrial processes rather than by the means of generating that energy. Such calculations show the scale of savings that are theoretically possible whilst still providing the same energy service.

However, the economics of such energy-saving improvements are not the sole considerations of regulators and consumers. A car that weighs less, is lower to the ground and has a lower coefficient of friction will not be as safe in an accident with a heavier, conventional car; it may be perceived as less stylish; and, most importantly it is currently more expensive to produce and manufacture. Other technologies such as Intelligent Transport Systems (ITS) which utilise microelectronics, satellite navigation, mobile communication and sensors can help with the logistics of traffic flow, reduce congestion, automate tolling, provide speed warnings and offer point-to-point mobility, thus offering opportunities to reduce journey emissions. Other advanced technologies such as regenerative breaking will also make important contributions. This raises a significant question of where technological innovation should be directed for the transport sector: towards fuel switching or towards demand reduction through improved efficiency or behaviour change.

Switching fuels from liquid petroleum to other fuels such as natural gas, diesel, biofuel or electricity provide some opportunity for emission reduction. If technology-based solutions are going to make any real impact on global decarbonisation on the scale required by the modelling results of earlier chapters, then personal vehicles will need to switch to advanced technologies such as electric vehicles, hybrid vehicles, and even hydrogen-based fuel cells. Implementation of advanced transport technologies, however, does have several drawbacks. Widespread installation of electric vehicles will only lead to reduced emissions if the power system that is used to charge vehicle batteries has also been radically decarbonised. It is still uncertain if the level of additional demand required by transport can be cost-effectively met by new low-carbon generation.

Although many of the more advanced technologies for personal vehicle transport appear out of reach given the short timeframe required to make substantial carbon reductions, at least at current rates of market penetration, there are other opportunities worth exploring that focus firstly on the service being provided by transport. Research has shown one of the best ways to reduce emissions is to change the mode of transport or to eliminate the need for transport altogether. The greatest reductions are shown to occur when a shift is made away from personal vehicles to cycling, walking or public transport. These switches however require changes in both mode choice attitudes and community spatial designs, each of which is influenced by economic and regulatory signals aimed at energy and carbon, but not yet fully determined by these.

Transport modes such as walking and cycling have significant potential for emission reductions; however such a strategy is only really viable for distances of between 2.5 km and 7.5 km for walking and cycling, respectively. In Europe 50% of all motorised trips are for distances less than 5 km so the impact of this strategy could be significant here. In countries such as the US and many developing nations, places of work, residence, markets and schools can be separated by much larger distances. Longer-term strategies of demand reduction for increased use of personal vehicles therefore are centred on changes in community design — and the regulatory and planning rules that govern their development — that aim for self-contained communities allowing people to live, work and access amenities within a small travel distance.

When all indicators of increasing global demand for transport are considered, it is evident that global emission targets will not be reached if present trends continue and the current mix of transport modes and technologies is maintained. The bottom–up analysis considered in this section suggests that policies must simultaneously create incentives for change in community design, material characteristics of vehicles, provision of mult-mode transport systems that allow consumers informed choices that include low-carbon alternatives to the personal vehicle, changes to the practice of shipping by heavy goods vehicles (HGVs) and decarbonisation of the energy supplied to whatever mode of transport a consumer chooses.

7.4. Industrial Sector

GHG mitigation potential from industry is usually treated on a sectoral basis. A major study covering a wide-range of sectors, analogous to the sectors in the analyses in earlier chapters of this book, is the one carried out by the Working Group III (WGIII) in the AR4 (IPCC, 2007). It takes a more bottom–up approach, however, by categorising the technical improvements that would lead to GHG emission reductions as sector-wide options, process-specific options and operating procedures. Each of them contains a set of options applicable to different sectors and their unique material and energy needs.

Their report focuses specifically on the following sectors: iron and steel, non-ferrous metals, chemicals and fertilisers, petroleum refining, minerals (cement, lime, glass and ceramics) and pulp and paper, which together accounted for about 85% of the global industrial sector's primary energy use in 2004. It noted that most of the industrial production of these sectors is processed in developing countries. Even if some of their facilities use international best-practice technologies and procedures, there is still notable room for efficiency and procedural improvements.

The mitigation potential of those sectors is based on three sources of CO_2: "(1) the use of fossil fuels for energy, either directly by industry for heat and power generation or indirectly in the generation of purchased electricity and steam; (2) non-energy uses of fossil fuels in chemical processing and metal smelting; and (3) non-fossil fuel sources, for example cement and lime manufacture" (Bernstein *et al.*, 2007).

Data from several studies are harmonised to the IPCC's A1B and B2 scenarios and the mitigation potential of each industry in 2030 is reported in Table 7.8 of their report. The mitigation potential for the industrial sector under the A1B scenario is 3.0–6.3 Gt CO_{2eq}/yr at a cost of <100 US$/t$CO_{2eq}$, and 2.0–5.1 Gt CO_{2eq}/yr under the B2 scenario. The mitigation range and costs are compared to a similar study led by the IPCC (2007) which provides ranges for mitigation potential in 2030 for nine groups of industrial technologies totalling about 2.5–3.0 Gt CO_{2eq}/yr, with mitigation costs estimated at <25 US$/t$CO_{2eq}$ (2004US$). IEA's estimate of mitigation potential falls in the previous range although their estimate of mitigation cost is significantly lower. This result can be due to the fact that the IEA study is based on the MARKAL model which provides an integrated optimisation of market allocation of different technologies across the different sectors (as opposed to the agglomeration and homogenisation of data in the WGIII), delivering cheaper options due to synergies within and between sectors.

There are other studies evaluating low-carbon industrial pathways and their associated emissions and costs. However, they do not assess industrial emissions in an integrated way, but rather examine only components of the global industrial system one at a time (so synergies are missed). Hughes and Strachan (2010) review 21 UK and international low-carbon scenarios. They find that most studies address the mitigation potential in the energy sector or the building sector generally, but do not examine technologies and material/energy demand characteristics of specific industrial products. Furthermore, only five studies address the social integration and feasibility of the changes in technology and consumer patterns. On the economic side, only six studies cover the associated costs of the mitigation potential when technological change causes changes in the processing of material and energy in economic systems. Only one addresses natural impacts such as environmental degradation. Inter-relations between technological change, policy, social acceptability and industrial change are not caught in any of the models described above and therefore they do not provide detailed bottom–up information on how the technologies can be deployed even if suitable economic and regulatory signals are applied. More research is required, therefore, in understanding these inter-relations and using them to inform the suite of policies and regulations required to place the global economy onto the low-carbon pathways noted in earlier chapters of this book.

The consumption side of industrial systems in regard to end-use of goods is another aspect that is not analysed in the WGIII. According to the European Environment Agency (EEA, 2010), “it is consumption and production patterns in developed countries, with developing countries catching up rapidly, that are the key drivers of global environmental problems”. This is due to the emission of pollution and waste during both the production and disposal phase of products. Inherent in such studies is a cradle-to-grave assumption in which goods are produced, used and then disposed of rather than being recycled back into industrial processes.

A major challenge in decarbonising industrial systems is therefore to develop procedures to understand how supply chains or production layers in an economy can improve the recycling and reuse of materials with high levels of embodied energy and carbon. In 2005, post-consumer waste contributed approximately 1300 $MtCO_{2eq}$ (IPCC, 2007), representing 5% of total GHG emissions. The largest source of emissions was landfill methane (CH_4), followed by wastewater CH_4 and nitrous oxide (N_2O); in addition, minor emissions of CO_2 result from incineration of waste containing fossil carbon, especially plastics and synthetic textiles. Given increases in waste production in both developed and developing economies, the total global economic mitigation potential for reducing landfill CH_4 emissions in 2030 is estimated to be above 1000 $MtCO_{2eq}$ at costs below 100 US$/$tCO_{2eq}$/yr. This represents up to a third of the projected industrial mitigation potential. Most of this potential is achievable at negative to low costs: 20–30% of projected emissions for 2030 can be reduced at negative cost and 30–50% at costs <20 US$/$tCO_{2eq}$/yr. Waste incineration and other thermal processes generate lower GHG emissions than landfilling, i.e. around 40 $MtCO_{2eq}$/yr. Those results are however the upper limits of a study from Monni *et al.* (2006) which uses strong assumptions such as that 15% of waste is disposed to landfills by 2010. Even if some developed countries adopted landfill ban policies in the late 1990s/early 2000 (EEA, 2007), such a global assumption was highly unrealistic to be applied at world level. This approach is already being implemented in Europe, where the EU Landfill Directive (1999/31/EC) aims for a gradual reduction of biodegradable waste up to 65% being diverted from landfills in 2020. Landfilling remains, however, the most common practice in developing countries.

Reduction of waste by changing industrial practices to cradle-to-cradle processing of materials, remains a significant strategy for decarbonising the industrial system of production and consumption. Recycling is mentioned in the waste management chapter in the WGIII (IPCC, 2007) and is presented as having significant potential for indirect reductions of GHG emissions according to WRAP (2006) and Smith *et al.* (2001). That potential has been confirmed by the 2010 update of WRAP (2010), where over 200 Life Cycle Analyses have been compared for different substance streams. The key impact categories used for the assessment of the different waste management options were depletion of natural resources, GHG emissions, cumulative energy demand and water consumption.

Recycling is also a key sector for dematerialisation of industrial processes, i.e. reducing the absolute material and energy intensity of the economy, which contributes to GHG mitigation by decreasing waste generation, reducing raw material consumption and reducing demand for energy and fossil fuels associated with the cycle of production. However, current dematerialisation indicators focus on industrial inputs, i.e. how many resources are saved to produce a similar product. A more constructive and informative approach — at least in regard to decarbonisation of the global economy — would be measures of how much waste is avoided throughout the life span of a product. As stated by Herman *et al.* (1990), "less is not necessarily less from an environmental point of view. If smaller and lighter products are also inferior in quality, then more units would be produced, and the net result could be a greater amount of waste generated in both production and consumption. From an environmental viewpoint, therefore, dematerialization should perhaps be defined as the change in the amount of waste generated per unit of industrial products."

A potential policy approach to closing the material and energy cycles of industrial systems, hence reducing emissions, is Extended Producer Responsibility (EPR) policies, in which producers of goods must recycle the products at the end of their lifetime. This provides incentives to the producers to both extend the lifetime of products and to make the products easily disassembled for recycling. Large-scale implementation of such policies is currently affecting the waste electric and electronic equipment (WEEE) and end-of-life light vehicles sector in Europe. The directives 2002/96/EC and 2000/53/EC specify that a producer takes responsibility for the waste management of the product and set minimal recycling and recovery targets. Such policies have a great potential to mitigate environmental degradation and GHG emission by reducing waste and increasing material and energy efficiencies of products through optimised design for recycling.

Mitigation potential in large industrial systems such as iron tends to be underestimated because non-destructive recycling is not included in the analyses of technological potential. Non-destructive recycling in this case means to recycle the material without re-smelting it, saving most of the energy required to reuse the material. Policies aimed at technology diffusion do not capture such mitigation measures. According to Allwood and Cullen (2012), GHG emissions in the steel sector could be halved in 2050 from best-practice results if non-destructive recycling is also implemented.

Therefore, even if not quantified, the mitigation potential for industry is larger than that stated in the AR4 and other low-carbon emission reports (see Hughes and Strachan, 2010). The main causes are, on the one hand, an underestimation of the technological improvement and their integration across and within sectors and, on the other, the lack of data on recycling and failing to recognise its central role in articulating dematerialisation targets. Development of effective policies and instruments for decarbonising the industrial sector depend on resolving these gaps in knowledge and focusing greater attention onto closing the cycles of material flow through industrial systems. Such policies and instruments will target the design of industrial systems to create incentives for cradle-to-cradle approaches rather than focusing solely on the energy and carbon intensity of particular components of the system.

7.5. The Role of MNEs

The macroeconomic analyses of earlier chapters parsed out the economy by considering the actions of the global economy divided into economic sectors. Those sectors interact through financial transactions, and so policies that influence one sector also influence others connected in both forward and backward linkages. Another path through these same interactions is to consider the roles of specific firms that may cut across several sectors, and who are part of and drive supply chains that spread over many layers of production and consumption. We turn now to a view of the global economy that considers how decarbonisation might be driven through the practices of these firms rather than solely through policies focused on nations or aggregated economic sectors.

Large enterprises, and in particular MNEs, have a fundamental role in the transition towards a low-carbon economy: they are a major part of both the problem and the solution of decarbonisation (UNCTAD, 2010). MNEs produce significant direct GHG emissions across their international operations. For example, the 2010 Global 500 Report by the Carbon Disclosure Project (CDP) — a not-for-profit organisation holding the world's largest database of primary corporate climate change information — recorded that the total direct emissions from the 367 enterprises that responded to the CDP's disclosure request (sent to the world's 500 largest public companies in the FTSE Global Equity Index Series) were 3.4 Gt CO_{2eq}, or approximately 11% of total global CO_2 emissions (PriceWaterhouseCoopers, 2010). In their role as "systems integrators", MNEs depend upon and coordinate to varying degrees the activities of many other enterprises along their supply chains both upstream (meaning companies that supply materials and services to the firm of interest) and downstream (meaning the consumers of the firm's products) (Nolan, 2008).

These activities entail further significant indirect emissions. For example, a recent UNEP report produced by Trucost (2010), estimated that the global top 3,000 publically listed companies caused 32% (US$ 1.44 trillion) of global GHG emission externality costs in 2008, based on a social cost of carbon of US$ 85/ton CO_{2eq}, with almost half of the externalities analysed coming from supply chains, i.e. from indirect upstream or downstream emission sources (Trucost, 2010). MNEs are in a prime position to generate and disseminate technology, influence their supply chains and finance investments for GHG emission abatement, whether via foreign direct investment, reform of patent ownership rights, exertion of pressure on suppliers or through the research and development of improved production processes and cleaner products that help reduce use-phase and end-of-life emissions (Dunning and Lundan, 2008). Their influence extends to creating demand for cradle-to-cradle production methods described in Section 7.3.

There is a disconnection between the nation-states that negotiate and agree mitigation targets on a geographic basis and the enterprises that control the actual sources of emissions where these targets will have to be met on the ground. Ultimately, nation-states pass on agreed targets to their domestic industries, but the leading enterprises in almost all national industries transcend national boundaries. These same enterprises also have internal policies that in many cases have been as ambitious and influential in reducing emissions, as have public sector efforts.

Despite this, policy elements to harness MNE contributions remain largely absent from climate change policies (UNCTAD, 2010). The lack of global consensus on environmental policy means that nation-states have to resort to national and or regional emission reduction targets. Given the dynamic nature of global markets, MNEs have the option to avoid these targets by relocating their operations and supply chains — a phenomenon of "carbon leakage" (Carbon Trust, 2010). However, even in the absence of a binding international climate agreement, leading MNEs are recognising that "it makes business sense to be ahead of the curve, to improve the energy efficiency of their operations and to seize emerging opportunities" (United Nations Environment Programme, 2010). MNEs are increasingly calling upon countries and their negotiators to reach concrete and ambitious agreements, often beyond what is currently on the table (United Nations Environment Programme, 2010). Institutional (e.g. pension and insurance funds) and private investors are also recognising the financial risks to which their portfolio investments are becoming exposed due to climate impacts (Trucost, 2010).

These forward-looking enterprises have come to see many different types of risks to their business in relation to climate change (Newell and Paterson, 2010): risks of incurring costs from later regulation if they are not prepared for it; supply disruption risks due to climate impacts (for example, changing weather patterns and flooding); risks to their reputation (and thus to potential markets and investment opportunities) if they are seen as unresponsive to changing consumer demands and employee, shareholder and investor expectations; risks of legal liabilities if they are seen to cause damages to others by refusing to cap their emissions; and risks of losing out on new low-carbon market opportunities. In becoming aware of the impact climate change and the transition to a low-carbon economy will have on their activities, large enterprises are realising that if they are to adequately understand their carbon footprint and the risks that they face, they must extend their focus from their own production processes to include their supply chains (Kearney, 2010). For example, up to 80% of a retailer's carbon footprint and between 40–60% for consumer goods manufacturers can be found in upstream supply chains due to indirect emissions associated with processing, packaging and transportation of goods. PepsiCo UK and Ireland discovered that only 30% of the carbon footprint of its Walkers products come from their own internal manufacturing processes (CDP, 2009).

In response to these pressures, a growing number of MNEs now measure and voluntarily disclose their GHG emissions and are beginning to incorporate carbon management into their core business strategy (PriceWaterhouseCoopers, 2010). Many MNEs are also requesting that their suppliers report and manage their GHG emissions (Kearney, 2010). However, the information being reported often lacks comparability, and reports of emissions from foreign affiliates and by value chains is often missing. In response, the UNCTAD *World Investment Report* calls for "a reliable internationally harmonized approach to measuring and reporting corporate climate change-related emissions" noting that this is "vital for the effective implementation and assessment of climate change policies … the internalization of climate risk into capital markets, and the monitoring of GHG emissions and clean technology diffusion throughout [MNEs'] value chains" (UNCTAD, 2010).

While we are still some way off UNCTAD's vision, we can begin to understand the roles of large enterprises relating to direct and indirect upstream emissions by considering their aggregate activity, along with smaller actors, within global industry sectors. Such understanding depends on the development of methods to better characterise how emissions of GHGs are generated at different stages of production in the global economy. Figure 7.5 presents results from a recent study investigating flows of embodied emissions through the global production system (Skelton, 2011). The study employs a multi-regional input–output model constructed from 2004 global economic and (energy use) CO_2 emissions data from the Global Trade Analysis Project (GTAP). The figure shows the evolution of sector production and consumption attributions of CO_2 emissions through the final four production layers of the global production system as a proportion of each sector's total consumption attribution.

The final production layer (PL^0) involves the processing of finished goods and services supplied to global final demand (i.e., household consumers, government and capital investment); PL^1 involves the processing of intermediate products supplied as inputs to PL^0; PL^2 involves the processing of intermediate products supplied as inputs to PL^1, and so forth. A sector's final production attribution is simply a measure of its total direct (i.e., Scope 1) emissions. Intermediate production attributions are similarly a measure of direct emissions, but only those that have been released up to a particular production layer. A sector's final consumption attribution is a measure of emissions, originating from any sector, that

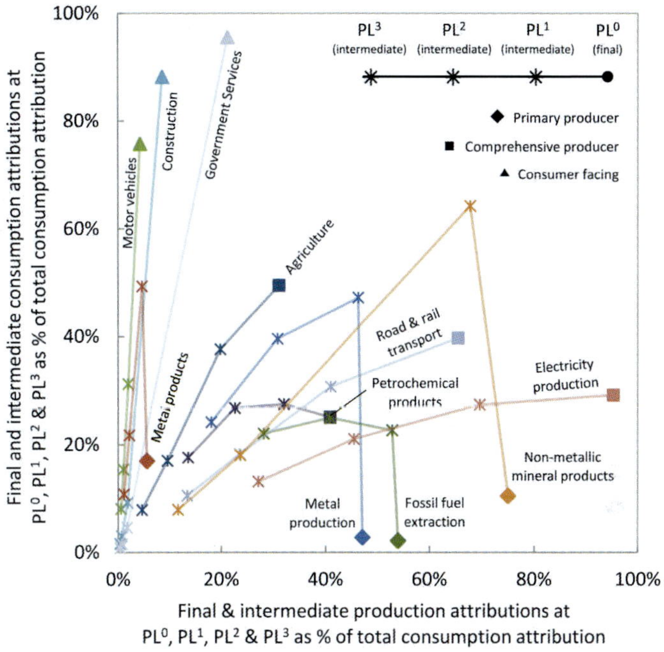

Figure 7.5. Evolution of sector consumption and production attributions of CO_2 emissions from PL^3 to PL^0 normalised as a percentage of sector total consumption attribution.

have become embodied in the sector's output of products sold to global final demand. Similarly, intermediate consumption attributions measure the emissions embodied in the intermediate products sold by a sector at a particular production layer. Finally, the total consumption attribution of a sector provides a measure of its upstream carbon footprint: all the emissions embodied in the sector's total output of products.

Conceptually we can envisage final demand purchases of finished products triggering cascades of intermediate product purchases that flow upstream through the sequential layers of the production system, with firms acting as the conduits through which these flows occur. Within each industry, at each layer, the processing of these intermediate products entails the release of emissions into the atmosphere. Similarly, we can envisage cumulative flows of embodied emissions running in the opposite direction, from the depths of the production system to the point of consumption of finished goods and services, picking up additional direct emissions along the way as intermediate products are processed.

In Figure 7.5, solid markers at the end of each line give a sector's final consumption attribution (read off the *y*-axis) and final production attribution (read off the *x*-axis) as a percentage of total consumption attribution. Data points marked with a star indicate intermediate consumption and production attributions at PL^1, PL^2, and PL^3, with PL^3 at the start of the line, again as a percentage of total consumption attribution. For example, for the metal production sector, the final production attribution is 1.56 Gt CO_2 or 47%, and final consumption attribution is 0.09 or 3% of the sector's total consumption attribution of 3.3 Gt CO_2. Intermediate production attribution at PL^1 is 1.53 Gt CO_2 or 46%, and intermediate consumption attribution at PL^1 is 1.56 Gt CO_2 or 47% of total consumption attribution. Intermediate production attribution at PL^2 is 1.02 Gt CO_2 or 31%, and intermediate consumption attribution at PL^2 is 1.31 Gt CO_2 or 40% of total consumption attribution. Finally, intermediate production attribution at PL^3 is 0.60 Gt CO_2 or 18%, and intermediate consumption attribution at PL^3 is 0.8 Gt CO_2 or 24% of total consumption attribution (Skelton, 2011).

Considering Figure 7.5, we find that for the sectors labelled motor vehicles, construction and government services, over 75% of their total consumption attribution can be accounted for by their final consumption attribution. For electricity production and non-metallic mineral products, the same can be said for their final production attribution. However, for all other sectors, neither final consumption nor production attributions account for more than 75% of their total consumption attribution. The most extreme case shown is that of metal products, where its final consumption attribution accounts for 19%, and its final production attribution only 8%, of its total consumption attribution. This suggests that the analysis of a sector's emissions impact in terms of only final product and consumption attributions runs the risk of underestimating the scale of the sector's overall impact and supports the position of leading-edge enterprises that are now seeking to evaluate their indirect carbon footprints.

In general, three broad groups of sectors, and hence constituent large enterprises, can be identified and are indicated on Figure 7.5 by different solid markers at the end of each line:

(a) The "primary producer" industries, such as non-metallic mineral products and metal production, which are characterised by a sudden drop in their consumption attribution and minimal increase in their production attribution at PL^0, as they tend not to supply few products to final demand. Enterprises within these industries are particularly vulnerable to the risk of incurring costs from regulation of direct emissions. In addition they may also face investment uncertainties as shareholders and investors seek to reduce their own risks.

(b) The "comprehensive producer" sectors, such as road and rail transport and electricity production, which are characterised by steadily increasing production and consumption attributions across all layers. These sectors tend to supply goods and services to both final demand and the intermediate demand of a wide range of other sectors. Enterprises within these sectors will face risks similar to the "primary producers" but in addition they are also open to reputational risks relating to consumer final demand. Furthermore, as suppliers of goods to all corners of the economy they face the greatest risk from losing out on new market opportunities as other actors seek to reduce risks — by reducing emissions — from their supply chains.

(c) And finally the "consumer facing" sectors, such as construction and motor vehicles, which are characterised by a low final production attribution and a sudden jump in their consumption attribution at PL^0. These sectors tend to only supply final demand with finished goods and services that require the processing of typically emission-intensive inputs. Enterprises within these sectors will face particular risks to their reputation with respect to consumers, potentially losing out on new consumer markets, if they are unable or unwilling to reduce the embodied emissions from their upstream supply chains. If unresponsive "primary and comprehensive producers" seek to pass on incurred costs from direct emissions to higher-order production layers in the economy (as might take place with consumer-based policies), then consumer facing enterprises will face increased costs of intermediate goods, which, in turn, will have to be passed on to their consumers, thus running the risk of losing existing markets or driving down demand.

Different MNEs will face different types of risks and thus their corporate strategies for dealing with them will vary considerably; however, it is evident that no matter what sectors they are involved with there is no escaping the need to reduce either direct or indirect emissions given the on-going trend for stricter emissions regulations and consumer, employee, shareholder and investor expectations. While the quantitative investigation of the role of large national and multinational enterprises in an emissions constrained economy is in its infancy, it remains clear that these actors will have considerable influence over the manner and rate at which the global economy transitions towards a low-carbon state. The creation of climate policy elements to harness the power of MNEs and the development of robust corporate emission-accounting approaches remain two major challenges within the current international climate agenda.

7.6. Conclusions

This chapter has examined material and energy use in the global economy using a bottom–up approach that reflects the complexities of material and energy decisions in specific sectors of the economy. The intent has not been to replace the modelling of earlier chapters in the book — indeed the material provided in the present chapter has not been placed into a modelling framework. Instead, the material presented here brings a nuanced understanding of the challenges facing specific sectors, supply chains and layers of industrial production, and the implications for policies aimed at reducing these challenges. It must be noted that opportunities for energy demand reduction within any given sector are, alone, insufficient for meeting the required level of global emission reduction. All sectors will need to pursue important demand reduction programmes simultaneously to meet the required emission reduction necessary. What is more, optimal solutions cannot be found without reasonable consideration given to the interdependencies between and within sectors. Likewise, energy demand cannot be adequately envisioned in isolation from energy supply and the availability of low-carbon supply options.

The central lesson is similar to that of the chapter on political feasibility for policies. The energy-economy-environment model at the global level at the heart of this book (E3MG) provides powerful insights into the decarbonisation of the global economy *if* economic and regulatory signals have the desired effects. The present chapter provides both a cautionary note that the rate of success of these policies may be lower than desired if we do not understand the limitations in viable technology and behavioural options of specific sectors, and a positive note that the rate of success can be increased if the broad, economy-wide policies identified through E3MG are married to more targeted policies that reflect the limitations to innovation and change in industrial systems and the built environment.

References

Allwood, J. and Cullen, J. (2012). *Sustainable Materials with both Eyes Open*, UIT Cambridge Ltd., Cambridge.

Association of British Insurers (2009). The Financial Risk of Climate Change. Available online at: ipcc-wg2.gov/njlite_download.php?id=6608. Accessed 6 October 2013.

Carbon Trust (2010). Tackling Carbon Leakage: Sector-specific Solutions for a World of Unequal Prices. Available online at: http://www.carbontrust.com/media/84908/ctc767-tackling-carbon-leakage.pdf. Accessed 6 October 2013.

CDP (2009). Raising Awareness: The First Step to Managing Climate Change in the Supply Chain. Carbon Disclosure Project Industry Viewpoints. Available online at: https://www.cdproject.net/en-us/WhatWeDo/Pages/Industry-viewpoints-first-step-to-managing-climate-change-in-the-supply-chain.aspx. Accessed 6 October 2013.

de la Rue du Can, S. and Price, L. (2008). Sectoral Trends in Global Energy Use and Greenhouse Gas Emissions, *Energy Policy*, 36, 1386–1403.

DEFRA (2010). Behavioural Economics and Energy Using Products: Scoping Research on Discounting Behaviour and Consumer Reference Points. March. Available online at: http://center.sustainability.duke.edu/sites/default/files/documents/consumers_%20and_behavior_uk.pdf. Accessed 6 October 2013.

Dunning, J. and Lundan, S. (2008). *Multinational Enterprises and the Global Economy*, Edward Elgar Publishing, London.

EEA (2001). *Environmental Signals 2001, European Environment Agency*, Copenhagen.

EEA (2007). *The Road from Landfilling to Recycling: Common Destination, Different routes*, European Environment Agency, Copenhagen.

EEA (2010). *SOER 2010 — Thematic assessment — Consumption and the environment*, European Environment Agency, Copenhagen.

Hope, C. (2008). Discount Rates, Equity Weights and the Social Cost of Carbon, *Energy Economics*, 30, 1011–1019.

Houser, T. (2009). Energy Efficiency in Buildings: A Global Economic Perspective. Available online at: http://www.iie.com/publications/interstitial.cfm?ResearchID=1198. Accessed 21 March 2013.

Hughes, N. and Strachan, N. (2010). Methodological Review of UK and International Low Carbon Scenarios, *Energy Policy*, 38, 6056–6065.

Iddon, C. and Firth, S. (2013). Embodied and Operational Energy for New-build Housing: A Case Study of Construction Methods in the UK, *Energy and Buildings*, 67, 479–488.

IEA (2002). World Energy Outlook. Available online at: http://www.ingentaconnect.com/content/oecd/1608019x/2002/00002002/00000016/6102271e. Accessed 6 October 2014.

IEA (2008a). IEA Worldwide Trends in Energy Use and Efficiency, Available online at: http://www.iea.org/publications/freepublications/publication/Indicators_2008.pdf. Accessed 6 October 2013.

IEA (2008b). World Energy Outlook. Available online at: http://www.worldenergyoutlook.org/media/weowebsite/2008-1994/weo2008.pdf. Accessed 6 October 2013.

IEA (2010). Energy Technology Perspectives. Available online at: http://www.iea.org/publications/freepublications/publication/etp2010.pdf. Accessed 6 October 2013.

IEA (2013). CO_2 Emissions from Fuel Combustion. Available online at: http://www.iea.org/publications/freepublications/publication/CO_2EmissionsFromFuelCombustionHighlights2013.pdf. Accessed 6 October 2013.

IPCC (2007). *Climate Change 2007: Mitigation. Contribution of Working Group III to the Fourth Assessment Report of the Intergovernmental Panel on Climate Change*, Metz, B., Davidson, O., Bosch, P., Dave, R. and Meyer, L. (eds), Cambridge University Press, Cambridge.

Kearney, A. (2010). *Carbon Disclosure Project: Supply Chain Report 2010*, A.T. Kearney, London.

Kelly, S., Pollitt, M. and Crawford-Brown, D. (2012). Building Performance Evaluation and Certification in the UK: Is SAP Fit for Purpose?, *Renewable and Sustainable Energy Reviews*, 16, 6861–6878.

Krzyżanowski, M., Kuna-Dibbert, B. and Schneider, J. (2005). *Health Effects of Transport-related Air Pollution*, World Health Organization, Geneva.

Levine, M. and Urge-Vorsatz, D. (2008). Residential and Commercial Buildings. Available online at: http://www.ipcc.ch/pdf/assessment-report/ar4/wg3/ar4-wg3-chapter6.pdf. Accessed 6 October 2013.

McCorkell, A. (2011). Rise in Flood Risk Could Make One Million Homes Uninsurable, *The Independent*, 9 January. Available online at: http://www.independent.co.uk/environment/climate-change/rise-in-flood-risk-could-make-one-million-homes-uninsurable-2179746.html. Accessed 6 October 2013.

McCormick, K. and Neij, L. (2009). *Experience of Policy Instruments for Energy Efficiency in Buildings in the Nordic Countries*, International Institute for Industrial Environmental Economics (IIIEE), Lund University.

Newell, P. and Paterson, M. (2010). *Climate Capitalism: Global Warming and the Transformation of the Global Economy*, Cambridge University Press, Cambridge.

Nolan, P. (2008). *Capitalism and Freedom: The Contradictory Character of Globalisation*, Anthem Press, London.

OECD (2003). *Environmentally Sustainable Buildings: Challenges and Policies*, Organisation for Economic Co-operation and Development, Paris.

PriceWaterhouseCoopers (2010). *Carbon Disclosure Project 2010: Global 500 Report*, PriceWaterhouseCoopers, London.

Ribeiro, S. and Kobayashi, S. (2008). Transport and Infrastructure. Available online at: http://www.ipcc.ch/pdf/assessment-report/ar4/wg3/ar4-wg3-chapter5.pdf. Accessed 6 October 2013.

Skelton, A., Guan, D., Peters, G. and Crawford-Brown, D. (2011) Mapping Flows of Embodied Emissions in the Global Production System, *Environmental Science and Technology*, 45, 10516–10523.

Trucost (2010). *Universal Ownership: Why Environmental Externalities Matter to Institutional Investors*, United Nations Environment Programme Finance Initiative (UNEP FI), Principles for Responsible Investment (PRI).

UNCTAD (2010). *World Investment Report 2010: Investing in a Low Carbon Economy*, UN Publications, Geneva.

United Nations Environment Programme (2008). Huovila, P. (ed.). *Buildings and Climate Change*, UNEP/Earthprint.

United Nations Environment Programme (2010). Press Releases September 2010: Global Business leaders to Support Transit to a Low Carbon Future. Available online at: http://hqweb.unep.org/Documents.Multilingual/Default.asp?DocumentID=647&ArticleID=6762&l=en&t=long. Accessed 10 November 2010.

Wang, X., Chen, D. and Ren, Z. (2011). Global Warming and its Implication to Emission Reduction Strategies for Residential Buildings, *Building and Environment*, 46, 871–883.

WBCSD (2004). *Mobility 2030: Meeting the Challenges to Sustainability: Executive Summary 2004, World Business Council for Sustainable Development*, Conches, Geneva.

WBCSD (2009). *Mobility for Development, World Business Council for Sustainable Development*, Conches, Geneva.

WGBC (2010). *Tackling Global Climate Change: Meeting Local Priorities*, World Green Business Council Special Report. London, UK.

Chapter 8

From Theory to Practice: Climate Policy and Political Feasibility

Sonja Klinsky and Michael Grubb

"A comprehensive strategy addressing all aspects of the problem and reflecting environmental, economic and social costs and benefits is necessary."

IPCC First Assessment Overview (1990)

"… there appears to be a nearly inverse relationship between those policies that policy analysts tend to endorse as holding the greatest promise to reduce emissions in a cost-effective manner and the political feasibility of respective policy options."

Rabe (2008: p. 106)

8.1. Introduction

When the IPCC (Intergovernmental Panel on Climate Change) released its First Assessment Report on climate change, the tone was serious but calm. The assessment stated that global climate change was occurring and that anthropogenic emissions of greenhouse gases (GHGs) were contributing to this change. Initial studies outlined some of the many possible negative implications of climate change, and the report indicated that a range of policy responses should be explored. These policy options included everything from sustainable forestry and the diffusion of energy efficiency, to consideration of tariffs and inclusion of environmental costs in energy prices. While not panicked, this assessment made a clear argument that diverse policy options needed to be evaluated and pursued within a global framework, and that the evaluation of these options had to include environmental, economic and social costs (IPCC, 1990).

Simultaneously, 1990 also marked the first steps towards the formation of the United Nations Framework Convention on Climate Change (UNFCCC). By 1994, the convention had been signed and ratified by enough nations to come into force, and in 1997 the Kyoto Protocol emerged. In less than a decade climate change had gone from being a relatively unknown issue to the subject of a major UN convention and legally binding international obligations to address GHG emissions.

Both the IPCC and UNFCCC processes called for systemic, integrated evaluations of the environmental, social and economic aspects of climate change and the possible responses to climate change. The need for solid analyses spurred the development of economic models that attempted to identify the implications of climate change, and climate change mitigation policies, from multiple perspectives. The growth of these models is clearly observed: over 750 emission scenarios were included in the IPCC's literature review of models during its Fourth Assessment Report (Metz *et al.*, 2007). In addition, as discussed elsewhere in this book, these models and the modelling groups responsible for their development have generated insights about a wealth of possible strategies for emission reductions.[1]

A consistent message from these studies has been that reducing emissions will not be cost-free but these costs are likely to be very modest in the context of global economic growth. Most studies cluster around costs of 1–2% of GDP by mid-century — deferring a few months to a year or two of foregone economic growth. The actual costs could vary more widely depending upon participation, innovation, policies, and the broader relationship between mitigation, revenues and investment — with some lines of modelling suggesting the potential for net economic benefits

[1]The IPCC Working Group III (Mitigation) reports contain comprehensive analyses of mitigation options and scenarios, with its Fourth Assessment Report published in 2007 (IPCC, 2007) and the Fifth Assessment Report due to be published in early 2014.

associated with accelerating energy efficiency and investment and reducing dependence on fossil fuels, as detailed elsewhere in this book.

However, if the first decade of international discussion appeared to move relatively quickly from problem identification to policy design, over the next decade policy progress slowed significantly and absolute emissions swelled. The US, then the world's largest emitter, failed to ratify the Kyoto Protocol and their emissions continued to grow (although these have started falling more recently). Russia's economy started to recover from its collapse following the fall of the USSR. China started a burst of economic growth that translated directly into a rapidly changing emission profile, and is now has the highest national annual GHG emissions.

Despite the increase of scientific concern and global debate, a global policy response has not been forthcoming. The Kyoto Protocol was supposed to have been either extended or replaced with another binding international agreement by 2012 but this has not been accomplished. Instead, the Durban Platform for Enhanced Action that emerged from the UNFCCC in 2011 starts negotiations for a future agreement with a deadline for completion of 2015. International agreements continue to insist that they are aiming to avoid a 2°C increase in global average temperature. However, the Durban Platform does not yet contain any strong commitments to reduce emissions, and it is widely expected that negotiations towards a global agreement will be extremely difficult. Analysis suggests there is *at least* a 5–6 Gt gap between even the reductions "pledged" by 2020, and the 2°C goal, which would imply global emissions peaking before that date and being no higher than 44 Gt in 2020. This also assumes that countries will actually meet their pledged reductions (UNEP, 2011).

Scientific evidence compiled by the IPCC is becoming ever more definite that emission reductions are a scientific necessity. Economic modelling, as discussed in this book, has routinely suggested that the economic costs of emission mitigation actions would be well within a range that could be absorbed without devastating consequences. The global rhetoric of the UNFCCC clearly recognises the importance of mitigation and adaptation actions, and yet progress has been limited; no agreement setting legally binding constraints on emissions has been reached since the adoption of the 1997 Kyoto Protocol. Evidence from economic modelling suggests that it is not the economic cost that is preventing sufficient action, it is the politics.

Importantly, some policy progress has been made. Large international goals have been repeatedly missed, but multiple efforts have been undertaken and have resulted in a variety of successes. Efforts to establish carbon pricing have been (or are being) pursued in Europe, North America, Japan, New Zealand, Australia, South Korea and China. Simultaneously, energy efficiency and renewable policies have been developed across diverse regions. These have included policies such as the US fuel efficiency improvements, the EU energy efficiency standards, or India's perform, act, trade system for industrial energy efficiency. Meanwhile, cities (Betsill and Bulkeley, 2007; Krause, 2011) and non-government actors, including industries (Betsill and Hoffmann, 2011; Meckling, 2011), have also initiated emission reduction efforts.

Acknowledging these policy efforts in the context of on-going debate about how best to pursue, design and implement climate policy is important for three reasons. First, existing climate policies provide the fundamental economic and political baseline against which all new policies will be developed. While some current policies may not be considered optimal or ideal, they are the real-life backdrop for further policy initiatives and will necessarily contribute to the way that future policies are evaluated by decision-makers. Failing to take the prior situation into consideration during analysis may overlook important information about the political or economic feasibility of policy options. It cannot be ignored that the current situation is characterised by a policy mix that includes a plethora of regulations, a diversity of market-based mechanisms in a non-global market and multiple policies aimed at technological change and strategic industrial development.

Second, despite on-going efforts at international coordination, the current global climate policy context is composed of multiple mitigation efforts, some of which are formally connected to one another while some continue to stand alone. This diversity of mitigation action is also spread across multiple policy contexts including city- and state-level governments and within the private sectors (Aulisi *et al.*, 2007; Betsill and

Bulkeley, 2006; Meckling, 2011; Rabe, 2007). As some scholars have pointed out, we cannot equate *governance* with *government*, and climate policy may best be understood as a form of governance in which decision-making is shared by many actors (Betsill and Bulkeley, 2006). From this perspective, we are not in a world characterised by a single decision-maker, or even an organised hierarchy of nested decision-makers. Instead, current climate policy decision-making emerges from the interactions of multiple players at multiple scales, many of them outside the formal institutions of government. Any attempt at understanding how climate policy has, or could, evolve must take this dispersion of decision-making into account.

Third, the political component of policy making cannot be ignored — especially in a world in which decision-making is necessarily shared among many players. It is very unlikely that any policy will emerge in the same form in which it exists in a theoretical space. Instead, thought must be invested into how policies will be shaped by the political spaces in which they will be developed and implemented. The diversity of current climate policies can be a boon to those interested in the political or economic evaluation of likely policy outcomes: if the argument that political inaction is the primary reason for the lack of policy movement in the climate arena is correct, the fates of previous policy proposals can provide insight into some of the political dynamics likely to be encountered by current and future climate policy attempts.

This chapter aims to summarise the key lessons about political feasibility that have emerged in the climate policy space. Our argument is that we can gain valuable insights into the ranges of political considerations that have shaped climate policies, and that these insights hold lessons both for those engaged in supporting future efforts and for scholars involved in modelling policy options and outcomes. These lessons can then guide efforts to place the world onto the trajectories of decarbonisation envisioned elsewhere in this book.

We first briefly discuss the concept of political feasibility, and then provide overviews of a cross-section of existing climate policies. Each of the policy sections outlines the development of a different high-profile policy approach, paying particular attention to factors that led to the policy's current manifestation. A comprehensive review of the plethora of climate policies undertaken globally is beyond the scope of the chapter so we have chosen to highlight two major categories of policy response. The first section discusses carbon pricing through examples in both Europe and North America. The second section addresses the use of standard setting through the example of UK and EU energy efficiency regulations and the development of North American fuel emission standards.

8.2. Political Feasibility

It has long been recognised that accurately assessing the political feasibility of policy proposals is an inexact art. Politics do not comply willingly with the demands of precise equations. As May argues, "it is fruitless to develop a predictive model that forecasts the political feasibility of policy proposals. The variability in proposals, settings, interests, groups, and the like are substantial enough to inhibit such modeling efforts" (May, 1986: p. 121). May's argument does not mean that the question of political feasibility should be ignored: just because it cannot be definitively predicted does not mean it is not crucially important. As discussed above, politics has been a central component of the varied abilities of governments to successfully develop climate policies. The core challenges facing those interested in supporting policy developments centre on finding ways of taking policy from theory into the actual policy world. Thinking carefully about what is and is not politically feasible is essential because policies that cannot be supported and pursued in a political space will not be implemented. If core factors that have shaped policy feasibility in real climate policy contexts thus far can be identified, these may assist in the design and implementation of future policy responses.

Predicting political feasibility is difficult, if not impossible, in part because different models of political processes stress particular factors of importance. For instance, some emphasise the role of institutions and the power politics in play within them. From this perspective assessments of political feasibility focus on the formal (and informal) institutions that shape decision-making (see, for instance, Paavola (2007)). Such analyses might focus on the legal or legislative systems that have contributed to the shape that specific policies have taken, or the ways in which multiple institutions have interacted to produce distinct political contexts. In addition, such analyses may include examinations of the political capital policy makers are willing to expend in order to ensure a particular policy makes it through the legislature or other political institutions (Meltsner, 1972). In order to assess political feasibility within this framework, knowledge about specific individuals and political parties at the time in question is needed in addition to an understanding of the formal institutional processes involved.

Other scholars have focused analysis on the interests of stakeholders beyond policy making institutions. These analyses have highlighted the likely distributive implications of policy shifts to identify likely winners and losers, and potential for coalitions (either for or against the policy proposal) among these groups (May, 1986; Meltsner, 1972). For instance, what constraints and resources are relevant for different players within a given policy process? How might different groups perceive opportunities for coalition building, support, or resistance to a particular policy shift? Importantly, in political systems and of particular relevance in collective action contexts, these analyses have focused on the dynamics by which certain stakeholders are able to generate political support (Olson, 1965). From this perspective the question of political feasibility is essentially one of interest politics; is a particular policy likely to engender support or resistance among interest groups with multiple types of influence and power? An assessment of political feasibility from an interest group-oriented frame requires consideration of the potential implications of the policy proposal across different actors or groups of actors.

Yet others have argued that it is core ideas, and attention to these ideas, that may drive policy in some situations. For instance, it has been suggested that there are cycles of public interest or concern about issues that then allow these to become a subject of political action (Downs, 1972). Similarly, ideas or concerns can be seen as nuclei around which a variety of stakeholders, including academics, policy makers and others may coalesce, leading to common ways of understanding the problem and the development of particular policy responses (Gough and Shackley, 2001; Haas, 1989). From this perspective the core issue or idea and how it is perceived by policy makers and others stakeholders is a central component of the likelihood that policies designed to address the issue are pursued and deemed politically feasible.

All of these ways of thinking about political processes are simplified archetypes designed to allow researchers partial theoretical windows into complex decision-making contexts. In reality all of these factors are likely to be involved in any given situation (Cohen *et al.*, 1972). As Kingdon argues, there can be many factors involved in allowing a policy response to emerge from a political process, and quite often the timing of these policy windows is both crucial and fleeting (Kingdon, 1984). Recognising the impossibility of developing a comprehensive and parsimonious theory of political feasibility in light of inevitable complexity, May (1986) proposes that analysts focus on developing specialised knowledge unique to the particular situation at hand about the players and context of policy debates, and then on using this knowledge to think through some of the common factors at play in political processes. In this chapter we do not attempt to derive a generalisable theory of political feasibility. Instead, this chapter examines how a broad range of existing climate policies have fared in order to develop the type of specialised knowledge that might help us think through challenges that may be common for climate change policies. We then extend these thought experiments to consider their potential implications for the kinds of approaches we use, and questions we might ask, when using economic models to assess climate change policy options.

8.3. Carbon Pricing

Establishing a carbon price has become a leading approach for climate change mitigation and has particular support from within the economics community. The rationale is simple, elegant, and well established in economic theory. GHG emissions are currently negative externalities, thus internalising the costs of emissions by imposing a carbon price will incentivise actors to reduce their emissions. Establishing a carbon tax or creating a cap-and-trade mechanism are two strategies for doing this.

As has been discussed in this book and elsewhere, there are myriad possible versions of cap-and-trade and carbon tax systems and there are long-standing debates about the economic costs and benefits of these (Grubb, 2014; Tietenberg, 2006). This chapter explores one key commonality of the two: a jurisdiction must make political commitments to some form of GHG emission limits in order to implement either cap-and-trade or carbon tax systems. The creation of either system is not only a technical mechanism for incentivising emission reductions but is a deeply political process. The following subsections summarise the political aspects of both carbon tax and cap-and-trade systems within Europe and North America.

8.3.1. *North American carbon taxes*

It has become rote within North America that carbon taxes are almost politically impossible. Following the emphatic defeat of President Clinton's energy tax in 1994, very few politicians were interested in pursuing carbon or energy taxes. However, in 2008 the province of British Columbia (BC) in Canada implemented the world's first universal consumer end-use carbon tax. Emboldened by the example, the Liberal Party of Canada featured a Canada-wide carbon tax in their 2008 federal election platform. The BC carbon tax has survived so far, despite some political difficulties, but the Canadian attempt was a spectacular disaster. The Liberal leader, Stephane Dion, lost both the election and his position at the head of the party as a result, and although still considering carbon taxes good policy, has subsequently called any effort to establish them an act of "political suicide" (Dion quoted in Harrison, 2012: p. 398).

8.3.2. *Clinton's energy tax*

In 1993, facing a serious federal deficit, President Clinton proposed a broad-based energy tax. In its initial incarnation it covered almost all fuel sources based on the heat content of the fuel. The primary impetus of the bill was to generate an alternative source of revenue, and to address on-going concerns about US energy independence by promoting energy conservation and the development of alternative technologies. Recognition that an energy tax would have environmental benefits was important but secondary, and the tax was explicitly not a carbon tax, largely due to foreseen problems with the coal industry (see Erlandson (1994) for a fuller discussion).

Despite receiving support from many academics and being seen as a relatively palatable solution to the pressing deficit problem, the energy tax almost immediately ran into problems. Regional and industry-specific lobby groups quickly objected, and used well-established arguments about job and competitiveness losses to "push buttons to ignite public passions" (Lee, 1994: p. 78). The administration acceded to some early lobby groups which resulted in a flood of exceptions tied to regional disputes, particular fuels, and specific industries. Eventually the exceptions largely gutted the measure and eroded support for it even within the Democrat caucus that had initially proposed it (Erlandson, 1994). Simultaneously, the political scene was becoming more polarised as Republicans and Democrats clashed on a number of large issues, including the budget. Political polarisation prevented the tax from achieving any bipartisan support, and made it vulnerable to any Democrats who did not support it (Horvitz, 1993). This vulnerability was eventually taken advantage of by Democrats from energy-exporting regions concerned that it would result in specific, regional costs (Erlandson, 1994). In roughly a year the energy tax went from being heralded as a solution to one of the most pressing political and economic challenges facing America to a politically expensive failure.

A number of lessons have emerged from reflections on this case. First, a classic public policy dilemma is that people respond more acutely to losses than to gains (Kahneman and Tversky, 1979), and the potential winners of any action tend to remain silent while those likely to bear costs mobilise quickly (Olson, 1965). Environmental benefits alone are unlikely to provide regulators with sufficient public support to face down lobby groups willing to use the familiar arguments of job and competitiveness losses, in part because these benefits are diffuse and difficult to identify. Lee (1994) argues that in order to be politically feasible the benefits of any tax must be clear and compelling, and those who are likely to benefit must be politically engaged.

Second, because of the diversity of ways in which energy is involved in people's lives, changes to energy prices easily raise political dispute. In the US and Canada these disputes will almost always have a regional dimension due to the geographical differences in the production and use of energy. However, regulators may have choices about how they engage with lobby demands. Erlandson argues that a key reason the BTU tax became a failure was that the administration was too quick to make deep and specific concessions, resulting in a bill that could not even claim legitimacy based on across the board changes in energy prices (Erlandson, 1994).

Third, popular wisdom has suggested that energy or carbon taxes in the US are simply not possible due to the depth of debate about taxation itself, even without the added economic and political dimensions inherent in energy taxes. However, a range of other fees related to energy have been implemented in North America. As Rabe points out, 18 American states have some form of very modest energy taxation, but frame these as "service charges" or "public benefit fees" (Rabe, 2008). This framing is hardly surprising and may be a crucial factor in their political feasibility. As a recent study indicated, many members of the US public, particularly Republicans, react more negatively to carbon taxes than carbon fees, even when the fees and the taxes are identical in every way except their name (Hardisty *et al.*, 2009). Stealth taxation raises a variety of moral questions about policy leadership, but also draws attention to the fundamental political difficulties of carbon pricing in America.

8.3.3. *BC carbon tax*

The BC Carbon Tax was announced under Premier Gordon Campbell in 2007, and it has been argued that his personal concern about climate change was a significant driver of its success (see Harrison (2012) for a more detailed discussion). Campbell established a Climate Action Secretariat (CAS) responsible for provincial climate policy (including both the Western Climate Initiative (WCI) and the carbon tax). The CAS was located within his office and reported directly to him. Simultaneously he established a high-level multistakeholder climate change advisory committee which he personally chaired.

The BC carbon tax started at US$10 per ton of CO_2, and was set to rise by US$5 per ton until it reached US$30 in 2012. The tax was also designed to be "revenue neutral" as the revenues generated through the tax allowed for individual and corporate income tax cuts. In addition, in the first year of the tax each BC citizen received a highly visible one-time

US$100 cheque as a carbon tax rebate (British Columbia, 2008). While this policy design received approval from many academics, it proved a difficult choice to explain to the public and industry: there is a distinct difference concerning revenue neutrality from the perspective of the state and that from the perspective of the individual. As Harrison points out, while the "transfer of wealth from business to the state is irrelevant from the perspective of economic efficiency.... it is highly relevant from the perspective of the actor whose wealth is being transferred" (Harrison, 2010: p. 512). In addition, the tax exacerbated regional tensions as northern and rural residents actively argued that the tax was "unfair" due to their limited transportation choices and additional need for heating fuel. Despite studies that demonstrated that these residents actually did not use substantially more fuel than southern or urban dwellers, they persisted until the government agreed to special rural and northern rebates. As others have argued, these disputes had less to do with the actual amount of tax, and more about long-standing regional disputes between north and south, and perceptions in the north that their needs were routinely overlooked in provincial politics (Peet and Harrison, 2012).

An additional political dispute emerged during the election year immediately following the tax's introduction when the opposition party ran an aggressive "Ax the Tax" campaign despite having initially supported the idea of carbon taxes. This campaign argued that the tax would hit "ordinary citizens" when it was "big polluters" who were really the problem, and suggested that the province should pursue a cap-and-trade system instead of a tax, which they implied would not impose any costs on citizens.

Despite these political challenges the BC Liberals maintained power after the 2009 election, and continued the tax. The Liberals lost substantial support because of the tax but had a more popular economic platform than the slightly left-leaning opposition, and in the years directly after the economic crisis these economic arguments were centrally important (Harrison, 2012). Compared to the level of public debate about the carbon tax, within BC the WCI cap-and-trade system went almost publicly unnoticed. This is hardly surprising — the WCI was a technical policy rendered largely invisible to individual citizens as the costs would be hidden. For industry, the co-development of the WCI and the tax meant that they did not necessarily have to worry about the carbon tax after the first year. Entities covered within the WCI would be exempt from the carbon tax when the cap-and-trade system started. The dual process of development may have helped erode industry resistance to the tax, as large industry expected to be able to use the WCI to meet obligations instead and the expected WCI permit prices were lower than the tax (British Columbia, 2008; Western Climate Initiative, 2010).

8.3.4. *Canadian federal carbon tax*

It could be argued that broader ideologies — such as support for social services and belief in government as the appropriate body to provide these — are linked to public acceptance of carbon taxes. It is no accident that carbon taxes have been adopted mainly in countries where citizens largely trust their governments and have a culture of relatively high taxation to support strong social services in health and education, such as Scandinavia. However, even within these, a study of the carbon taxes adopted in four Scandinavian countries in the early 1990s found that only one comes close to the theoretical ideal of fixed-rate taxation across different sectors, and all vary according to specific national political characteristics (Anderson, 2004).

Within Canada, widespread support for public health care and other social services was used as a framework for the carbon tax proposal. Federally, Dion's proposed carbon tax looked similar to the BC version. The "Green Shift" carbon tax started at 10 per ton of CO_2, and would be raised incrementally until it hit US$40 per ton four years later. Although most fuels were included, gasoline was excluded due to arguments that it was already taxed at the national level. The tax was combined with commercial and individual income cuts, a child tax credit (argued to be essential because families with children were likely to have higher energy costs), higher employment tax credits aimed at low-income working families, and the development of a federal fund earmarked for research and development into green technologies (Liberal Party of Canada, 2008). As with BC, at the aggregate level the tax would be revenue neutral.

However to run an election on a tax increase as the 2008 economic crisis unfolded proved politically untenable, despite the potential cuts to other taxes. Despite Dion's insistence that the public would appreciate good policy for its own sake, and framing the carbon tax as a politically courageous move that placed faith in the Canadian public (Liberal Party of Canada, 2008), the public voted otherwise. The opposing parties immediately attacked the tax, arguing that the only effect of the tax would be in "shifting of money out of people's wallets into Mr. Dion's coffers" (Kenny quoted in Fitzpatrick, 2008). In a period in which economic stability was considered crucial, arguments about the revenue neutrality of the tax fell flat. Framing the "Green Shift" as the "Green Shaft" (Morris, 2008) and insisting that it would result in massive increases in the cost of living for middle- and low-income Canadians, the Conservative Party of Canada won the election without even having to present a coherent climate change policy of their own. Dion lost the election and his position as head of the party, despite proposing a carbon tax based on core principles suggested by many academics.

8.3.5. *North American cap-and-trade system*

Some have pointed to the centrality of the US in getting "flexible mechanisms" into the Kyoto Protocol (Convery, 2009; Ellerman *et al.*, 2010). The idea behind these mechanisms, most clearly operationalised through the development of the clean development mechanism (CDM) and joint implementation, was that states could claim credit for emission reductions achieved outside their borders through trading obligations for investments in countries in which emission reductions could be achieved at lower prices. The flexible mechanisms essentially entail the creation of a carbon market in which a price is placed on GHG emissions through the emission cap set for covered states, and entities are allowed to buy and sell credits as they seek to minimise their costs.

During the Kyoto negotiations the US was reflecting on early results from one of the largest experiments ever conducted on the use of market mechanisms to address an environmental "bad". Starting in 1990, the US had designed and implemented a trading system for acid rain pollutants, in particular SOx and NOx emissions. At this time, this system appeared to have achieved significant reductions in pollutants at costs that were lower than might otherwise have been achieved through straight regulation (Ellerman 2000). These initial successes suggested that there might be potential for a similar system to provide benefits in the climate policy space internationally, and the US insisted that such mechanisms be included in an international climate agreement.

Of course, these developments are often pointed to ironically because the US never ratified Kyoto, and has never participated in the flexible mechanisms. Canada did ratify Kyoto, started but did not continue to use flexible mechanisms, and made international headlines in 2011 when it was the first state to officially withdraw from Kyoto to avoid liability for millions of dollars of compliance penalties.

Not only have both Canada and the US chosen not to participate in international emissions trading schemes, but they also have been unable to establish national systems. The US tried to implement a federal cap-and-trade system immediately following the election of President Obama. The Waxman–Markey Bill proposed a national cap-and-trade system designed to encompass the bulk of American GHG emissions. At the start of the programme about 80% of the permits were freely allocated, although not solely to industry. The Waxman–Markey Bill was designed with the potentially regressive impacts of cap-and-trade policy in mind and included elements designed to augment this. State governments, electricity providers, and gas distributors were also allocated permits but were required to use the value or these permits to reduce costs to households or ratepayers, or to invest in clean technologies (Pew Center, 2009). However, these efforts were insufficient in the face of deeply divided politics right at the time of the economic crisis. Despite getting through the Democrat-controlled House of Representatives, party-line debates in the face of economic concerns prevented it from receiving sufficient votes in the Senate and the system was abandoned.

Meanwhile, following the Conservative Party's 2006 win, Canada has consistently declared that it will not pursue cap-and-trade policies without an American agreement and has subsequently shown no interest in pursuing federal carbon pricing. For the brief months in 2009 when it looked like the US would be able to implement a national policy, Canada also indicated that it might consider some form of national cap-and-trade system. However, this policy direction was entirely dependent on the US efforts and stopped immediately following the defeat in the US Senate.

Instead of national systems, within North America the bulk of policy innovation on emissions trading systems has happened at the sub-national level (see, for instance, Rabe (2007, 2008)). Two important sub-national emissions trading systems have developed in North America. The Regional Greenhouse Gas Initiative (RGGI) cap-and-trade scheme among power plants in the northeast US was started in 2005 (RGGI, 2005) and the WCI was started in 2007.

8.3.6. *Regional Greenhouse Gas Initiative*

The RGGI is a mandatory emissions trading system covering power plants in a group of northeastern US states. When the first Memorandum of Understanding (MOU) was signed, RGGI included Connecticut, Delaware, Maine, New Hampshire, New Jersey, New York, and Vermont (RGGI 2005). By 2007 it also included Maryland and Rhode Island. New Jersey withdrew in 2011 (New Jersey, 2011), although it has continued to face pressure to rejoin.

The RGGI consists of a regional emission cap on GHG emissions from power plants in the participating jurisdictions. This cap is modest and aims for emission stabilisation between 2009 and 2014 and a 10% decrease from this level by 2018. Within the system offsets can be used for 3.3% of the permits required for compliance although this limit can be increased to 5 and 10% if the 12-month rolling average market price of permits goes above US$7 or US$10, respectively. Importantly, RGGI auctions almost all of the allowances (roughly 90%). Even though the price of allowances has remained extremely low (below US$2 per ton of CO_2), these auctions have generated significant revenue for the participating jurisdictions. In 2011 it was estimated that the regional auctioning revenue generated by 2010 was almost US$800 million (RGGI, 2011).

The RGGI appeared during the same federal climate policy vacuum that gave rise to the WCI. However, unlike the WCI, RGGI grew out of an already established network of jurisdictions in the US northeast and the neighbouring Canadian provinces. As Selin and Vandeveer point out, the leaders of these jurisdictions have met annually since the 1970s and the environmental bureaucracies within them have had to engage with a range of previous coordination challenges — including work on acid rain, mercury pollution and ozone regulations (Selin and Vandeveer, 2005). One of the interesting features of RGGI is the use of a "Model Rule" (RGGI, 2008), which may have been facilitated by these long-term relationships. Although each state must pass its own legislation, considerable energy was invested in crafting a model rule which states — both those already involved and those who might want to join — could use as a template for their legislation. The model rule could facilitate policy coordination and, considering the substantial administrative effort involved in crafting such legislation, could lower the transaction barriers to entry for other jurisdictions.

In addition to helping with policy coordination, the pre-existing relationships among RGGI members, along with their close economic ties, may have helped address concerns from lobby groups within individual states. Also, few of these states have strong fossil industries although they do have large manufacturing sectors. It has been argued that the basic political economy of the northeastern states was a factor in its development (Byrne *et al.*, 2007). Most of these states have limited indigenous energy resources, making energy security and pricing a long-term issue of interest for them and reducing the power of fossil fuel lobbies. However, internal lobbying from the fossil fuel industry did play a role in shaping the initial programme, as it was the likely cause of the delayed entry of Massachusetts and Rhode Island (Wall, 2007). The importance of the coal industry to Maryland was also a crucial factor in its decision-making process, although modelling suggested that RGGI would have little negative economic effect and could have some small net benefits (Ruth *et al.*, 2008).

The degree of RGGI's success has been the topic of on-going debate. On one hand, the programme has resulted in limited GHG reductions due to its modest cap and the broader economic downturn which lowered business-as-usual (BAU) projections for the region (Stavins, 2012). Simultaneously, these factors have combined to keep RGGI permit prices extremely low; during 2010 prices remained well below US$2 per ton (Potomac Economics, 2012). It is unlikely that this price alone would be able to trigger innovation and reinvestment in low-carbon technologies.

On the other hand, although low prices may not have substantially aided emission reductions, the use of revenues to fund energy efficiency investments may have also contributed to emission reductions in some states (Schneider and Elliott, 2012). Because such a large proportion of permits are auctioned, almost US$800 million has been generated in revenue. This revenue has been used for a variety of goals including energy efficiency (52%), support of renewable technologies (11%) and relief of energy costs (14%). In addition, a substantial amount of this revenue (17%) has been used to reduce general state deficits (RGGI, 2011).

These debates about success have been echoed by a series of political challenges within RGGI states. In 2011 New Jersey withdrew, although not without internal debate as the governor vetoed a Senate decision to rejoin RGGI. RGGI has provided states with alternate sources of revenue that they have been able to use for a variety of actions including deficit reductions, making it less attractive to pull out. However, the diversity of uses for which revenues have been used has also raised political challenges. For example, in 2012 New Hampshire almost withdrew, but opted to remain in RGGI and adopted limitations on revenue use. Republicans had argued that RGGI was essentially a tax, an argument that caused the legislature to threaten to withdraw from the programme unless revenues were used for rebates only instead of energy efficiency grants and similar programmes (Spolar, 2012). Similarly, New York landed in a Supreme Court case when the state was sued by claimants arguing that RGGI was essentially an illegal, hidden tax. The court ruled against the claimants, not on grounds that it was not an illegal tax, but on grounds that the suit had been filed too long after the legislation had been passed and that the claimants had not suffered distinct, identifiable harms from the state's participation (Navarro, 2012).

Both of these cases highlight the political challenges of implementing such systems and point to the political difficulties faced in particular by carbon taxes, or any systems that look like carbon taxes in the American context. These political challenges — and the role that the generation and use of revenues have played in them — also highlight the potential risks and benefits of tying climate policy to other policy issues. For some, RGGI's ability to generate revenue for energy efficiency is a chief success of the programme. For others RGGI's revenue generation capacity is merely an inappropriate and non-transparent tax grab that does nothing *per se* to reduce GHG emissions, the official goal of the policy. Systems that are not directly tied to single policy goals, or that could have overlapping foci, can present political risks, especially in polarised contexts in which multiple concepts of good governance and appropriate regulation coexist.

8.3.7. *Western Climate Initiative*

The WCI is the most ambitious North American cap-and-trade system under development and is composed of US states and Canadian provinces. By 2008 it looked like it would be the second largest system in the world, and many felt it would be a natural testing ground or initial platform for federal US and Canadian systems. At its peak, the WCI had 11 partners committed to cap-and-trade policies: Arizona, California, Montana, Utah, New Mexico, Washington, Oregon, British Columbia, Manitoba, Ontario and Quebec. Another 16 jurisdictions had observer status. In this system all installations emitting over 25,000 ton of CO_2 per year would be included in the scheme, except for emissions from forestry, waste and agriculture. Compared to RGGI or the EU ETS (the EU Emissions Trading Scheme), the coverage of the WCI is extensive and was designed so that when the second phase started in 2015, almost 90% of the GHG emissions in the partner jurisdictions would be covered. Of all the existing cap-and-trade systems, the design of the WCI is the closest to the proposed but defeated US federal cap-and-trade system.

However, by 2011 most states had decided not to participate, the provinces were on the fence, and only California and the province of Quebec were on track to start trading in 2012. This section provides a brief discussion of some of the key factors that both supported its initial development, and that contributed to the multiple jurisdictions' decisions not to continue involvement (see Klinsky (2012) for further discussion).

The WCI was started when two initiatives, one between California, Washington and Oregon, and the other between Arizona and New Mexico, merged. Shortly after these early partners formed the WCI the four Canadian provinces, Utah and Montana joined in. For the early partners of the WCI, the scheme had several key appeals. First, the WCI structure was extremely decentralised. Jurisdictions remained in charge of setting their own caps and designing their own legislation. Once individual trading systems had been set up they would then be linked. This structure reduced the extent to which jurisdictions had to surrender authority within their own boundaries.

Second, and related, the WCI offered a way for jurisdictions to attempt to influence national policies in ways that would ultimate benefit them. As Governor Schwarzenegger stated, the WCI "sets the stage for a regional cap and trade program, which will provide a powerful framework for developing a national cap and trade program" (Schwarzenegger quoated in Gregoire *et al.*, 2007).

Third, coordination provided regulators arguments about the benefits of a regional market and protection from domestic lobby groups' concerns about competitiveness. Jurisdictions repeatedly signalled that "collective effort" was crucial in enabling action and that "a broad geographic scope [would] reduce overall compliance costs and help mitigate leakage risks" (WCI, 2010).

Finally, the WCI presented small initial costs and potentially significant economic opportunities. None of the earliest-adopting jurisdictions had large fossil fuel industries but several had renewable energy potential. British Columbia, Quebec, Manitoba, Washington and Oregon all had hydroelectric resources, and in Arizona, New Mexico and California the potential for solar power was substantial. For most of these jurisdictions transport remained the largest source of emissions, but transport was not to be included until 2015 which allowed these potential stakeholder costs to be delayed.

Several strategies were used by the WCI to cement the partnership and to create incentives to bring less climate policy inclined neighbours into the scheme. First, an enormous amount of effort was invested in coordinating guidelines for emission inventories, offset protocols, and rules about auction and market oversight. In addition, learning from the EU ETS, it was recognised that a consistent and secure trading platform capable of tracking permits across jurisdictions was important in order to provide widespread confidence in the market.

Second, several neighbouring states were not traditionally involved in climate policy but were seeking economic opportunities, and were vulnerable to energy policy changes. For example, Montana exports roughly 40% of its electricity to its neighbours (Montana Climate Change

Advisory Committee, 2007), much of it coal powered, and Utah has been interested in expanding its fossil industry. The First Jurisdictional Deliverer Rule meant that any electricity being imported into the WCI had to be covered by permits at the border. In both Utah and Montana initial involvement in the WCI was heavily framed as a prudent financial move and economic opportunity.

The WCI dramatically changed shape in the years following the 2008 economic crisis. In 2009 Washington and Oregon both decided against cap-and-trade systems due to concerns about employment and policy costs. In addition, the crisis had eroded support for market based mechanisms in general; as the speaker of the Washington legislature explained, "with what's happened in the last year, one would wonder about the wisdom of the markets" (Chopp quoted in Cornwall, 2009). This decision was also fuelled by concerns about "political favouritism and accounting tricks" (Washington, 2011b) in the EU ETS.

Simultaneously, the US was undergoing political polarisation as the Republican Party moved to the right with the growth of the Tea Party movement (Williamson *et al.*, 2011). Cap-and-trade considerations had been included in Senator McCain's bid for presidential office in 2008, but by early 2009 the Republican Party had changed position and were attempting to block President Obama's proposed federal cap-and-trade scheme. The WCI became consumed in much broader political debates which also prevented the development of a federal cap-and-trade scheme in the US.

During this period of time debates about climate science were reignited in the US. In 2010 only 53% of American felt there was solid evidence of global warming, the lowest level since 1997 (Newport, 2010). Broad mistrust of climate science was reflected in several states. For instance, the Utah legislature passed a resolution which withdrew Utah from cap-and-trade and, citing the "climategate" scandal, urged against any federal regulation of GHGs until "climate data and global warming science are substantiated" (Utah, 2010).

To date only California and Quebec have proceeded with cap-and-trade regulations, although the other provinces have remained nominally involved. Challenges have emerged due to the disparity in size between the partners, and work creating a unified market platform has persisted, partly in the hope that other jurisdictions will join. California's progress was slowed by a court case brought forward by environmental justice groups concerned that the trading component of the policy would allow Californian industry to avoid local abatement through the purchase of geographically distant offsets, a situation which would not benefit local air pollution. The case was settled but draws attention to the importance of non-industry stakeholders in these systems.

8.3.8. *European carbon pricing*

It is a full quarter of a century since carbon pricing came onto the European political agenda. It did so from the combination of two factors: a decisive shift in favour of market liberalisation as an economic philosophy (embodied in the Single European Act in 1986), and a rising tide of concern about the international and indeed global environment. The latter originally focused on acid rain and ozone depletion, but climate change became the dominant issue in 1988, with the Toronto climate conference of that year setting a broad goal to reduce industrialised-country CO_2 emissions by 20% by 2005. By 1990 most of the major Member States had adopted CO_2 emission targets, and the debate turned to how to implement them.

The European Community, as it then was (now the EU), first committed to stabilisation of GHG emissions in 1990, just before the Second World Climate Conference. A series of proposals followed in 1992, including one for a directive on a community-wide combined carbon and energy tax. This was proposed to start in 1993 at a relatively low level, but ramp up towards 2000. The proposal met considerable opposition from industry and also from individual Member States. To deal with this, substantial exemptions for energy-intensive industries were proposed, along with a move to tax electricity as per output rather than as per input fuels.[2] In the run-up to the Earth Summit of 1992, the tax was also made conditional on the introduction of similar measures in the other countries of the OECD (Organisation for Economic Co-operation and Development).

[2]The French government wanted a pure carbon tax in power generation; the German government wanted it on an energy basis, to avoid hitting its coal industry too hard.

With such commitments not forthcoming, the EU carbon tax proposal struggled on for some years. A further attempt was made in 1994 with even greater concessions, including maintaining a low level of taxation for three years until a review could be conducted, greater exemptions for both energy-intensive industries and also exemptions for countries with relatively low levels of emissions. These further concessions weakened the emission reduction impact of the tax even further, but were still not sufficient to gain the level of support required, and at the end of 1994 attempts at an EU-wide carbon tax were abandoned in favour of attempts to harmonise existing energy taxes across the Community.[3]

These efforts finally culminated in 2003 with a directive on energy taxation. This aimed to remove distortions due to differences in rates between fuels in different Member States, and to encourage greater energy efficiency. After more than a decade of struggle, this — together with a clutch of Scandinavian carbon taxes enacted around 1990 — was the result of Europe's first efforts to price carbon.

8.3.9. *The EU Emissions Trading Scheme*

The mantra in Europe had been that the EU was generally better disposed towards taxation, whereas emissions trading was seen as a US invention of debatable effectiveness or morality, since it allowed polluters to either get free allowances rather than pay, or to "trade pollution".

It was thus a surprise to many when a rapid political debate in the late 1990s led to a Green Paper by the European Commission on emissions

---

[3] Despite the political failures to enact an EU-wide carbon tax there are individual countries which have enacted such instruments. Beginning with Finland in 1990, which introduced a carbon tax levied on all energy products except transport fuels, the four Scandinavian countries (along with the Netherlands) have introduced carbon taxation. Even here, where there are significant environmental constituencies, the taxes adopted have run into significant difficulties. In all four countries the taxes are beset by exemptions and differentiated rates that mean that they are very different on paper from the uniform economy-wide tax that theory dictates. The levels of carbon tax vary between and within the countries, with the level of taxes varying by up to a factor of five in both cases. Energy-intensive industries and electricity producers have managed to negotiate significant exemptions in most of these tax regimes, playing on competitiveness concerns.

trading in 2000, which moved the tone of the debate from "should there be" to "how could there be" such a trading system.

The first discussions of the possibility of an EU-level trading system had begun in the wake of the Kyoto agreement of December 1997, but it was the realisation in the post-Kyoto implementation debates that the UNFCCC would be unable to create strong enforcement mechanisms that pushed the EU more strongly to consider an EU-level arrangement, which could constitute the EU's primary instrument for implementing its Kyoto commitments and fall under the purview of the European Court of Justice.

A major advantage that became apparent with the move to an emissions trading scheme over a tax came from the political structure of the EU decision-making process. To implement an EU-wide tax requires unanimity between all Member States, effectively giving each a veto. Given the political difficulties the tax encountered, a unanimous decision was extremely unlikely. An emissions trading scheme, on the other hand, fell within the realm of environmental policy rather than taxation and as such only required "qualified majority voting", making it impossible for a single country (or even a small group) to block it.

The cautious support of industry bodies such as the Union of Industrial and Employers Confederations of Europe, who had previously opposed carbon tax proposals, was vital in allowing discussions to make progress.[4] But the decisive event was the massive political boost given when President Bush rejected the Kyoto Protocol in March 2001. Amidst a flurry of unilateralist US actions, the rest of the world looked to Europe to hold the Kyoto agreement together and thus uphold basic principles of multilateral cooperation, whilst the Bush administration suggested Europe had no idea how to meet its targets.

This raised the issue to the level of heads of state across Europe, and the EU ETS became both the symbol and substance of Europe's response to these political challenges. A draft directive was quickly assembled and submitted to the Council of Ministers and the European Parliament. After a year or so of discussions between the Parliament and key stakeholders, both industrial and political, the Emissions Trading Directive was

[4] UNICE — now renamed Business Europe, The Confederation of European Business.

formally issued in October 2003, with the system to begin with a three-year phase from 2005 to 2007.

The EU itself had been 50 years in creation, and despite being founded mainly on energy-related treaties, had been struggling for most of that time to establish a common energy policy.[5] An integrated system for pricing carbon had been created almost from scratch in just three. The most powerful opposing bloc — the German chemical industry — was exempted, whilst most other actual or potential opposition was simply swept aside — or never even had time to realise what was happening.

Thus, the world's principal system for pricing carbon, designed amongst 15 countries but expanding to include 12 new Member States before it even started operating, was born in a set of political circumstances that are unlikely ever to be repeated. It also built upon the 50 years of institutional and legal development of the EU itself. There are many lessons — but most of those about how it came to be are unlikely to be directly transferrable to efforts elsewhere.

8.3.10. *Key design features of the EU ETS*

In many ways the EU ETS is remarkably close to the theoretical ideal of a cap-and-trade system, setting legally binding emission allowances for all the regulated entities, which can then trade them across the entire system. The trading price emerges from this market-based activity, in theory according to the differential cost of compliance. The allowances are freely tradable not only with installations, but also with financial intermediaries and indeed individuals able to register to buy and trade them.

The focus of the EU ETS is to control CO_2 emissions from power generation and industry, riding to some extent on the pre-existing legal and monitoring infrastructure of industrial emission control. It covers approximately 11,500 energy intensive installations across the 30

[5] The European Coal and Steel Community, and Euroatom. Yet in the Single European Act, energy remained explicitly a preserve of Member State legal competence. The EU had also struggled for well over a decade to introduce effective directives on liberalising its energy markets, a process that did not even approach completion until the Energy Third Package directive of 2010.

countries, which represent just over 40% of total EU emissions. The majority of emissions come from combustion installations (mostly power stations), with process emissions from cement, iron and steel, and refineries also important elements.

Emissions from domestic consumers, transport and agriculture are not covered under the EU ETS.[6] This was a source of some criticism, particularly as proposals for more ambitious schemes developed in the US, Australia and New Zealand. The narrow focus of the EU ETS was however one of the keys to its implementation. Excluding potentially difficult sectors such as transport and agriculture hugely simplified the system.

This contrasts with the vast sweep of the US Waxman–Markey Bill, or the agricultural complexities that helped to bedevil the Australian Carbon Pollution Reduction System.[7] The US proposal also tried to cover emissions "upstream" (e.g. to include the oil flowing through refineries) to give it greater coverage. However moves upstream helped to raise fears that the cap-and-trade systems were essentially taxes in disguise. Upstream systems mandate that fossil fuel producers are required to surrender allowances for the carbon content in the fuels that they sell and are in contrast to *direct emission* systems such as the EU ETS, which requires

[6] Part of the rationale was that transport, especially road transport, is already heavily taxed in many EU countries and the implementation of a carbon price would be marginal compared to the current tax levels. The administrative dimension of emission allowances for individual vehicles were also prohibitive, and there are fewer substitution possibilities to justify the effort. Policies around the domestic sector were considered to be much closer to the responsibilities of Member States, and a separate Directive was negotiated establishing emission targets for the non-ETS sectors. This all explains why the scope of the EU ETS is focused on industry and electricity, and so far has only been expanded to aviation without current ambitions to include other transportation and buildings activities.

[7] The final version of the Waxman–Markey Bill that passed the House of Representatives ran to almost 1,500 pages to accommodate the various sectors and provide for compensation to several constituencies. It also attempted to cover emissions from transport by bringing oil producers into the system. The Australian Carbon Pollution Reduction System (first discussed in 2007 but delayed in 2010), had proposed to bring in agricultural emissions from 2015. In addition to huge complexity, the wide scopes of these systems helped to create larger constituencies lobbying against them.

installations to surrender allowances for the carbon dioxide emissions that they actually emit.[8]

The most contentious area of EU ETS design was, not surprisingly, in the allocation of emission allowances. Since these were clearly likely to be very valuable assets, Member States made it plain at the outset that the allocation of allowances should be the prerogative of Member States, a principle that was embodied through the process of National Allocation Plans. In practice, almost all Member States started out with generous allowances to most installations, many with complex rules for allocating free allowances to new entrants, and withdrawing allowances from closing installations. The initial economic theory fed to the Commission had been that allocation did not matter to the system — which would deliver the cap set efficiently irrespective of how allowances were distributed. This is probably the area in which both theory and practice was to prove most problematic.

In addition to the trade of emission allowances, installations can purchase credits from outside the system to set against their emissions. The ETS Linking Directive was put forward in July 2003 and came into effect in October 2004, to allow use of the Kyoto offset mechanisms and synchronised to the Kyoto system.

All this required a huge investment of intellectual and political effort over a relatively limited time span, during which the EU was also handling expansion to include the new Member States, and grappling with the transatlantic politics of climate change (and much else). Probably the smartest feature of the EU ETS was that it was designed to evolve, through sequential phases of development which would enable learning. Indeed, it turned out that the system still had a lot to learn.

8.3.10.1. *Phase I: 2005–2007*

The first phase of the EU ETS from 2005 to 2007 was always perceived as a learning step to build experience for Phase II (2008–2012). Along with the establishment of a carbon price, it served to establish principles

[8] This also contrasts with hypothetical downstream systems in which final consumers who are required to surrender allowances for the carbon content of the goods they consume.

and capacity within the EU Commission, industry and Member State governments, as well as revealing many issues in data and allocation. As indicated, the cap and the distribution of allowances were decided by Member States in their National Allocation Plans (NAPs). Almost all of these freely allocated to firms with most allocations based upon historical emission levels.

History may judge 2006 as the defining year for the EU ETS. It started with prices for Phase I (2005–2007) carbon emission allowances trading at levels higher than anyone predicted, and governments confidently issuing draft NAPs for how they intended to allocate allowances for Phase II, the Kyoto period of 2008–2012. The year ended with Phase I prices sinking close to zero, and several countries threatening to take legal action to overturn the European Commission's rejection of almost all the submitted NAPs as inadequate. It was certainly a year of vast learning — as befits the middle of the first learning period of any major new system.

The evolution of carbon prices since the inception of the ETS is shown in Figure 8.1. Concerns from some commentators about overall shortage in Phase I allocations proved groundless, when in May 2006 the release of data on verified emissions for 2005 showed a substantial surplus. The price halved overnight, and as the situation clarified over subsequent months, it sank further. The final tally showed that emissions in 2005 were about 100 Mt (5%) below the allocated amount, and shortly after the new year Phase I allowances became essentially worthless as it became clear that Phase I was in overall surplus; the rules for Phase I had prohibited banking forward into Phase II to prevent any major problems spilling over into the critical Phase II.

8.3.10.2. *Phase II: The allocation battle*

It was against this roller-coaster backdrop that countries sought to define emission allowances for Phase II, the Kyoto period of 2008–2012. There was a great deal at stake. The EU ETS in Phase II was central to meeting Kyoto Protocol targets, but Phase I had shown the huge potential financial value of emission allowances — against a background of prices exceeding €20 per tonne of CO_2. It was plain that governments were allocating

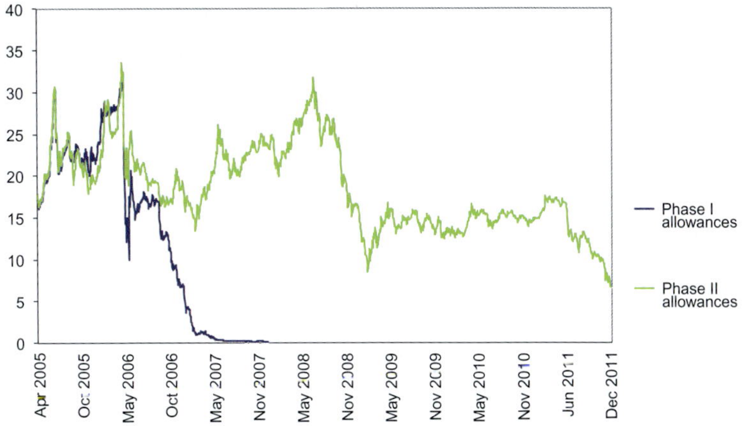

Figure 8.1. EU ETS prices 2005–2011 (per tonne CO_2).

assets worth probably more than €200 bn in total. Not surprisingly, they were subject to huge lobbying pressures.

The European Commission was empowered to reject NAPs if they did not meet certain criteria laid out in the directive, relating to the avoidance of surplus allocations and consistency with Kyoto targets.[9] The NAPs initially proposed for Phase II mostly offered modest cutbacks relative to projections of sharply rising emissions — but in aggregate, would have resulted in an increase of around 5% relative to the verified levels of 2005. This was not only inconsistent with the EU's aggregate target, it would also have left a precariously thin margin below BAU emission projections — the Commission feared this could render the EU ETS almost impotent for the whole of Phase II, requiring hardly any real abatement. On 29 November 2006 the Commission rejected all but one of the allocation plans submitted so far.[10]

[9] The data on verified 2005 emissions were published only six weeks before the official deadline for submitting proposed Phase II NAPs to the European Commission — clearly insufficient to consider a wholesale revision of allocation approaches. The decline halted in the first half of this decade as these factors petered out, and rising gas prices pushed power generators back towards coal.

[10] A total of ten, after the French government withdrew its plan a few days before; the Commission rejected all but the UK's as inadequate.

In fact the Commission went further than this. It clarified its interpretation of the directive in terms of specific total allocations that would be deemed acceptable, linking this to EU projections and Kyoto targets.[11] By announcing decisions on such a big group of countries simultaneously, the Commission raised the stakes enormously. Any country that challenged its ruling would essentially be disputing the underlying interpretation of the Directive, as applied consistently across all countries; and would thereby open the floodgates for all to appeal. This would have locked up the EU ETS in legal disputes from which it would probably never have recovered — certainly not in time to be of much use to investors wanting to know the rules for Phase II.

The day after the Commission ruling, the German economics minister issued a statement saying that Germany would take the European Commission to court. However, it was spring of 2007 and Chancellor Merkel was due to host the G8 summit that year, and had placed climate change high on the agenda. To chair that Summit whilst suing the Commission to emit more was hardly an attractive proposition. Germany backed down, and so did the other western European Member States.

As threats of legal challenges slowly dissipated, and as the deadline for appeals on the first round of decisions passed, the Commission turned to the other allocation plans. Those for eastern Europe were also mostly

[11] The Commission criteria linked allowed allocations to two main factors. The first was an explicit numerical formula that total allocations could not exceed 2005 levels multiplied by projected economic growth, corrected for trends in energy intensity (energy per unit of economic output). Moreover, the economic growth projections and energy intensity corrections were taken from international (EU) sources, not those that Member States themselves presented. The second was a correction to ensure that allocations were consistent with Kyoto targets, after taking account of other aspects of Member State implementation plans including provisions for purchase of international Kyoto credits. The total for NAPs was capped so it exceeded neither the level of BAU emissions nor the level of emissions that would be compatible with a Member State's achievement of its Kyoto target as determined under the Burden Sharing agreement. This was determined by a straight line from 1990 emission levels to the Kyoto target. Of the 15 Member States at the time, ten were determined to need to cut emissions below BAU levels and those were capped by the latter of these two conditions, with the remaining five allowed to produce NAPs up to BAU levels.

rejected, this time not on grounds of Kyoto compatibility, but on grounds of implausibly inflated baselines. Legal dispute on this dragged out for many years, but the Commission had already won the decisive battle. In aggregate, their decision cut total allocations in Europe by 10% as compared with the submitted and draft plans, turning a proposed aggregate *increase* of 5% from 2005 levels into confirmed allocations 5% *below* 2005 levels. The final allocations totalled almost exactly ten billion tonnes of CO_2 over the period — two billion tonnes annually.

8.3.10.3. *Into Phase III*

There were numerous crucial outcomes from the Phase II battles and learning experiences. The warnings of economists that power generators would accumulate windfall profits under the scheme proved unambiguously correct, establishing a political context which enabled its next phase to finally move (in that sector) to the economic ideal of auctioning allowances in full. Politically even more fundamental was that the allocation battle marked a seismic political shift; Member States were exhausted by the allocation process and the Commission had been forced to step in and impose semi-harmonised allocation rules. The Commission boldly followed through its position to the logical conclusion, and proposed that for Phase III all allocation rules should be set centrally, to be applied across all Member States. The Member States, who had so forcefully protected their prerogative over allocation only five years earlier, acquiesced with barely a murmur. The sheer impossibility of doing facility-by-facility negotiation helped also to establish the principle that the industrialised sectors received allocations "benchmarked" to best practice, further removing distortions and tightening the incentives. Thus Phase III finally brought a logical, coherent design that seemed to respect the best economic principles.

The massive irony is, of course, that all this turned out to be wholly inadequate. Moreover, with energy prices escalating globally and the financial crisis erupting in 2008, carbon prices that had been projected to rise steadily from €20–30 per tonne of CO_2 actually fell to €15 per ton of CO_2 after the Copenhagen conference; as the subsequent eurozone crisis unfolded, emissions capped by the EU ETS fell further and dramatically.

By the end of Phase II in 2012, the system over which so much political blood had been spilt to strengthen it, had accumulated more than 2 billion tonnes of surplus allowances.

Moreover, the overall allocation for Phase III had been set by an agreement to achieve at least 20% reductions by 2020, but 30% in the event of an effective global deal. That deal never materialised, leaving the Phase III allocations substantially weaker than the environmental community had hoped for. Two billion tonnes of allowances from Phase II banked forward are sufficient to cover the entirety of expected shortfall right through to the end of Phase III in 2020. In late 2011, prices sank below €10 per tonne of CO_2 and have stayed well below that level ever since. The only thing that prevents them sinking further is a growing clamour for intervention to tighten the system yet again.

As a way of establishing a carbon price, emissions trading in Europe had indeed proved easier to implement than its ill-fated carbon tax. Getting it right, and adequate, has not.

8.4. Regulations and Standard Setting

A third archetypal policy strategy for GHG reductions is the use of regulations to set standards. As with the other strategies discussed in this chapter, and in the rest of the book, there can be myriad specific designs of regulatory standards. This section looks at how two particularly important real-world examples of GHG emission standards have been development and implemented. The example here explores the development of fuel efficiency standards in California and the US.

When the Obama administration went to the Copenhagen COP in 2009 it did so with a stated policy and reserve one. A federal cap-and-trade system was the stated policy and has not materialised. The bird in the hand was new national fuel emission standards which by default have become central components of national US and Canadian climate policies.

Transport is currently responsible for roughly 30% of the US's GHG emissions, and between 1990 and 2006 accounted for almost half of the increase in total GHG emissions in the US (Department of Transportation, 2010). Following the oil crisis and amidst concerns about energy security,

President Carter had pushed for fuel efficiency standards to be raised from 18 miles per gallon for light passenger vehicles to 35 miles a gallon in 1979. However, this policy direction was immediately halted by the election of President Regan who took a strong stance against any regulation. For almost three decades US fleet-averaged fuel efficiency standards stagnated around 27 miles per gallon (mpg) for cars, and only 21 mpg for larger vehicles (National Highway Traffic Safety Administration, 2011). Facing a powerful and large automobile manufacturing industry, Canada had long since used voluntary agreements with manufacturers to regulate efficiency, and piggybacked these agreements on the US regulations (Perl and Dunn, 2007).

The convoluted story of US fuel efficiency and emission standards revolves around three key components: the legal capacity for California to regulate air quality independently of the federal administration, the ability (or inability) of the Environmental Protection Agency (EPA) to regulate carbon dioxide, and the balance of power among key stakeholders.

In 1963 the Clean Air Act (CAA) formalised federal control over ambient air quality. California, having already been suffering from poor air quality, was the only state with pre-existing standards regulating emissions from motor vehicles. After a lengthy dispute between California and the federal administration, the resulting CAA text stipulated that federal air quality regulations pre-empted state rules, but that California *alone* could qualify for a waiver from federal control as long as the state regulations were at least as stringent as the federal ones. In 1977 the CAA was amended to allow all other states to follow *either* the federal *or* the Californian standards (Davis, 2009). This accumulation of legal rules has contributed to the powerful position California holds in US climate change policy.

Facing the continued lack of federal movement on climate change, California developed a range of state climate policies. One early component of this was the Pavley Bill which established limits on GHG exhaust pipe emissions based on the size of vehicle. This bill did not target fuel efficiency directly, but the emissions themselves, leaving the door open for emission reductions through upstream interventions as well as efficiency increases. Strict party line voting allowed the bill to squeak through the California legislature with several amendments (Perl and Dunn, 2007), but it still required an EPA waiver. California applied for the waiver in 2005. The only formal opposition to the waiver came from the automotive industry, but by 2007 California had still not received a response and Governor Schwarzenegger threatened to sue the EPA for administrative delay (Davis, 2009). Finally in late 2007, under the Bush administration and against internal EPA counsel (Wannier, 2010), the EPA took the unprecedented step of denying California's waiver.

Legally, the EPA may deny California's waiver requests for only three reasons: if California has been "arbitrary or capricious" in declaring that its rules are less stringent than the federal ones; if California does not need the rules for "compelling and extraordinary conditions"; or if California's rules and enforcement would be inconsistent with the EPA's authority to regulate emissions of an air pollutant (United States of America, 1963). The argument taken by the EPA in 2007 was that California did not have compelling reasons because a new national system had been implemented (Freeman, 2011). Immediately before the denial of the waiver, the Bush administration announced a new target for national fuel economy standards of 35 miles per gallon by 2050, which allowed an argument to be made that California no longer had due cause for concern about the lack of sufficient national regulation (Berck *et al.*, 2010). However, this new target was far weaker and slower than the standards being pursued by California. In his 2008 presidential platform Barack Obama promised to revisit the EPA's waiver denial, and upon his request (Obama, 2009) the EPA reconsidered and granted California's waiver (Environmental Protection Agency, 2009). California, and any other state that wanted to follow it, could now pursue regulations to lower vehicular GHG emissions.

However, the legitimacy of regulating CO_2 as a pollutant under the EPA at all was a consistent area of legal and political dispute nationally, and had the capacity to undermine both the Californian process and national efforts to reduce emissions. When the CAA was developed climate change was not a major concern, and CO_2 was not included in the list of priority air pollutants considered to endanger human welfare and so was not regulated. In 1990 some amendments made it possible to address stratospheric ozone depletion, but the rulings were ambiguous when it came to non-ozone depleting GHGs. In 2003, again under the Bush administration and again against internal legal council (Davis, 2009), the EPA decided that CO_2 emissions did not fit the category of a pollutant under the CCA, did not constitute a danger to the public and that therefore the EPA lacked a federal mandate to regulate it.

After a series of legal battles the question landed in the Supreme Court which decided in 2007 that the EPA did have the authority to regulate CO_2. In addition, it decided that scientific uncertainty over whether CO_2 was linked to climate change was the only legitimate reason the EPA could use for not regulating CO_2. Internally the EPA started preparations for regulation but the Bush administration overruled this process and, citing the need for a legislative response to climate change and the importance of a strong economy, decided not to pursue regulations (Freeman, 2011). Upon his election President Obama requested the EPA revisit their endangerment decision. The EPA subsequently decided that exhaust pipe CO_2 emissions did present a danger to public well-being, and that it would be regulated as a pollutant (Environmental Protection Agency, 2009). Due to the structure of the CCA, once CO_2 was recognised as a pollutant emerging from motor vehicles, it was automatically included in the lists of pollutants for all other sources as well, opening up a range of options for stationary and non-stationary source regulations (see Burtraw *et al.* (2011) for discussion of these).

By 2009 the core legal pieces were in place not only for California to pursue fuel emission standards, but for the entire country to do so. However, fuel economy remained a politically contentious issue that was repeatedly resulting in bitter legal disputes. The inclusion of CO_2 as an EPA pollutant was legally contentious both before and after the 2007 Supreme Court decision that the EPA could regulate it (Freeman, 2011; Wall, 2007). These legal battles resonated with deep divisions about climate science and appropriate regulation that were part of the increasingly polarised political climate policy landscape (Klinsky, 2012). During this time multiple states passed bills urging the federal government to revisit and reverse the EPA endangerment decision that had declared CO_2 a threat to public well-being (Bravender, 2010). In addition, several states passed bills arguing that there was insufficient scientific proof of climate change to warrant any legislative action (see, for example, Montana Legislature (2011) and Utah (2010)). These bills stopped climate policy development in their own states and attempted to prevent any federal legislation from emerging as well.

The direct lobbying power of the automobile and oil industries was an important component of these debates. The automobile manufacturing industry had steadily used the court system to hinder California from developing its own regulations (Berck *et al.*, 2010; Wall, 2007; Wannier, 2010) and had been the only group formally opposing California's initial request for its EPA waiver. It is striking that when applying to the EPA to make a decision about its waiver, the attorney general of California explicitly requested that the EPA do this themselves instead of deferring to the Bush administration based on his recognition that both President Bush and Vice President Cheney were deeply connected to the oil industry (Davis, 2009).

Once the Obama administration had secured clear legal authority to regulate CO_2, fuel efficiency, standards were formalised relatively quickly. The President's Office of Energy and Climate Change facilitated cooperation amongst the key agencies involved at the federal level (Freeman, 2011), and President Obama himself announced the new fuel standards in a Rose Garden press meeting. In addition, significant stakeholder involvement was used to get the automobile industry on board. All of the major automobile industry stakeholders signed letters of commitment, promising that they would not sue or use legal means to prevent the new fuel standards. These letters were signed before the formal release of the new standards which indicates either a level of trust between these industries and the regulatory bodies (Freeman, 2011), and/or a significant reduction in industry lobbying power. California had a strong bargaining chip in pressuring for a national policy: other states could follow Californian regulations. Therefore, if California developed different regulations from the federal system, car manufacturers could be forced to produce different cars for different states with the US, a costly and complicated possibility. It is also important to note that these letters of commitment were drafted as the federal government poured billions of dollars of public money into bailing out the failing automobile industry, and "they had reason to make nice" (Banerjee, 2011). Strikingly, the only other time the US federal government had increased fuel efficiency standards was following the Carter administration's massive financial bailout of Chrysler, while it looked as if both Ford and GM would be requiring assistance as well (Perl and Dunn, 2007).[12]

Finally, California agreed to accept the US federal standards as sufficiently stringent to be acceptable in their system without sacrificing their right to future waivers from federal regulation. In 2011 President Obama announced a new fuel efficiency standard which aims to have a fleet average of 54.5 mpg for cars and light-duty trucks by Model Year 2025 (Environmental Protection Agency, 2011). The regulations were complicated, and have been criticised for failing to incentivise smaller cars because efficiency is determined based on the footprint of the vehicle, but for the first time in 30 years the US had a nationally comprehensive fuel efficiency standard that could result in emission reductions. Initial analysis suggests that the emission reductions possible under these regulations may be comparable to the level expected under the proposed federal cap-and-trade scheme, although likely at a higher cost (Burtraw *et al.*, 2011). As many had predicted, Canada almost immediately followed suit and indicated its intentions of following the US lead (McCarthy, 2010; Ministry of Environment, Government of Canada, 2010). Despite the widespread efforts invested in emissions trading and other climate policies, the new fuel standards in both countries are the most comprehensive climate policy tool currently available in North America.

8.5. Discussion

This chapter provided brief overviews of a range of climate change policies in Europe and North America in order to identify political factors that have shaped their development. Five central categories of lessons are highlighted in this chapter and are discussed below. These categories are: policy evolution, policy framing, non-state actors, collective action and institutional structures. Each provides insights into the challenges that lie ahead for achieving the policy and political support required to place the global economy onto the trajectories discussed in the other chapters of this book.

[12] As discussed earlier, these movements towards fuel efficiency were reversed immediately upon Carter's defeat to Regan and the long stagnation of US fuel efficiency standards began.

8.5.1. *Policy evolution*

One of the central lessons from this comparison is that climate policies have not emerged fully formed from any single jurisdiction's policy design crucible, but are instead dynamic. All of the policies discussed in this chapter evolved from previous policy attempts and through interactions with other policies. For example, neither the WCI nor RGGI would have likely survived without the collapse of the federal cap-and-trade bill in the US, but both were shaped by lessons from the EU ETS. The Canadian federal government would have been unlikely to attempt a carbon tax had BC not already developed one, although the federal failure may impede future carbon tax attempts just as the failed Clinton energy tax has done.

Policy learning is not just transferred from one policy to another, but evolves internally as learning occurs about what works and what doesn't. The EU ETS did not start with the level of centralisation and sophistication evident in Phase III; it was a process. Similarly, even though the BC carbon tax started with significant revenue recycling through income tax reductions, this design looked insufficient to ensure its success. Subsequently further changes to income tax were required in order to address regional distributive concerns. In the case of BC income, tax rebates are now dependent on carbon tax revenue, which means that it may prove to be difficult to remove even if public sentiments change. Not only does this evolution process mean that policies do not look like textbook cases, but it also highlights the "stickiness" of policies. Once they are implemented they may be harder than expected to eliminate.

Recognition of the dynamic process of policy development has several implications. First, once a policy is in place it may be "sticky", leading to a certain degree of path-dependency. The long-term political implications of any policy — including those that are only partially successful initially — can be hard to assess. This stickiness could theoretically provide policy makers with greater leeway to ratchet policies up over time, although thus far only the EU ETS has demonstrated this process. Second, no policy is going to be enacted in the form in which it theoretically originates. Finally, in most cases jurisdictions have pursued multiple climate policies simultaneously. Not only do policies influence each other sequentially, but it is entirely likely that multiple policies are interacting within a single jurisdiction at the same time. Recognition of the stickiness, less-than-ideal policy structure, and policy interactions may be issues to consider both in efforts to support policy and to model policy implications.

8.5.2. *Policy framing*

A second observation emerging from the cases in this chapter is that policies cannot be examined without understanding how they fit with much broader issues. In some situations background ideological debates about regulation and the role of government have shaped perceptions of policy options. Several states pulled out of the WCI in part due to deep concerns about "excessive" regulation. In contrast, it is possible that Scandinavian countries had a greater ability to implement early carbon taxes due to relatively widespread acceptance and trust of government taxation.

Unfortunately for analysts and policy makers, overarching framings can shift. For example, the economic crisis of 2008–2009 played havoc with North American climate policy attempts. In the case of the WCI, collective action under climate policy and green growth was perceived as a positive pull into the scheme. However, the economic crisis sharpened public and state economic concerns, and changed some states' perceptions of the trustworthiness of markets in general. In Canada, the economically strong platform of the BC Liberals allowed them to maintain power despite public tensions over the carbon tax, but the federal Liberals lost their tax scheme and the general election due to the stronger economic rhetoric of the opposing Conservative Party.

The fluidity of overarching frames and their ability to shift the fates of a particular policy proposal highlight the dangers of attempts to tie climate policy to other issues. A clear example of this is seen in the debate about revenue use in RGGI, raising the question of whether RGGI is a climate policy or a means of reducing state deficits. The need for state revenue may be a pull into policies but, depending on ideological stances, state generation of revenue can also erode support for policies perceived as "hiding" taxation behind climate goals.

The tight connection between energy, climate policy and economic vitality make economic framings likely to be particularly central to climate policy debates. From a modelling perspective this highlights the importance of underlying assumptions about economic conditions, especially when communicating model results to possible users.

8.5.3. *Non-state actors*

One of the central observations of this chapter echoes those who have been arguing that climate policy must be seen as a form of multilevel governance (Betsill and Bulkeley, 2006; Rabe, 2007). A key lesson from all climate policies is that many actors — including non-state actors — are involved in decision-making. In none of the cases in this chapter was policy crafted by a single centralised decision-maker, although the centrality of particular decision-makers (and levels of decision-makers) varied by policy. Public opinion and electoral power were crucial factors in the North American carbon tax debates, essentially resulting in a situation in which millions of people were implicated in decision-making. In contrast, the bulk of the US fuel efficiency process involved primarily those within government, the automotive lobby groups and the courts. In the fuel efficiency case decision-making was thus shared among a much smaller number of actors and remained dynamic. Similarly, the EU ETS, WCI and RGGI involved negotiations primarily among jurisdictions and key industries — a smaller cohort than represented by voters in general but still wider than government alone.

It is not just that many actors are involved in decision-making, but that the needs or desires of these actors may not be aligned, or may have different components. Policies which involve all voters — such as carbon taxes — may need to be more carefully aligned with the ways in which people self-identify individual well-being. One of the key lessons emerging from the Clinton energy tax fiasco was the difficulty of communicating clear benefits to individuals. Similarly, identifying the types of costs and benefits relevant to decision-making is important even for policies which involve fewer decision-makers, such as cap-and-trade or industrial policies. The core concerns of industry and regulators are likely to differ, as are the types of evidence they are going to rely upon because they may use fundamentally different decision-making strategies. For example, while aggregate optimisation of decision-making may be of interest to central regulatory authorities, it is unlikely to be at the centre of industry-specific interests.

The lobbying power of large industries is clearly a major component of many, if not most, policy processes although it does not have to prevent climate policy. One observation is that timing is crucial. It is unlikely merely coincidental that the only times the US government has successfully pursued fuel efficiency improvements have been immediately following government bailouts of the automotive industry. It is also possible that the sheer speed of EU ETS development limited the extent to which industry groups became involved in explicit lobbying against the policy.

However, the consistency of the policy message is important. For example, as the EU ETS policy has become further established industries have become more involved in trying to influence how the regulations will look, although there is limited pressure to get rid of it entirely. Similarly, many key industries in the WCI became involved during the consultation process once it became apparent that some form of cap-and-trade was a likely outcome. Industries facing changes in large investment strategies are unlikely to shift trajectory for fleeting policy interests; if regulators are serious about climate policy they must be consistent in their messaging.

Finally, a balance must be struck between accommodating specific industry needs and maintaining across-the-board policy measures. In the US fuel efficiency standards, the WCI and the EU ETS, industries were involved in crafting the agreements and had their concerns taken seriously, but were not entirely exempt from regulation. In contrast, the Clinton energy tax was largely gutted with exemptions for many industries. These exemptions eroded broad support for the tax and resulted in an unpassable bill.

Lobbying is likely an inevitable challenge for climate policy but several potential strategies may exist for managing it. First, there may be timing windows that facilitate policy development, either due to decreased lobbying power or through rapid policy development. Second, the policy patterns noted above would suggest that sending consistent messages about the inevitability of some form of climate policy may be an important factor in facilitating constructive engagement across industry sectors.

Third, while some concessions may be required, these must be balanced with across-the-board measures to ensure that the underlying arguments for regulation are not excessively weakened.

These observations have potential implications for modelling as well. A central feature of the engagement of multiple parties is that while aggregate optimisation information may be used at some point in a political process, many of those involved in decision-making may make limited use of such information. For instance, evidence regarding aggregate savings may be of limited value for regulators facing significant lobbying pressure, especially from industries or publics concerned with competitiveness losses or distributive impacts. This raises questions about the types of information that are likely to be desired or useful during policy making processes that are dynamic and involve multiple actors. If a centralised decision-maker is not going to use model results, who is, and what kinds of indicators or results are likely to be of most assistance? How are costs and benefits selected for assessment, measured and communicated?

8.5.4. *Collective action*

Climate policies must ultimately be made within localised jurisdictions but have inevitable connections far outside specific jurisdictional boundaries. Even when policies are not explicitly linked by overarching institutional relationships, policy action in one jurisdiction is influenced by that in others. The logic of collective action is crucial in climate policy decision-making at (and across) all scales.

The EU ETS epitomises several of the dynamics of collective action that are useful, possibly necessary, for the development of climate policy. First, it originated in a context in which jurisdictions already had a history of cooperative action, and for whom a shared sense of identity was important. One of the differences between RGGI and the WCI is that RGGI states already had close relationships which may have facilitated the relatively easy and rapid cooperation that emerged through its development. Collective action may be possible among jurisdictions without historical ties but may be slower and more difficult.

Second, the background political economy of jurisdictions is likely to affect how actors form coalitions and cooperate. It is unsurprising that within the EU, Poland has maintained reservations about the EU ETS. Similarly, the structure of the WCI was strongly influenced by the divisions between jurisdictions without their own fossil energy reserves (such as California, Quebec or Ontario) and those that have maintained, and want to expand, fossil fuel industries (such as Montana and Utah). The WCI was able to create an association while economic times were good and while it looked like such an association would protect long-term economic interests, but as soon as the economic and federal policy landscapes shifted the collective action of the WCI was insufficiently grounded and it disintegrated.

Third, the momentum of collective efforts can be crucial. Regulators may use arguments about the inevitability of climate policy and the need for first-actor advantages to address industry concerns. However, such arguments become less believable as collective action weakens. It is possible that efforts to maintain momentum can be important. For example, RGGI's use of a model rule allowed regulators an avenue for rapid policy development that was explicitly tied to policies in neighbouring jurisdictions. Similarly, states involved in the WCI were explicit that such involvement was predicated on collective action. As the collective action of the WCI eroded it became impossible for states not deeply committed to climate policy, such as Montana or Utah, to continue involvement and the trading system unravelled.

Finally, no policy exists in isolation and movements in other jurisdictions can affect how policies are perceived and pursued (or not pursued). The EU ETS is formally built upon the EU's obligations through Kyoto, providing a clear example of how the global sense of collective action translated into specific policy obligations and directions within a jurisdiction. However, this lesson cuts both ways. Washington state's decision to withdraw from the WCI officially noted the EU ETS's problems with fraud and price instability.

8.5.5. *Institutional structure*

One of the core threads of political feasibility research has emphasised is the centrality of institutions, and the relevance of this to climate policy can hardly be overstated. Policies depend on the core institutional structures within their jurisdictions. For example, the EU had tried to implement carbon taxes but failed due to EU rules which allowed states veto power on financial decisions. Instead, cap-and-trade did not require unanimity but only a two-third majority, which allowed it to be pursued. At the same time, an EU-wide cap-and-trade system was possible only because of the deep connections built over years of European integration and negotiation.

As an institution, the court has become particularly important in the US. The central role of the court system was blatantly evident in the development of US fuel efficiency. Both lobby groups and environmental organisations actively used the courts to influence and ultimately direct climate policy. The on-going debates about revenue use in RGGI are also being conducted through the courts. It was ultimately legal rules dating back over 50 years that facilitated California's ability to pursue fuel emission standards.

Institutions might provide a basic landscape for policy but they are not immutable. The influence of the presidential administration on the way in which fuel standards in the US were shaped is striking: the attitude of the EPA was entirely different under the Bush and Obama administrations. The BC carbon tax was also strongly influenced by high-level support by Premier Campbell. In both the US fuel efficiency and BC carbon tax cases, leadership created new institutions with high-level responsibilities that played central roles in facilitating and coordinating climate policy actions. Similarly, both RGGI and the WCI resulted in the creation of entirely new institutions responsible for supporting these trading systems. These institutions are now part of the policy backdrop against which new policy options will be evaluated. Institutions can be changed, reinterpreted and created but they provide a fundamental framework in which policies will operate and cannot be ignored.

References

Aulisi, A., Larsen, J., Pershing, J. and Posner, P. (2007). *Climate Policy in the State Laboratory: How States Influence Federal Regulation and the Implications for Climate Change Policy in the United States*, World Resources Institute, Washington, DC.

Banerjee, N. (2011). The Road to New Fuel Economy Standards Was Not Smooth, *Los Angeles Times*, November 18. Available online at: http://articles.latimes.com/print/2011/nov/18/business/la-fi-fuel-economy-deal-20111119. Accessed 2 June 2013.

Berck, P., Brännlund, R. and Spindell Berck, C. (2010). Green Regulations in California and Sweden, *Journal of Natural Resources Policy Research*, 3, 49–61.

Betsill, M. and Bulkeley, H. (2006). Cities and the Multilevel Governance of Global Climate Change, *Global Governance*, 12, 141–159.

Betsill, M. and Bulkeley, H. (2007). Looking Back and Thinking Ahead: A Decade of Cities and Climate Change Research, *Local Environment*, 12, 447–456.

Betsill, M., and Hoffmann. M. (2011). The Contours of "Cap and Trade": The Evolution of Emissions Trading Systems for Greenhouse Gases, *Review of Policy Research*, 28, 83–106.

Bravender, R. (2010). State Legislators Ramp up Campaigns Against EPA Climate Rules, *The New York Times*, 29 March. Available online at: http://www.nytimes.com/gwire/2010/03/29/29greenwire-state-legislators-ramp-up-campaigns-against-ep-73429.html?pagewanted=all. Accessed 2 June 2013.

British Columbia (2008). Carbon Tax Act. SBC 2008 [Chapter 40]. Available at: http://faolex.fao.org/docs/html/bc85495.htm. Accessed 2 June 2013.

Burtraw, D., Fraas, A. and Richardson, N. (2011). Policy Monitor — Greenhouse Gas Regulation Under the Clean Air Act: A Guide for Economists, *Review of Environmental Economics and Policy*, 5, 293–313.

Byrne, J., Hughes, K., Rickerson, W. and Kurdgelashvili, L. (2007). American Policy Conflict in the Greenhouse: Divergent Trends in Federal, Regional, State, and Local Green Energy and Climate Change Policy, *Energy Policy*, 35, 4555–4573.

Cohen, M., March, J. and Olsen, J. (1972). A Garbage Can Model of Organizational Choice, *Administrative Science Quarterly*, 17, 1–25.

Convery, F. (2009). Origins and Development of the EU ETS, *Environmental and Resource Economics*, 43, 391–412.

Davis, K. (2009). The Road to Clean Air Is Paved with Many Obstacles, *Fordham Environmental Law Review*, 19, 39–101.

Department of Transportation (2010). *Transportation's Role in Reducing U.S. Greenhouse Gas Emissions: Report to Congress*, US Department of Transportation, Washington, DC.

Downs, A. (1972). Up and Down with Ecology: The "Issue-attention Cycle", *Public Interest*, 28, 38–50.

Ellerman, A. (2000). *Markets for Clean Air: the U.S. Acid Rain Program*, Cambridge University Press, Cambridge.

Ellerman, A., Convery, F. and de Perthuis, C. (2010). *Pricing Carbon*, Cambridge University Press, Cambridge.

Environmental Protection Agency (2009). *California State Motor Vehicle Pollution Control Standards; Notice of Decision Granting a Waiver of Clean Air Act Preemption for California's 2009 and Subsequent Model Year Greenhouse Gas Emission Standards for New Motor Vehicles*, Federal Registry, FR-2009-07-08, Washington, DC.

Environmental Protection Agency (2011). *2017 and Later Model Year Light-Duty Vehicle Greenhouse Gas Emissions and Corporate Average Fuel Economy Standards*, Federal Registry, Washington, DC.

Erlandson, D. (1994). Btu Tax Experience: What Happened and Why It Happened, *Pace Environmental Law Review*, 12, 173.

Fitzpatrick, M. (2008). Liberal Carbon Tax Plan Good for Planet, Wallet: Dion, *National Post*, 19 June. Available online at: http://www.nationalpost.com/news/story.html?id=599498. Accessed 2 June 2013.

Freeman J. (2011). The Obama Administration's National Auto Policy: Lessons from the "Car Deal", *Harvard Environmental Law Review*, 35, 343–374.

Gough, C. and Shackley, S. (2001). The Respectable Politics of Climate Change: The Epistemic Communities and NGOs, *International Affairs*, 77, 329–346.

Grubb, M. (2014). *Planetary Economics: Energy, Climate Change and the three Domains of Sustainable Development*. Routledge, Oxford.

Haas, P. (1989). Do Regimes Matter? Epistemic Communities and Mediterranean Pollution Control, *International Organization*, 43, 377–403.

Hardisty, D., Johnson, E. and Weber, E. (2009). A Dirty Word or a Dirty World? Attribute Framing, Political Affiliation, and Query Theory, *Psychological Science*, 21, 86–92.

Harrison, K. (2010). The Comparative Politics of Carbon Taxation, *Annual Review of Law and Social Science*, 6, 507–529.

Harrison K. (2012). A Tale of Two Taxes: The Fate of Environmental Tax Reform in Canada, *Review of Policy Research*, 29, 383–407.

Horvitz, P. (1993). Clinton Retreats on Energy Tax in Fight over Budget, *The New York Times*, 9 June. Available online at: http://www.nytimes.com/1993/06/09/news/09iht-plan_1.html. Accessed 2 June 2013.

IPCC (1990). *Intergovernmental Panel on Climate Change First Assessment Report Overview*, Intergovernmental Panel on Climate Change, United Nations, Geneva.

IPCC (2007). *Climate Change 2007: Mitigation. Contribution of Working Group III to the Fourth Assessment Report of the Intergovernmental Panel on Climate Change*, Metz, B, Davidson, P. R., Bosch, R. and Meyer, L. A. (eds), Cambridge University Press, Cambridge.

Kingdon, J. (1984). *Agendas, Alternatives, and Public Policies*, Little, Brown, Boston.

Krause, R. (2011). An Assessment of the Greenhouse Gas Reducing Activities Being Implemented in US Cities, *Local Environment*, 16, 193–211.

Lee, H. (1994). Political Economy of Energy Taxes: An Assessment of the Opportunities and Obstacles, *The Pace Environmental Law Review*, 12, 77.

Liberal Party of Canada (2008). The Green Shift. Available online at: http://www.poltext.org/sites/poltext.org/files/plateformes/ca2008lib_plt_eng._05012009_111617.pdf. Accessed 26 March 2014.

May, P. (1986). Politics and Policy Analysis, *Political Science Quarterly*, 101, 109–125.

McCarthy, S. (2010). Canada, U.S. Announce Plans for Fuel-efficiency Regulations for Heavy Trucks, *The Globe and Mail*, 9 August. Available online at: http://www.theglobeandmail.com/news/politics/canada-us-announce-plans-for-fuel-efficiency-regulations-for-heavy-trucks/article589878/. Accessed 2 June 2013.

Meckling, J. (2011). *Carbon Coalitions Business, Climate Politics, and the Rise of Emissions Trading*, MIT Press, Cambridge, MA.

Meltsner, A. (1972). Political Feasibility and Policy Analysis, *Public Administration Review*, 32(6), 859–867.

Ministry of Environment, Government of Canada (2010). *Passenger Automobile and Light Truck Greenhouse Gas Emission Regulations*, Amendment to the Canadian Environmental Protection Act 1999, SOR/2010-201, September 23, 2010.

Montana Legislature (2011). *House Joint Resolution No. 18: A Joint Resolution of the Senate and the House of Representatives of the State of Montana Urging Montana's Governor to Withdraw Montana from the Western Climate Initiative*, HJ0018.01. Available online at: http://leg.mt.gov/bills/2011/houjoint/HJ0018_1.pdf. Accessed 28 March 2014.

Morris, C. (2008). Liberal Green Shift Is "Green Shaft," Says Harper, *The Toronto Star*, 14 August. Available online at: http://www.thestar.com/news/canada/article/478887--liberal-green-shift-is-green-shaft-says-harper. Accessed 2 June 2013.

National Highway Traffic Safety Administration (2011). *Summary of Fuel Performance*, Department of Transportation, Washington, DC.

Navarro, M. (2012). Challenge to New York's Carbon Trading Fails, *The New York Times*, 14 June. Available online at: http://green.blogs.nytimes.com/2012/06/14/challenge-to-carbon-trading-fails. Accessed 2 June 2013.

Obama, B. (2009). *Presidential Memorandum: State of California Request for Waiver Under 42 U.S.C. 7543(b), the Clean Air Act*, The White House, Washington, DC. Available online at: http://www.whitehouse.gov/the-press-office/Presidential-Memorandum-EPA-Waiver/. Accessed 2 June 2013.

Olson, M. (1965). *The Logic of Collective Action: Public Goods and the Theory of Groups*, Harvard University Press, Cambridge, MA.

Paavola, J. (2007). Institutions and Environmental Governance: A Reconceptualization, *Ecological Economics*, 63, 93–103.

Peet, C. and Harrison, K. (2012). Historical Legacies and Policy Reform: Diverse Regional Reactions to British Columbia's Carbon Tax, *BC Studies*, 173, 97–122.

Perl, A. and Dunn, J. (2007). Reframing Automobile Fuel Economy Policy in North America: The Politics of Punctuating a Policy Equilibrium, *Transport Reviews*, 27, 1–35.

Pew Center (2009). Pew Center Summary of H.R. 2454: American Clean Energy and Security Act of 2009 (Waxman-Markey), Pew Center. Available online at: http://www.c2es.org/docUploads/Waxman-Markey%20summary_FINAL_7.31.pdf. Accessed 2 June 2013.

Potomac Economics (2012). Annual Report on the Market for RGGI CO_2 Allowances: 2011, Regional Greenhouse Gas Initiative. Available online at: http://www.rggi.org/docs/MM_2011_Annual_Report.pdf. Accessed 2 June 2013.

Rabe, B. (2008). States on Steroids: The Intergovernmental Odyssey of American Climate Policy, *Review of Policy Research*, 25, 105–128.

Rabe, B. (2007). Beyond Kyoto: Climate Change Policy in Multilevel Governance Systems, *Governance*, 20, 423–444.

Rabe, B. (2008). Regional Greenhouse Gas Initiative Model Rule, Regional Greenhouse Gas Initiative. Available online at: http://www.rggi.org/docs/Model%20Rule%20Revised%2012.31.08.pdf. Accessed 2 June 2013.

Rabe, B. (2011). Investment of Proceeds From RGGI CO_2 Allowances, Regional Greenhouse Gas Initiative. Available online at: http://www.rggi.org/docs/Investment_of_RGGI_Allowance_Proceeds.pdf. Accessed 2 June 2013.

Ruth, M., Gabriel, S., Palmer, K., Burtraw, D., Paul, A., Chen, Y., Hobbs, B., Irani, D., Michael, J., Ross, K., Conklin, R. and Miller, J. (2008). Economic and Energy Impacts from Participation in the Regional Greenhouse Gas Initiative: A Case Study of the State of Maryland, *Energy Policy*, 36, 2279–2289.

Schneider, J. and Elliott, M. (2012). Benefits of the Regional Greenhouse Gas Initiative, Environment New Jersey Research and Policy Centre. Available online at: http://www.environmentnewjerseycenter.org/sites/environment/files/reports/Benefits%20of%20RGGI.pdf. Accessed 2 June 2013.

Selin, H. and Vandeveer, S. (2005). Canadian-U.S. Environmental Cooperation: Climate Change Networks and Regional Action, *American Review of Canadian Studies*, 35, 353–378.

Spolar, M. (2012). Bill That Reforms RGGI Becomes Law. Concord Monitor. Available online at: http://www.concordmonitor.com/article/338194/bill-that-reforms-rggi-becomes-law?SESS3a3f6a2f017b76baab785194a2a5630b=google. Accessed 2 June 2013.

Stavins, R. (2012). Low Prices a Problem? Making Sense of Misleading Talk About Cap-and-Trade in Europe and the USA, Harvard Kennedy School Belfer Center. Available online at: http://www.robertstavinsblog.org/2012/04/25/low-prices-a-problem-making-sense-of-misleading-talk-about-cap-and-trade-in-europe-and-the-usa. Accessed 2 June 2013.

Tietenberg, T. (2006). *Emissions Trading: Principles and Practice*, Resources for the Future, Washington, DC.

UNEP (2011). Bridging the Emissions Gap: A Synthesis Report, United Nations Environment Programme. Available online at: http://www.unep.org/pdf/UNEP_bridging_gap.pdf. Accessed 2 June 2013.

United States of America (1963). Clean Air Act. Descriptions of the Act may be found at http://environmentallaw.uslegal.com/federal-laws/clean-air-act/. Accessed 28 March 2014.

Utah (2010). Climate Change House Joint Resolution N. 12. Available at: http://le.utah.gov/~2010/htmdoc/hbillhtm/hjr012.htm. Accessed 28 March 2014.

Wall, M. (2007). The Regional Greenhouse Gas Initiative and California Assembly Bill 1493: Filling the American Greenhouse Gas Regulation Void, *University of Richmond Law Review*, 41, 567–588.

Wannier, G. (2010). Examining California's Right to Regulate Vehicle Greenhouse Gas Emissions, Climate Law Blog: Center for Climate Change Law, Columbia Law School. Available online at: http://blogs.law.columbia.edu/climatechange/2010/12/14/examining-california%E2%80%99s-right-to-regulate-vehicle-greenhouse-gas-emissions/. Accessed 2 June 2013.

Western Climate Initiative (2010). Updated Economic Analysis of the WCI Regional Cap-and-Trade. Available online at: http://www.westernclimateinitiative.org/document-archives/Economic-Modeling-Team-Documents/Updated-Economic-Analysis-of-the-WCI-Regional-Cap-and-Trade-Program. Accessed 2 June 2013.

Chapter 9

Co-impacts of a Decarbonised Economy

Douglas Crawford-Brown and Ann Thompson

9.1. Introduction to Co-impacts

Justifications for decarbonising the global economy usually focus on the reduction in greenhouse gas (GHG) emissions as described in Chapter 1. These reductions in emissions are then related to the predicted reduction in atmospheric GHGs, and the resulting change in climate change risks (Markandya and Chiabai, 2009), e.g. human health effects due to higher summer temperatures, ecosystem loss or asset damage through increased storm frequency and/or intensity. The political will to carry out mitigation and adaptation policies then rests on a belief in the reliability of the science of climate change and of models predicting the effectiveness of policies at reducing GHG emissions. As discussed in Chapter 8, this political will can face significant hurdles when the sole justification for policy is reduced climate risks.

There are, however, many other reasons why an institution or nation might move along a trajectory of decarbonisation towards the targets described in Chapter 1. There might be concern for the air pollution effects of burning fossil fuels, such as risks of damage to human health from particulates and agriculture from low-level ozone. There might be concern for energy security, reducing reliance on foreign energy sources or reducing fuel poverty. There might be a desire for lower energy bills through reduced energy consumption. There might be a government target for economic development, with investment in green buildings, energy systems and/or infrastructure providing a stimulus for this development. Experience shows that moving a policy such as decarbonisation forward is most viable when there are multiple reasons for doing so, ensuring the widest range of participating actors and protection against deep regrets should any one policy justification prove in hindsight to be unwarranted.

It is therefore natural to ask whether decarbonisation might bring with it impacts that affect the ability of a nation or institution to reach other economic, energy and/or environmental policy goals, apart from the reduction in GHG emissions. These impacts might be beneficial, helping to reach these other policy goals, or they might present new problems that impede reaching climate change goals (a process of maladaptation). An example of the former is less exposure to fluctuating oil prices when much more energy supply comes from renewable sources. An example of the latter is increased pressure on land and water resources — two key indicators of sustainability — when bioenergy becomes a significant aspect of a policy of decarbonisation.

Consideration of these co-impacts is important for several reasons:

- If they are co-benefits, they can be used to marshal additional political support for policies of decarbonisation.
- Inclusion of co-impacts provides a more complete understanding of the multiple effects of a policy, improving the ability to inform multi-attribute decisions.
- Assessment of co-impacts potentially provides a more balanced perspective on the net costs and benefits of a policy to and within society.
- Co-impacts may reveal a multiplier effect of policies in the economy, better characterising the beneficiaries of those policies.
- Identification of co-impacts provides insights into where hurdles might be encountered in implementing policies of decarbonisation, and how they might be overcome.
- Early recognition of adverse co-impacts can avoid situations of backlash if the impacts become evident in the future.

This chapter assesses co-impacts in regard to climate change and decarbonisation policies using the examples of global public health and natural resource use (specifically land and water).

Since the assessment of co-impacts can add appreciably to the effort involved in policy assessment, it is natural to ask whether the additional information brings significant new insights to the overall societal effect of a policy. As a partial answer to that, consider the case of a study of the impact of carbon taxes on the economy of Bangkok, Thailand (Li, 2002). In that study, the impact of a carbon tax on GHG emissions and national gross domestic product (GDP) was first assessed using an economic model without co-benefits associated with health; here, it was an economic model based on a social accounting matrix without the addition of econometric relationships (and so a different approach to economic modelling than is found elsewhere in this book). The policy was re-assessed taking into account the co-reduction of emissions of particulate matter (PM) when strategies that reduce GHGs through reduced energy consumption driven by a carbon tax are introduced.

Using methods described below, the projected reduction in PM emissions was converted into reductions in ambient air exposures, and then into reductions in health effects from PM and finally into the reduction in health care costs and time lost by the workforce. The results are shown in Figure 9.1, indicating that consideration of co-benefits of reduced PM emissions not only changed the magnitude of predicted GDP impacts, but reversed the sign of those impacts. What appeared to be a climate change policy creating a loss of GDP without considering the co-benefits of PM reduction became an impact that was positive on GDP growth when these co-benefits were considered.

A word of caution should be added concerning the results in Figure 9.1. Bangkok is one of the world's megacities, with very high ambient air PM concentration and a large number of people exposed near roadways. As a result, policies that reduce PM emissions there can have a dramatic effect on the incidence of disease, health care costs and worker productivity. The same dramatic impact when including co-benefits would not be expected in cities with significantly lower levels of PM, such as most major Western cities. Still, the point remains: consideration of

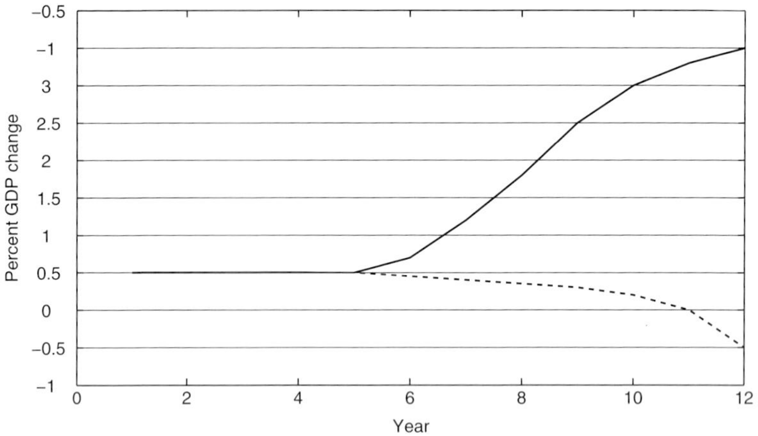

Figure 9.1. Predicted impact of a carbon tax on the economy of Bangkok without considering co-benefits of PM reduction (dashed line) and with co-benefits (solid line). Note that inclusion of co-benefits indicates a growth in GDP, rather than a decline. The *y*-axis is cumulative percent change in GDP relative to the baseline projection with no carbon tax.

co-impacts can at times completely alter the implications of a climate policy, improving political acceptance of the policy.

Figure 9.1 concerns only the influence of PM reductions, and only the impacts in Bangkok. Section 9.2 provides the details on calculating the health risk co-benefits, using again the example of PM but extending the analysis globally. There are, however, many more co-benefits than that from PM reduction, including other impacts on health as well as impacts on a variety of environmental and ecological indicators of sustainability. A good review of the literature on these has been provided by Bollen *et al.* (2009), the results of which are summarised here. Their review draws several key conclusions:

- Mitigation policies yield long-term benefits, but effects from co-benefits often are much more immediate and reliable, raising the willingness of decision-makers to implement policies in the short term. In fact, in the short-to-medium term, it will be largely the co-benefits that will be realised by a nation rather than the direct impacts related to climate change.

- Co-benefits could be achieved by other, more direct policies than climate change mitigation, and so the incentives to decision-makers depend critically on how well these other policies might have delivered on the co-benefits. A key issue here is the cost effectiveness of producing the same improvements in the environment and/or health through GHG reduction policies or through policies targeting, e.g. solely and specifically PM.
- Climate change and local air pollution policies can at times be in conflict. For example, aerosols can cool the atmosphere by blocking sunlight. Therefore, the reduction in PM that accompanies reduction in GHGs may in turn accelerate warming. It is crucial that these interactions be taken into account to ensure the net effect of a co-benefit is truly beneficial.
- The health care cost savings of these co-benefits range between US$10 and US$60 per ton of carbon dioxide abated. This value is similar to the carbon prices or marginal abatement costs (MAC) for many strategies of carbon dioxide reduction, at least for energy efficiency improvements, although well below the MAC values for some of the more expensive options such as solar power (using current levelised costs). As shown in a study by the OECD (Organisation for Economic Co-operation and Development) (Bollen *et al.*, 2009), the cost savings from the health co-benefits alone will be higher than the cost of abatement for a ton of carbon dioxide equivalent for five of the 15 regions they studied in the global economy. Even in those regions where this was not the case, inclusion of the co-benefits reduced the apparent costs of carbon abatement by more than 50% in five of the ten. A summary of their findings are presented in Figure 9.2.
- The same OECD study (Bollen *et al.*, 2009) estimated that the percentage reduction in premature mortality of all air pollutants combined was slightly below the percentage reduction in GHG emissions; an 80% reduction in GHG emissions produced approximately a 50% reduction in annual premature mortality (attributable to air pollution) as of 2050.
- Depending on existing concentrations of pollutants in ambient air, the co-reduction of non-GHG pollutants can bring about savings in air

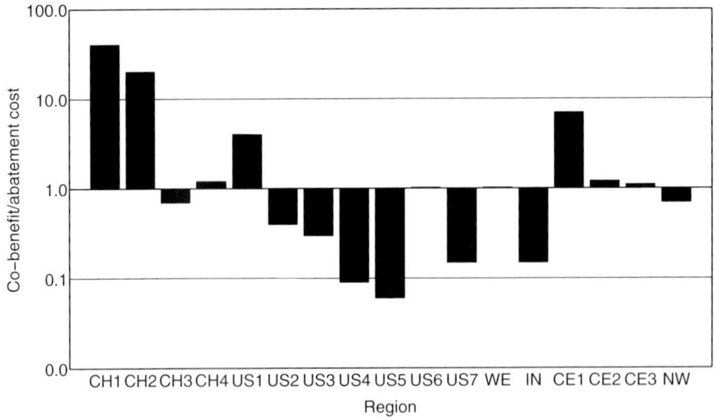

Figure 9.2. Summary of calculations of the ratio of cost savings of co-benefits over cost of carbon abatement measures examined, obtained from Bollen *et al.* (2009). Regions refer to China (CH), United States (US), Western Europe (WE), India (IN), Chile (CE) and Norway (NW). Values of more than 1.0 represent cases where cost savings exceed cost of carbon abatement.

pollution control costs, since a reduction in PM and other pollutants by GHG mitigation equates with a savings in reduction by other methods such as investments in scrubbers, baghouses, etc.

Co-benefits appear in a number of forms, depending on the source of the GHGs being reduced. The most significant pollutants impacted by policies that reduce reliance on fossil fuels are black carbon, PM, SOx (oxides of sulphur), NOx and N_2O (oxides of nitrogen), CH_4 (methane), O_3 (ozone) and NH_3 (ammonia). As just one example, a study by IIASA (International Institute for Applied Systems Analysis) and the European Commission (EC) Joint Research Centre (IIASA, 2004) indicates that a pathway to decarbonisation would produce a reduction of between 3 and 12 ppb (parts per billion) of O_3 in different regions of the world, against a current global background level of between 30 and 50 ppb. The resulting decrease of 10% to 25% for ozone associated with a decarbonisation pathway is in stark contrast to the increase in tropospheric ozone that will occur through 2050 without control policies, which is estimated to cause a cumulative loss globally of US$580 billion as a result of increased mortality and morbidity associated with ozone (Selin *et al.*, 2009).

Aside from their direct impacts on health, reduction of these air pollutants in concert with climate mitigation also:

- Reduces deposition of nutrients onto waterways, in turn reducing impacts on both water quality and eutrophication.
- Reduces acidification of ecosystems, in turn reducing the damage to species such as fish that might be part of the diet or an indicator of ecosystem vitality.
- Decreases damage to plant life (e.g. due to ozone), increasing the yield of crops, pasture and forests, all of which play an important role in the economic welfare of the poorest populations.
- Reduces indoor air pollution, especially in lesser-developed countries where the burning of dung or wood indoors is common; in those countries, indoor air pollution is a more significant determinant of health than is ambient air and so decarbonised energy sources for the home play a central role in public health.
- Decreases damage to buildings and other assets of the built environment such as infrastructure resulting from corrosion.

As a result, the climate benefits of GHG mitigation are bolstered by reference to co-impacts on a range of other environmental policy goals.

9.2. Co-benefits to Human Health

We turn now to the co-benefits of the air pollutant that is of primary concern globally in regard to human health, the reduction of which is a co-benefit of GHG reduction: particulate matter (PM, taken here to be PM-10, or PM with an aerodynamic diameter of 10 microns or less). The methodology is similar to that used in regulatory assessments of policies aimed at other reductions in air or water toxics. The modelling converts estimates of emission rates of air toxics into estimates of ambient air concentration, human health risk, incidence of disease and cost of illness (COI). Statistical values of life, often used to convert mortality risks to costs, are not employed here because such values are not true costs, there is no market for these (and hence they cannot be used in macroeconomic analyses) and their philosophical basis within public policy is contentious.

9.2.1. *Method of estimating co-benefits for PM reduction*

A simplified diagram of the relationships between economic systems, material/energy use, environmental quality and human health is shown in Figure 9.3. It can be seen that impacts on human health from economic activities have two components: (i) the use of energy and/or material in economic activity that generates waste, including emissions of air toxics, which in turn degrades environmental quality and reduces human health, and (ii) income or welfare, including per capita wealth, that influences the sensitivity of populations to environmental degradation such as pollutants in ambient air (poverty being a significant determinant of sensitivity to environmental pollutants, due in part to malnutrition and in part to lack of access to health care), with improved wealth generally decreasing sensitivity to environmental stress.

Calculations of the relationship between health and emissions of an air toxic such as PM are often executed at refined levels of spatial and temporal resolution for regulatory assessments (Crawford-Brown, 2001). This involves (i) creating the emissions inventory for each pollutant for

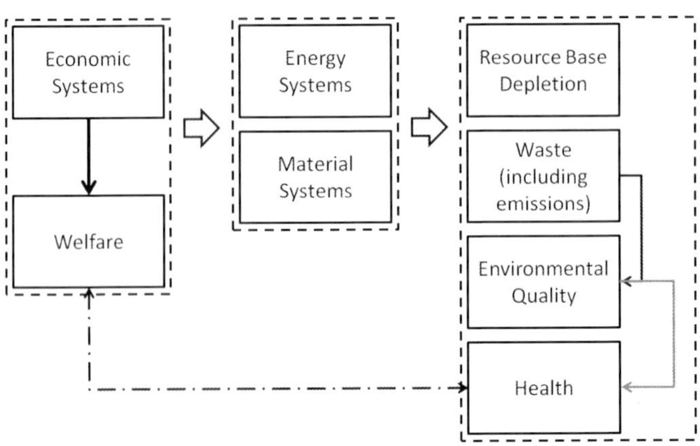

Figure 9.3. Schematic of the relationship between economic activity and human health. Impacts on health are mediated through the energy and material systems affected by the economic system, but there is also a direct effect of economic activity on welfare and, through this, health.

each source category (e.g. mobile, stationary, non-point); (ii) gridding the geographic area, often into census blocks or census tracts of several thousand people each; (iii) assigning each source a pollutant emission rate into the relevant geographic grid block; (iv) using atmospheric dispersion and chemistry models to calculate ambient air concentration; (v) applying "micro-environment" factors to account for differences between ambient air concentration and the concentration indoors, near roadways, etc; (vi) assigning individuals to specific grid blocks; (vii) developing activity patterns that move people between grid blocks (e.g. through travel) and that determine metabolic rates due to physical activities (e.g. breathing rates); (viii) calculating exposures in each grid block based on air concentration and activity; (ix) applying age-specific risk coefficients in each grid block to convert exposures to probability, frequency and/or severity of effect; (x) multiplying per capita incidence by population size to obtain estimates of annual or lifetime disease incidence; and (xi) applying COI values to convert incidence of morbidity and mortality into annual costs of health impacts.

This level of spatial and temporal detail is not feasible for global models of coupled energy-economy-environment policies such as E3MG (energy-economy-environment model at the global level), both because the computational resources required are quite high, justified only for regulatory actions, and because the data required are insufficient for the majority of nations and regions. As a result, the Human Health Module considered here uses an approximation employed elsewhere in the literature (Li, 2008; Li and Crawford-Brown, 2011) in which:

1. An economics model such as E3MG is used to estimate economic activity, energy demand, and energy provision due to a policy scenario. This is repeated for all regions of the global economy. The same is done for the baseline (no policy) scenario.
2. The material and energy use from Step 1 is used to estimate emissions of air toxics such as PM. By running these first two steps both with and without a climate change policy, the difference in emissions resulting from the policy can be calculated. The fractional change in PM emissions due to the policy is equal to the difference in PM

emissions between the "with policy" and "without policy" runs of the model, divided by the PM emissions from the "without policy" run. This is repeated for all regions in the global economy.

3. Empirical factors are developed to convert fractional changes in emissions (Step 2) into fractional changes in ambient air concentration. For PM, the empirical factor is approximately 0.5 (a 50% change in ambient air concentration associated with a 100% change in emissions) based on available global data. This empirical factor varies between regions depending on degree of industrialisation, density of vehicular traffic, mix of petrol and diesel use in vehicles, and existing air pollution control measures. For example, in Hyderabad, India, it is estimated to be 0.5 (Guttikunda and Aggarwal, 2009). The values are approximately 0.53 in several regions of China (Zhang *et al.*, 2004) and 0.4 in Korea (Heo *et al.*, 2009). Roughly similar results have been obtained in the southeastern US (Lee *et al.*, 2007). This chapter uses a central tendency value of 0.5.
4. The result of Step 3 is multiplied by region-specific estimates of baseline ambient air concentration to calculate the change in ambient air concentration resulting from a policy, in units of $\mu g/m^3$. This is the value of Δx in Equation 9.1, below. Where more than one country is in a region, the value used is the population-weighted average over the countries contained in that region. For PM, the assumption used here is that the contribution of a given source to ambient air concentration in a particular region is proportional to the contribution of that source to the total rate of PM emissions in that region, following a process of "source apportionment" (Lee *et al.*, 2007). This is repeated for all regions in the global economy.
5. Changes in ambient air concentration (Step 4) are converted to estimates of change in incidence of disease (number of cases) through application of risk coefficients taken from meta-analyses performed on the global literature concerning population sensitivities, using Equation 9.1, and total exposed population in each region. This is repeated for all regions in the global economy.
6. Finally, changes in COI are determined through application of COI factors for each illness, multiplying these by the results of Step 5 on numbers of cases avoided by the policy.

To translate the benefits of reduced health risks into economic terms, it has been suggested that the most appropriate economic valuation method is willingness to pay (WTP) (e.g. Freeman III, 1993; Hubbell *et al.*, 2005). However, the WTP method is not an appropriate measure of economic impact to be fed back into a macroeconomic model because WTP represents only the stated will to take an action in avoiding an illness, not the amount an individual actually pays in experiencing that illness. Hence, COI, which can be quantified from previous studies (e.g. Smith *et al.*, 1997; Stanford *et al.*, 1999), is used here to evaluate economic impact of morbidity (non-fatal) effects.

The illnesses considered here are respiratory and cardiovascular, as the majority of epidemiological studies conducted worldwide have focused on these effects for exposures to air toxics (USEPA, 2010). Only health endpoints for which the concentration–response (C–R) relationships have been adequately quantified through meta-analysis of worldwide epidemiological studies and that have incorporated control for confounding variables, such as smoking, are included. These studies have consistently shown that the relative risk (percentage change in the background incidence of illness per unit exposure) is stable across regions of the world, even if the absolute risk varies regionally due to changes in background incidence of illness. Based on the US Environmental Protection Agency's list of PM health effects (USEPA, 2008, 2010) and other relevant literature from worldwide studies (e.g. Cifuentes *et al.*, 2001; Ostro and Chestnut, 1998; Thanh and Lefevre, 2001), mortality outcomes and several morbidity outcomes are included here in calculating co-benefits from PM reduction.

The selected C–R functions for each health endpoint are used in health impact functions to calculate the changes in health outcomes (R) resulting from changes in ambient air concentration (C). The Human Health Module uses an exponential C–R function for health effects. Therefore, the health impact function is:

$$\Delta y = y_0 \times (e^{\beta.\Delta x} - 1) \tag{9.1}$$

(Deck *et al.*, 2001; Hubbell *et al.*, 2005; Ostro *et al.*, 2006; Sanhueza *et al.*, 2003) where Δy is the change in the number of cases of each health endpoint (i.e. mortality or morbidity); y_0 is the baseline incidence of cases, equal to the baseline per capita incidence (I) times the potentially affected population:

$$y_0 = I \times POP \tag{9.2}$$

β is the relative risk coefficient, equal to the fractional change in the background incidence of illness per unit change in ambient air concentration; and Δx is the estimated change in the ambient air concentration as a result of a policy being simulated. The baseline incidence by nation is taken for the Human Health Module from the WHO (World Health Organization, 2004).

For premature mortality, meta-analyses of epidemiological studies report an increase of 0.5–1% in daily respiratory and cardiovascular mortality per 10 μg/m^3 increase in daily average PM amongst exposed adults (USEPA, 1996; USEPA, 2008). Since differences in particle characteristics (e.g. chemical composition) and other factors that influence public health status can affect this risk coefficient, the results from studies in both developed and lesser-developed countries were stratified within the meta-analysis, using the US and Europe as the representative developed countries and countries of Southeast Asia as representative of lesser-developed countries. To date, two epidemiological studies on the associations between daily PM concentration and short-term mortality in Bangkok have been published by Ostro *et al.* (1999), and Vajanapoom *et al.* (2002). For a 10 μg/m^3 increase in daily PM, the former reported a 1.0% increase in daily mortality whereas the latter reported 0.77%. These results indicate a significant level of agreement in C–R estimates for acute mortality from Thailand, the US and Europe, despite very different levels of development. Since studies in industrialising countries show similar results (Chen *et al.*, 1998; Wells *et al.*, 1994), an average value of 1% increase in mortality per 10 μg/m^3 increase in daily PM is used here for all regions (i.e. β in Equation 9.1 is 0.001 per μg/m^3).

In contrast to epidemiological studies on mortality, studies on the association of morbidity and air pollution exposure are less comprehensive (Wang and Mauzerall, 2006). The relative risk ratio for chronic obstructive pulmonary disease and pneumonia has been found to be between 1.06 and 1.25 for a 50 μg/m^3 increase in daily PM. This translates into a relative risk coefficient for acute morbidity effects of approximately 1% to 5% per 10 μg/m^3 increase in daily PM, a factor of up to five times that of mortality. Studies in developing countries produce approximately the same range of values (Chen *et al.*, 2002; Hagler Bailly, 1998; Ostro, 1987; Xu *et al.*, 1995), so as in acute mortality effects, a common risk coefficient is used for all regions, set to the median value of 3% per 10 $\mu g/m^3$ increase in daily PM (i.e. β in Equation 9.1 is 0.003 per $\mu g/m^3$).

A significant issue of concern is that currently there is no COI study specific to many of the regions of the global economy. Since COI is significantly affected by income, values may not be readily transferred from a country to another without adjustment (Thanh and Lefevere, 2001). The simplest adjustment corrects only for the income difference between countries (Thanh and Lefevere, 2001). For example, based on the disparity in per capita incomes of the US (US$29,760) and Thailand (US$1,345) as in Crawford-Brown *et al* (2013), a ratio of 0.046 is derived.

However, studies also suggest that income elasticity of WTP or COI is less than one, which means that "other things being equal, WTP in low-income countries is lower than WTP in high-income countries, but proportionately less than the income differentials" (Thanh and Lefevere, 2001). Again using the US and Thailand as an example, Thanh and Lefevere (2001) suggested a ratio of 0.285 to be used to extrapolate US values to Thailand, rather than the value of approximately 0.046 (US$1,345/US$29,760) estimated solely on the basis of income differential. A World Bank report (Hagler Bailly, 1998) used a ratio of 0.2 to extrapolate US WTP values to Thailand. The studies employed in the World Bank report suggest that, for countries in which COI has not been measured directly, the value in lesser-developed countries should be equal to that in developed nations (taken here to be the US and EU), multiplied by the ratio of per capita GDP (lesser developed/developed). Where that ratio is below 0.2, it should be adjusted upwards to 0.2.

The economic value associated with a change in health outcome (V_i), is given by:

$$Vi = A_i \, x \, COI_i \, x \, SF \tag{9.3}$$

where A_i is the change in the incidence (annual number of cases) of a specific health endpoint i, COI_i is the cost of illness per case of effect *i* (for morbidity) and *SF* is the scaling factor equal to the ratio of per capita GDP in the region divided by per capita GDP in the reference developed nation (here, the US), with a minimum ratio of 0.2 as discussed previously.

The specific values for COI are taken from the USEPA Benefits Assessment (USEPA, 2008), using the 2020 values provided in that report. All are given as 2020 USD per statistical case of the effect. They are equal to: US$90,000 for non-fatal heart attack; US$16,600 for hospital admissions for chronic obstructive pulmonary disease; US$8,900 for hospital admissions for asthma; US$24,700 for hospital admissions for all cardiovascular combined; US$24,600 for hospital admissions for all respiratory effects combined; US$130 for work loss days.

The risk coefficients for morbidity effects specify only the increase in PM-related effects overall, and are not tied directly to these specific categories of health care expenditure. In Deck *et al.* (2001), the summary of studies examined produce a ratio of all respiratory effects over hospital admissions and extended health care (including cardiovascular and respiratory admissions) — as a result of PM exposures — equal to 0.018/0.0016 or approximately 10:1. From the cost information above, the average cost per case of hospital admission and/or significant health care is US$27,000. Since only 10% of effects are assumed to result in such health care costs, the average cost per effect used here is US$2,700 (alternatively, the COI value is left at US$27,000 and the number of effects multiplied by 0.1).

9.2.2. *Applying the methodology*

What are the implications of the co-benefits of global decarbonisation when such policies also reduce PM emissions? To explore this question, we consider policies that focus solely on reducing energy consumption from fossil fuels, in which a percentage change in GHG emissions is accompanied by the same percentage change in PM emissions. This is equivalent to assuming that the climate policy changes the total consumption of fuels by changing demand, but not the mix of fuels or the technologies in which they are used.

We then consider percentage change in GHG emissions that vary between 0 and 100%. Fractional reductions in GHG and hence PM emissions are assumed uniform in all nations globally (which differs from many of the results reported elsewhere in this book). The previously outlined approach was used in all nations, including the assumption that a fractional reduction in GHG emissions of 0.1 produces the same value of 0.1 fractional reduction in PM emissions from the modelled components of the economy, and the further assumption that a fractional reduction of 0.1 in PM emissions from the economy produces a fractional reduction of 0.05 in ambient air concentration (as described previously).

Published values of the ratio of percentage PM reduction to percentage carbon dioxide reduction vary from 0.77 (Cifuentes *et al.*, 2001) to 1.0 in transport resulting from fuel efficiency improvements (Caton and Constable, 2000) and as high as 140% (Chanhong *et al.*, 2004). Policies that cause a switch to biomass fuel, carrying carbon dioxide emissions reductions (by up to 80%), result in reductions of up to 75% when the switch is from coal-fired plants, but can increase PM emissions when the switch is from fuel oil, gas oil or natural gas (AEA, 2008). Hence the assumption in the current study that a percentage reduction in PM emissions equals that of the carbon dioxide emission reduction appears valid.

Figures 9.4, 9.5 and 9.6 show the reduced annual costs and number of cases of morbidity (all PM-related causes combined) and mortality

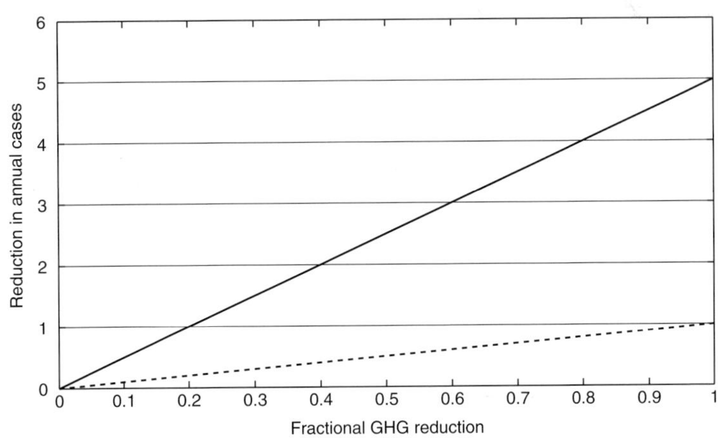

Figure 9.4. Reduction in annual PM-induced morbidity (respiratory and cardiovascular combined) globally for different levels of GHG emission reductions shown on the x-axis. The solid line is the case where all nations reduce GHG emissions; the dashed line is the case where only Annex I parties reduce them.

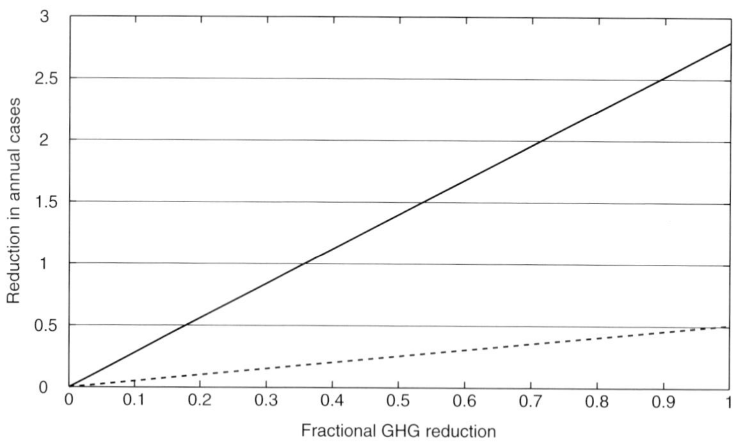

Figure 9.5. Reduction in annual PM-induced mortality (respiratory and cardiovascular combined) globally for different levels of GHG emission reductions shown on the x-axis. The solid line is the case where all nations reduce GHG emissions; the dashed line is the case where only Annex I parties reduce them.

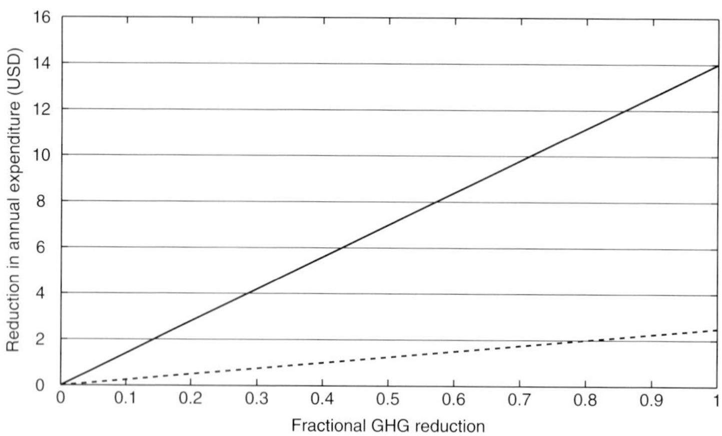

Figure 9.6. Reduction in annual cost of PM-induced illness from morbidity (respiratory and cardiovascular combined) globally for different levels of GHG emission reductions shown on the x-axis. The solid line is the case where all nations reduce GHG emissions; the dashed line is the case where only Annex I parties reduce them.

(number of cases only) globally as a function of the fractional reduction in GHG emissions. The reader should bear in mind that the target emissions reduction set by international discussions, as explored in Chapter 1, is in the neighbourhood of 0.6 to 0.8 by 2050. In Figure 9.4, the results shown as solid lines assume that climate policies cause emission reductions uniformly across all nations, with no distinction between Annex I and non-Annex I parties (and hence no distinction between developed and developing economies). The results shown as dashed lines assume that reduction policies are in place only in the Annex I, developed, parties. All results under both cases are for years after attainment of the GHG reduction levels shown on the x-axes.

9.2.3. *Policy implications*

The examination of co-benefits from PM reduction alone support the contention that co-impacts of climate change mitigation strategies can be a significant adjunct to policies focused solely on climate change itself. Consider as an example the results of Figure 9.5. The reduction in lives lost (mortality) associated with an 80% reduction in carbon dioxide emissions globally is approximately 220,000 per year across these nations. With a global population of 7 billion, this corresponds to a reduction in mean annual probability of death equal to 3.1×10^{-5}, or about three chances per year per 100,000 people. Over the course of a lifetime of 70 years, the reduction in mean lifetime probability of death is 2.2×10^{-3}.

Environmental regulations tend to focus on a "fuzzy bright line" in which a mean lifetime probability of death from a given pollutant of below 1×10^{-6} is considered *de minimus* ("beneath the interest of the law") and above 1×10^{-4} is considered a level requiring corrective action. Note that the co-benefits of GHG reduction examined here for PM are at least an order of magnitude above the upper level of this "fuzzy bright line" and so provide health benefits that would be considered a valid basis for regulatory controls on emissions of other air toxics. Clearly these co-benefits add significantly to the arguments for decarbonisation of the world's economy.

Short-term action on climate change is restricted to the Annex I parties under the UN Framework Convention on Climate Change (UNFCCC).

If attention is focused solely on these nations, arguing that they are the ones that must marshal the will to political action first, the results are not dissimilar to the global results above. The reduction in lives lost (mortality) associated with an 80% reduction in carbon dioxide emissions in only the Annex I parties is approximately 50,000 per year across these nations. With a population of 1.3 billion in the Annex I parties, this corresponds to a reduction in mean annual probability of death equal to 3.9×10^{-5}, or about four chances per year per 100,000 people. Over the course of a lifetime of 70 years, the reduction in mean lifetime probability of death is 2.7×10^{-3}. Again, these co-benefits would usually be considered adequate justification for regulation under risk-based approaches in the Annex I parties. The caveat here, both for Annex I parties and globally, is that any legal attempt to use the co-benefits in such a risk-based argument would invariably result in challenges to demonstrate that GHG emission reductions are the most effective means to obtain the co-benefits.

The calculation above applies only once the 80% reduction of GHG emissions has been achieved. A key reason for invoked co-benefits, however, is that these benefits are seen in the short term — certainly within the span of a career of a politician who must support climate policies. Hence we need to consider the timing of the reductions and not simply what the co-benefit will be after the nations have decarbonised by 80%, which might be as late as 2050 under the UNFCCC. Figure 9.7 examines the lifetime reduction in risk from the co-benefits of PM reduction at the levels of GHG reductions in Figures 9.4–9.6. Also shown on the figure is the 1×10^{-4} risk (as the horizontal dashed line), as well as the approximate year in which that level of risk reduction is obtained, assuming a linear decrease in emissions between 2010 and 2050 down to 80% reduction by 2050.

Note from Figure 9.7 that the threshold of a 1×10^{-4} risk reduction is reached almost immediately, suggesting that a risk-based approach to GHG emission reductions on the back of PM co-benefits apply essentially from 2012.

The results above do not include changes in socio-economic status (SES) as the economies of nations — especially in the developing nations, grow and raise the standard of living for their populations. The effect of SES on sensitivity to pollution is not well established at present, although

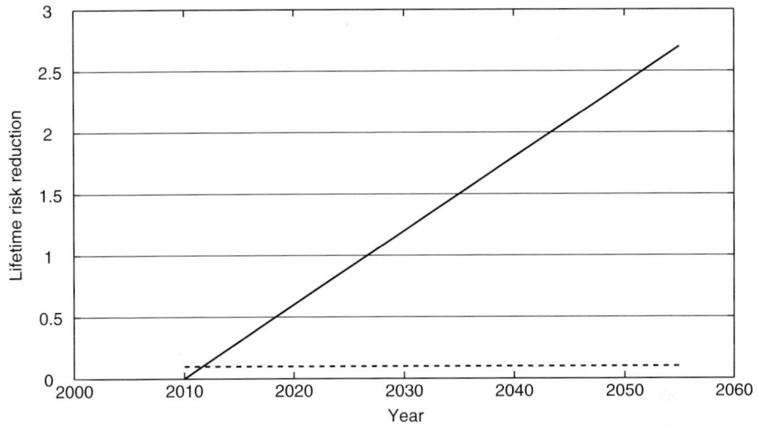

Figure 9.7. Reduction in lifetime excess risk of mortality from PM reduction as a function of time, assuming reductions in GHG emissions are approximately 2% per year.

Chen (2013) has produced an analysis based on the global literature, with findings showing that sensitivity for both morbidity and mortality effects decreases by a factor of between 1.5 and 2 as annual household income level increases up to approximately US$25,000, but then rises again with further increases in SES (presumably due to the onset of obesity and the cardiovascular diseases that brings). Such a pattern suggests that co-benefits could be reduced in the poorer nations as they improve their economies over time, but this effect will be in part offset in the middle-income nations as their improving economies push their populations into income levels associated with higher sensitivity. The net effect of these changes in SES as economic development proceeds globally cannot be estimated at present.

9.3. Water, Food and Energy

The summary of politics surrounding climate policy provided in Chapter 7 indicates that effective policies often require mobilising the support of multiple constituencies. These constituencies have different aims and will support climate policies if their own aims are furthered by what might begin as a climate policy but brings a variety of benefits. At the least, these additional aims must not be contradicted by, or come into conflict with, the proposed climate policy. No policy exists in isolation of others.

9.3.1. *Co-impacts of policies relating to land use and land-use change*

An emerging example of this mobilisation of the will to action is a set of policies aimed at reducing GHG emissions from land use and land-use change (LULUC) and keeping carbon sequestered in the biosphere. Aside from the benefits to the carbon cycle, such policies are tied politically to efforts simultaneously promoting poverty alleviation through sustainable livelihoods, increasing the resilience of crops to a changing climate, improving the market value of those crops and attracting investment capital from both donors and the private sector. We therefore consider here whether programmes such as REDD+ (Reducing Emissions from Deforestation and Degradation), climate smart agriculture or development of livelihoods based on growth of crops for bioenergy might carry net beneficial or detrimental co-impacts on non-climate aims (and whether these aims are supported or weakened by the climate policies).

The global population is projected to increase to 9 billion by 2050 (UNDESA, 2004), potentially requiring an increase in crop production by 60% or more to meet food demand (Bruinsma, 2009). Since agriculture contributes approximately 25% of the global emissions of GHGs through LULUC, there is increased focus on "climate smart" agricultural practices, which the Food and Agriculture Organization (FAO) defines as "agriculture that sustainably increases productivity, resilience (adaptation), reduces/removes GHGs (mitigation), and enhances achievement of national food security and development goals", and on keeping carbon stored in forests through programmes such as REDD+. The emissions from LULUC are primarily from developing nations, and the negative impacts of climate change on crop production are also concentrated in these regions. This has led to a need for programmes of international finance that allow for economic development of the poorest in these nations, raising populations at the "bottom of the economic pyramid" out of poverty while using LULUC policies and instruments to reduce GHG emissions and reduce the vulnerability of developing economies to climate impacts.

Both economic growth and climate policy in the developing nations therefore are increasingly linked through a focus on sustainable development. These linkages, usually called the food-water-energy nexus (World Economic Forum, 2011), raise a broad issue of land management or land economics, requiring a balancing of potentially competing uses to which land and other natural resources can be put in the face of climate change risks. We use this nexus here as a representative example of the potential co-impacts of decarbonisation through LULUC policies, and ask whether economic development, climate aims and the food-water-energy nexus are complementary or contradictory forces. We explore this question by considering three interconnected issues:

- The potential competition between land for fuel and land for food.
- The REDD+ programme to reduce emissions from conversion of forests to cropland.
- The introduction of climate smart agricultural practices in rural communities as a way to secure sustainable livelihoods.

To support a growing and more affluent population, the FAO has projected a 50% increase in the demand for food while the IEA forecasts a 40% increase in energy demand and IFRI anticipates a 30% increase in demand for water by 2030 (World Economic Forum, 2011). These increases will result in pressures on natural resources such as land and water and on natural services such as those provided by ecosystems, pressures which are already strong inhibitors of traditional agricultural practices in many parts of the world and will be compounded by the anticipated changes in hydrology and weather patterns resulting from climate change. These same increases in pressure have important implications for climate mitigation aims in regard to the potential for increased contributions to emissions from LULUC, reduced sequestration of carbon in forests (as these forests are converted to cropland) and increased reliance on bioenergy. Failure to address these issues by coordinating them with climate policy may result in increasing economic disparity and a high risk of geopolitical conflict as peoples and nations compete for finite natural resources. Were this conflict to emerge, it could derail climate policies.

9.3.2. *Biomass and bioenergy*

Turning first to biomass as a source of potentially low-carbon energy, there is increased reliance on bioenergy as part of the strategy for decarbonising energy provision in nations. Bioenergy from biomass is currently the largest global renewable energy resource being exploited (50.3 EJ/yr, 10.2% of global annual energy) (IEA, 2009a). Most of this bioenergy is woody biomass burnt to provide fuel for heating and cooking in the developing world, in a low-technology, inefficient, highly polluting method that has been in use for millennia. This use of biomass degrades indoor air conditions, causing many of the same health effects noted in Section 2 of this chapter.

Bioenergy feedstocks are a primary energy source which can be converted into high-quality secondary energy carriers such as electricity or liquid transport biofuels, which can feed directly into current energy systems, with applications in both centralised and decentralised settings. Biomass can be stored offering a constant and controllable output and a fill-in source of heat and power when intermittent renewable energy resources such as wind and solar are unavailable or insufficient to meet demand.

9.3.3. *The technical potential of biomass*

A broad range of bioenergy conversion technologies are commercially available or in pre-commercial stages as well as many emerging in demonstration and research and development phases. Based on this diverse range of feedstocks and technologies, the technical potential for biomass is estimated in the literature to be as high as 1500 EJ/yr by 2050 without taking into account sustainability, food-for-fuel or cost constraints (IEA, 2009a).

It is estimated that biomass can provide 100 EJ/yr to 300 EJ/yr of total global energy requirements in 2050 even under constraints of sustainability, as estimated by the IPCC scenario reviews of potential global future energy resource mix (IPCC, 2011). This represents between 10% and 30% of projected total world energy demand by 2050. The upper end of this range would require a two-to-three fold increase in bioenergy feedstocks worldwide, much of which might be expected to be supplied by programmes in developing nations. Since sustainably harvested biomass for energy offers an 80 to 90% GHG emission reductions compared to the fossil fuels that it can replace (IPCC, 2011), there will be increasing policy pressures to increase the supply of biomass for energy. However, those reductions can decrease dramatically or even disappear if the feedstock is not grown sustainably (Searchinger *et al.*, 2008).

9.3.4. *Growth of bioenergy and pressure on land and water resources*

This expected growth in bioenergy in order to reduce GHG emissions could result in pressure to devote a high proportion of the global land and water resources to bioenergy production at the expense of food production, especially in the developing nations. Currently only 1% of agricultural land is utilised for bioenergy production, but this would increase significantly under the more ambitious global bioenergy targets mentioned above (IEA, 2010). There is a danger that land needed to meet the projected demand for food production if used instead to meet the demand for bioenergy production will cause spikes in food prices or food shortages.

This could prove particularly problematic if bioenergy crops for export to developed regions are grown in developing and ecologically vulnerable regions. Competition over finite land resources required to meet both food and energy production also can have negative social impacts related to, for example, land rights. Managed correctly though, the production of bioenergy in developing regions has the potential to be a source of locally generated cheap energy and fuel particularly beneficial for rural populations (IEA, 2009a).

The availability of sufficient water in order to grow the forecasted volume of food, feed and fibre as well as bioenergy threatens to be a significantly limiting factor in meeting future demands for food and energy. Agriculture, primarily for food production, already consumes more than 70% of the total global water demand with the production of meat being particularly highly water-intensive due to the double demands of the volume of feed crops required in feeding and in the process of rearing livestock.

Dedicated bioenergy crops are anticipated to increase this demand with the IEA's prediction alone indicating that the 5% of global transport fuel in 2030 (estimated 3.2 million barrels a day) to be met by biofuels will require between an increase of 20–100% of the water demand for current agricultural needs (World Economic Forum, 2011). Clearly this level of water consumption requires trade-offs between the components of the food-energy-water nexus in order to meet growing demand for all three.

Managing these trade-offs is anticipated to be best met through a combination of market mechanisms and regulation.

9.3.5. *Reducing GHG emissions from land-use change*

Every land type contains a different level of stored carbon sequestered as soil organic carbon (SOC) and in vegetation, both above and below the ground. High-carbon stock land includes forests and low-carbon stock land includes desert and other sparsely vegetated areas. Increasing demand for land to produce food and energy crops could lead to pastures and grasslands being converted to croplands for annual arable crop agriculture with practices that release GHGs. Changes in land use in order to provide land directly for bioenergy production (land-use change; LUC) or the use of land for bioenergy production resulting in another type of land elsewhere replacing that of the land use displaced by bioenergy production (indirect land-use change; ILUC) can negate the GHG savings offered by bioenergy over their fossil fuel equivalents (IEA, 2009b).

GHGs are emitted from burning biomass during land clearing and in the change of land management practice altering the carbon stocks as deforestation and agricultural practices release the carbon stored in vegetation and SOC. Hence the REDD+ programme was introduced as a climate policy aimed at preventing stored carbon in forests from being released in clearing land for agricultural (including bioenergy) activity.

The intensive and continued tillage practices used in the production of arable crops have been calculated to cause enormous losses of SOC and nitrogen from soils as GHGs to the atmosphere. It is estimated that as much as 60% of SOC in temperate regions and 75% in the tropics has been depleted by intensive plough tilling, contributing around 23% of the total GHG concentration in the atmosphere (IPCC, 1996). Changes in agricultural practices such as no-tillage, can significantly reduce the SOCs lost and the GHGs emitted from soils. This has resulted in the introduction of the third issue discussed in this section: policies aimed at promoting climate smart agricultural practices, tied to development aid.

9.3.6. *Co-impacts from reducing GHG emissions from fertilisers*

Application of chemical fertilisers containing reactive nitrogen compounds has been shown to compromise soil and water quality. They release GHGs (e.g., CO_2, CH_4, N_2O, NOx), and increase tropospheric ozone and particulate matter, resulting in local air pollution and endangering surrounding ecosystems and biodiversity through leaching that damages the balance of surrounding species and the quality of surrounding bodies of water. The leaching of reactive nitrogen compounds from fertilisers into surrounding soils and waterways is not only a waste of energy in lost nitrogen but can also upset the balance of sensitive ecosystems that have evolved to exist in low-nitrogen conditions (European Nitrogen Assessment, 2011).

Excess reactive nitrogen compounds can also cause water pollution such as eutrophication, supporting excessive algal blooms in waterways, whose rapid growth, death and decomposition deplete the water of available oxygen, causing the death of other organisms such as fish. Future farming methods and applications of chemical fertilisers will have to be at a level to maintain land productivity but also to reduce the energy demands of crop production and losses through leaching. Key challenges include the development of technologies and agricultural practices to minimise nitrogen loss and increase efficiencies that include future arable crops that have the potential to fix their own nitrogen or can use nitrogen more efficiently resulting from the use of plant engineering, whether by GM, selective breeding or other technologies. Clearly climate smart agriculture, while introduced ostensibly as a climate policy, has implications far beyond climate change.

9.3.7. *Co-impacts from REDD+ and finance for low-carbon development*

The challenges in implementing REDD+, climate smart agricultural practices and bioenergy production have not primarily been technological. They have instead been related to problems of governance and finance. Introducing them at the scale required for effective global climate policy requires the transfer of finance from the developed to developing nations, including significant investment by the private sector. For example, the African Union's Comprehensive Africa Agriculture Development Programme (African Union, 2003) notes the importance of increasing agricultural growth and investment into the sector in order to achieve food security and poverty alleviation.

Such flow of finance could provide livelihoods for local populations, reducing the need for higher-carbon livelihoods. Hence for development organisations such as the Department for International Development (DFiD) in the UK, these programmes are seen as much more than climate policies; they have the primary aim of promoting sustainable livelihoods, alleviating poverty and increasing equity in developing nations. The challenges emerging are:

- Finance of these programmes has been hampered by a lack of clarity over land tenure. Where ownership of the land is unclear, private finance will not flow.
- The development programmes may serve the climate policy aims initially, but they also increase the disposable income of local populations. An unintended consequence of the improved wealth of the local populations is a shift to more consumptive and carbon-intensive lifestyles.
- Governance mechanisms are not in place in many countries to ensure aid, and the profits from shifting agricultural and forestry practices, appear in the hands of local communities and individual farmers rather than of governments. When this occurs, participation in REDD+, climate smart agricultural practices and bioenergy production are often seen as means to raise government income, not as ways

to reduce global climate risks, and so the allocation of funds on the ground is driven by the former aims and not the latter.

- This same issue of governance can result in widening gaps in income and a decline of equity, where donor funds and revenue from crops and bioenergy profit the upper classes in control of government decisions rather than the local communities and farmers who carry out the work.
- The measurement, reporting and verification (MRV) procedures needed to ensure these programmes produce beneficial climate effects are not yet in place at local levels. This exacerbates the problem of programmes attracting aid in support of climate policies but providing little or no benefit to those policies.
- Environmental regulations in developing nations are lagging behind the creation of aid programmes. As a result, the pressures on water, soil and air described earlier are not yet factored into many of the development programmes, which instead focus on economic development and climate policy aims.

9.3.8. *Conclusions on co-impacts relating to food, energy and water*

These are not insurmountable challenges. If properly executed, policies such as REDD+ could reduce the problems associated with the food-energy-water nexus while also bringing much needed economic development to the billions of poor at the bottom of the economic pyramid. Bioenergy can for example be produced sustainably using the inedible lignocellulosic portions of food crops and growing perennial lignocellulosic crops that typically do not require the application of chemical fertilisers. Dedicated bioenergy crops could be grown on marginal land unsuitable for arable agriculture that ideally requires little or no irrigation (to avoid competition with resources for food and water as well as preventing clearance of forests). Governance mechanisms could be put in place to ensure project finance and revenues flow to the local communities and farmers. Green economies could be created to allow improvement of living conditions and disposable income without necessarily increasing per capita GHG emissions associated with the lifestyles of developed nations.

Gaining support for climate policies will therefore require recognising these challenges and demonstrating that the policies support, or at least are not in conflict with, the other policies aimed at economic development, equity, human rights, environmental protection and resource conservation. Where this is accomplished, climate policies will find much broader political support, and finance will be both unlocked and flow to where it is most needed.

References

AEA (2008). Measurement and Modelling of Fine Particulate Emissions (PM10 & PM2.5) from Wood-burning Biomass Boilers. Report to the Scottish Government, 26 September.

African Union (2003). New Partnership for Africa's Development (NEPAD), Comprehensive Africa Agriculture Development Programme, Report by the African Union.

Bollen, J., Guay, B., Jamet, S. and Corfee-Morlot, J. (2009). Co-benefits of Climate Change Mitigation Policies: Literature Review and New Results. OECD Economics Department, Working Paper 693.

Bruinsma, J. (2009). The Resource Outlook to 2050, in Experts Meeting on 'How to Feed the World in 2050' (Rome), Food and Agricultural Organization. Available online at: ftp://ftp.fao.org/docrep/fao/012/ak542e/ak542e00.pdf.

Caton, R. and Constable, S. (2000). *Clearing the Air: A Preliminary Analysis of Air Quality Co-benefits from Reduced Greenhouse Gas Emissions in Canada*, David Suzuki Foundation, Vancouver, BC.

Chen, C., Chen, D., Green, C. and Wu, C. (2002). Benefits of Expanded use of Natural Gas for Pollutant Reduction and Health Improvement in Shanghai, *The Sinosphere Journal*, 5, 58–64.

Chen, P. (2013). *The Health Co-benefit of Global Decarbonisation, the Climate Change Act 2008 and the Effect of Income and Age*, MPhil dissertation, Department of Land Economy, University of Cambridge.

Chen, P., Lai, Y., Wang, J., Yang, C., Hwang, J., Kuo, H., Huang, S. and Chan, C. (1998). Adverse Effect of Air Pollution on Respiratory Health of Primary School Children in Taiwan, *Environmental Health Perspectives*, 106, 331–335.

Cifuentes, L., Borja-Aburto, V. H., Gouveia, N., Thurston, G. and Davis, D. L. (2001). Assessing the Health Benefits of Urban Air Pollution Reductions Associated with Climate Change Mitigation (2000–2020): Santiago, Sao Paulo, Mexico City, and New York City, *Environmental Health Perspectives*, 109(Suppl. 3), 419–425.

Crawford-Brown, D. (2001). *Mathematical Methods of Environmental Risk Modelling*, Kluwer Academic Publishers, Boston.

Crawford-Brown, D., Chen, P.-C., Shih, H.-C. and Chao, C.-W. (2013). Linking Climate Change Co-benefits Risk Reduction and Macroeconomic Modelling, *Journal of Environmental Management*, 125, 1–6.

Deck, L. B., Post, E. S., Smith, E., Wiener, M., Cunningham, K. and Richmond, H. (2001). Estimates of the Health Risk Reductions Associated with Attainment of Alternative Particulate Matter Standards in two U.S. cities, *Risk Analysis*, 21, 821–836.

European Nitrogen Assessment (2011). *Sources, Effects and Policy Perspectives*, Cambridge University Press, Cambridge.

Freeman III, A. (1993). *The Measurement of Environmental and Resource Values: Theory and Methods*, Resources for the Future, Washington, DC.

Guttikunda, S. and Aggarwal, R. (2009). Contribution of Vehicular Activity to Air Pollution in Hyderabad, India, *Indian Journal of Air Pollution Control*, 9, 37–46.

Hagler Bailly (1998). Final Report: Health Effects of Particulate Matter Air Pollution in Bangkok. A report prepared for Air Quality and Noise Management, Pollution Control Department, Bangkok, Thailand, March 1998.

Heo, J., Hopke, P. and Yi, S-M. (2009). Source Apportionment of PM2.5 in Seoul, Korea, *Atmospheric Chemistry and Physics*, 9, 4957–4971.

Hubbell, B., Hallberg, A., McCubbin, D. R. and Post, E. (2005). Health-related Benefits of Attaining the 8-Hr Ozone Standard, *Environmental Health Perspectives*, 113, 73–82.

IEA (2009a). *Bioenergy — A Sustainable and Reliable Energy Source: A Review of Status and Prospects*, IEA Bioenergy: ExCo: 2009:06.

IEA (2009b). *Bioenergy — The Impact of Indirect Land Use Change. Summary and Conclusions from the IEA Bioenergy ExCo63 Workshop*, IEA Bioenergy: ExCo: 2009:04

IEA (2010). *Bioenergy, Land Use Change and Climate Change Mitigation. Report for Policy Advisors and Policy Makers*, IEA Bioenergy: ExCo: 2010:03

IIASA (2004). Medium-term Mitigation for Long-term Stabilization: National Mitigation Potentials, Costs and Co-benefits. Presentation at the COP10, December 8, 2004.

IPCC (1996). *IPCC Second Assessment Climate Change 1995: A Report of the Intergovernmental Panel on Climate Change*, Cambridge University Press, Cambridge.

IPCC (2011). 'Bioenergy', *Special Report on Renewable Energy Sources and Climate Change Mitigation*, Edenhofer, O., Pichs-Madruga, R., Sokona, Y., Seyboth, K., Matschoss, P., Kadner, S., Zwickel, T., Eickemeier, P., Hansen, G., Schlömer, S. and von Stechow, C. (eds), Cambridge University Press, Cambridge, pp. 209–331.

Lee, S., Russell, A. and Baumann, K. (2007). Source Apportionment of Fine Particulate Matter in the Southeastern United States, *Journal of the Air and Waste Management Association*, 57, 1123–1135.

Li, J. (2002). Including the Feedback of Local Health Improvement in Assessing Costs and Benefits of GHG Reduction, *Review of Urban and Regional Development Studies*, 14, 282–304.

Li, Y. (2008). *Health Benefits of Traffic-Related Particulate Matter Control Policies: The Case of Bangkok, Thailand*, PhD dissertation, University of North Carolina at Chapel Hill, Chapel Hill, NC.

Li, Y. and Crawford-Brown, D. J. (2011). Assessing the Co-benefits of Greenhouse Gas Reduction: Health Benefits of Particulate Matter Related Inspection and Maintenance Programs in Bangkok, Thailand, *Science of the Total Environment*, 409, 1774–1785.

Markandya, A. and Chiabai, A. (2009). Valuing Climate Change Impacts on Human Health: Empirical Evidence from the Literature, *International Journal of Environmental Research and Public Health*, 6, 759–786.

Ostro, B. D. (1987). Air Pollution and Morbidity Revisited: A Specification test, *Journal of Environmental Economics and Management*, 14, 87–98.

Ostro, B. D. and Chestnut, L. (1998). Assessing the Health Benefits of Reducing Particulate Matter Air Pollution in the United States, *Environmental Research (Section A)*, 76, 94–106.

Ostro, B. D., Chestnut, L., Vichit-Badakan, N. and Laixuthai, A. (1999). The Impact of Particulate Matter on Daily Mortality in Bangkok, Thailand, *Journal of Air and Waste Management*, 49, 100–107.

Ostro, B. D., Tran, T. and Levy, J. I. (2006). The Health Benefits of Reduced Tropospheric Ozone in California, *Journal of Air and Waste Management*, 56, 1007–1021.

Searchinger, T., Heimlich, R., Houghton, R.A., Dong, F., Elobeid, A., Fabiosa, J., Tokgoz, Hayes, D. and Yu, T.-H. (2008). Use of U.S. Croplands for Biofuels

Increases Greenhouse Gases through Emissions from Land-use Change, *Science*, 319, 1238–1240.

Selin, N., Wu, S., Nam, K., Reilly, J., Paltsev, S., Prinn, R. and Webster, M. (2009). Global Health and Economic Impacts of Future Ozone Pollution, *Environmental Research Letters*, 4, 1–9.

Sanhueza, P., Reed, G. D., Davis, W. T. and Miller, T. L. (2003). An Environmental Decision-making Tool for Evaluating Ground-level Ozone-related Health Effects, *Journal of Air and Waste Management*, 53, 1448–1459.

Smith, D. H., Malone, D. C., Lawson, K. A., Okamoto, L. J., Battista, C. and Saunders, W. B. (1997). A National Estimate of the Economic Costs of Asthma, *American Journal of Respiratory and Critical Care Medicine*, 156, 787–793.

Stanford, R., McLaughlin, T. and Okamoto, L. J. (1999). The Cost of Asthma in the Emergency Department and Hospital, *American Journal of Respiratory and Critical Care Medicine*, 160, 211–215.

Thanh, B. D. and Lefevre, T. (2001). Assessing Health Benefits of Controlling Air Pollution from Power Generation: The Case of a Lignite-fired Power Plant in Thailand, *Environmental Management*, 27, 303–317.

UNDESA (2004). *World Population to 2300*, UN Department of Economic and Social Affairs, ST/ESA/SER.A/236.

USEPA (1996). *Air Quality Criteria for Particulate Matter*, Office of Research and Development, Office of Health and Environmental Assessment, EPA report no. EPA/600/P-95/001aF.

USEPA (2008). *Integrated Science Assessment for Particulate Matter*, Office of Research and Development. Washington, DC.

USEPA (2010). *Environmental Benefits Mapping and Analysis Program User's Manual*, Office of Air Quality Planning and Standards, Research Triangle Park, NC.

Vajanapoom, N., Shy, C. M., Neas, L. M. and Loomis, D. (2002). Associations of Particulate Matter and Daily Mortality in Bangkok, Thailand, *The Southeast Asian Journal of Tropical Medicine and Public Health*, 33, 389–399.

Wang, X. and Mauzerall, D. L. (2006). Evaluating Impacts of Air Pollution in China on Public Health: Implications for Future Air Pollution and Energy Policies, *Atmospheric Environment*, 40, 1706–1721.

Wells, G., Xu, X. and Johnson, T. (1994). 'Valuing the Health Effects of Air Pollution: Application to Industrial Energy Efficiency Projects in China', Report 23807: China: Issues and Options in Greenhouse Gas Emissions Control, no. 8, Washington DC: The World Bank.

World Economic Forum (2011). *Water Security: The Water-Food-Energy-Climate Nexus*, Island Press. Washington, DC.

World Health Organization (2004). Global Burden of Disease. Available online at: www.who.int/topics/global_burden_of_disease/en. Accessed 21 March 2013.

Xu, X., Li, B. and Huang, H. (1995). Air Pollution and Unscheduled Hospital Outpatient and Emergency Room Visits, *Environmental Health Perspectives*, 103, 286–289.

Zhang, Y., Zhu, X., Slanina, S., Shao, M., Zeng, L., Hu, M., Bergin, M. and Salmon, L. (2004). Aerosol Pollution in some Chinese Cities, *Journal of Pure and Applied Chemistry*, 76, 1227–1239.

Chapter 10

Conclusions

Terry Barker

10.1. The Basis from Climate Science for Global Decarbonisation

Mean global temperature rises from climate change are likely to pass the 2°C threshold, as agreed and repeatedly confirmed by world governments in successive Intergovernmental Panel on Climate Change (IPCC) meetings as the maximum risk acceptable, by 2060 or 2070 if current emissions continue. Reducing the probability of this occurring before 2100 to levels of confidence above 30% requires limiting accumulated emissions between 2010 and 2050 to the neighbourhood of 1000–1500 Gt CO_2, about 5% to 30% below 2014 levels.

There is significant uncertainty in the temperature increase by any point in time, with the uncertainty characterised by a factor of 2. Uncertainties in precipitation change are even larger, and uncertainties in damage due to climate change are larger still. Climate policies therefore are made in an arena of significant uncertainty. If uncertainty is dealt with through the precautionary principle, this suggests the cumulative emissions mentioned above should be closer to 1000 than 1500 Gt CO_2.

There is increasing focus on adaptation in addition to mitigation in reducing the risks of climate change, as it becomes increasingly less evident that the 2°C limit will be achieved. However, this book is dedicated to showing pathways forward that would allow the world to retain this limit as an aspiration.

All climate policies must reflect the need for economic growth in the poorest nations, with recognition that there are vulnerable nations that will experience the greatest impacts of climate change yet currently have insufficient capacity to respond through adaptation. This requires some form of both deeper mitigation by developed nations and transfer of adaptive capacity to these vulnerable nations.

10.2. Policies and Measures for Decarbonisation

There are many ways to reduce greenhouse gas (GHG) emissions. Voluntary abatement, such as corporations operating an internal cap-and-trade scheme, or "save energy" campaigns are useful in raising awareness but are too fragmented, unreliable and uncoordinated given the scale of the problem. Broadly based, international, government action is required. Government policies divide into regulations affecting products and processes, i.e. command-and-control legislation, and those affecting prices and incentives, so-called market-based instruments. Each group has its advantages. Command-and-control is appropriate when market-based instruments are ineffective either because markets do not work or work too slowly, e.g. when coal plants for power generation become inefficient through reductions in gas prices, but continue to operate at a loss. Market-based instruments can provide a diffused and continuous economic pressure to reduce use of carbon-based fuels via the carbon price, imposed by a tax or an emissions trading scheme (ETS). In addition they can give incentives for investment in low-carbon equipment via feed-in tariffs or other subsidies or allowances.

The case for international coordination arises not just for efficiency but for fairness. It can reduce the risks of "free-riding" i.e. countries taking advantage of others' international action and the risks of loss of competitiveness and "carbon leakage" from unilateral action. Against these concerns there is the first-mover advantage, in that technological innovation in low-carbon equipment and procedures can lead to increases in competitiveness. There are also advantages in domestic action such as reductions in local air pollution from non-GHG emissions.

Since various policies and measures complement each other and the circumstances of countries vary, a different mix of policies is appropriate depending on country, existing policies and international agreements.

10.3. Scenarios for Assessing Decarbonisation

The basis of the economic assessment of feasibility is the scenarios approach, prevalent in the literature on climate change mitigation. All the scenarios adopt the conventional assumption that they exclude the effects of climate change on the economy. These are likely to be relatively small in relation to the expected growth of the economy, as the IPCC Fifth Assessment Report (AR5) makes clear. These scenarios are not forecasts but plausible outcomes depending on assumptions about (a) critical but uncertain factors, such as world oil prices, and (b) policies and measures by governments to reduce GHG emissions.

Two sets of scenarios are developed in Chapters 4 and 5 using energy-environment-economy model at the global level (E3MG). One set includes variants of a reference scenario, with annual projects at a detailed 20-region level, of population, the economy and air emissions from human activity, but with largely business-as-usual economic and environmental policies. The other set includes the Decarbonisation Scenario and is developed from the first set assuming the same population projections but including a mix of policies and measures to decarbonise the world economy. The Decarbonisation Scenario in the model solution projects accumulated emissions, 2010 to 2050, of 1036 Gt CO_2. This outcome is close to the lower end of the 1000–1500 Gt CO_2 carbon budget discussed above as justified by the precautionary principle. The outcome is between the AR5 Representative Concentration Pathways (RCPs) of RCP2.6, a mitigation scenario that easily achieves the 2°C target by 2100 and RCP4.5, which misses it. The Reference Scenario is between the RCP4.5 and RCP6.0, implying substantial global warming.

For the scenarios to be plausible, we have chosen to model the policies actually used and proposed by governments as summarised by the International Energy Agency (IEA). The projections in the scenarios are adapted to follow the broad outlines of the IEA's 2010 World Energy Outlook (WEO). The Reference Scenario follows the IEA's Current Policies Scenario and the Decarbonisation Scenario follows the IEA's 450 Scenario. However, we have extended the scenarios to 2050, deepened the policies in terms of extra investments and tax rates and widened them to cover more countries in order to reduce emissions sufficiently to meet the target.

10.4. The Economic Feasibility of Decarbonisation

One of the key findings for economic plausibility is that with the right mix of policies, the Decarbonisation Scenario shows slightly higher gross domestic product (GDP) growth, at 2.7% p.a., compared to 2.6% p.a. between 2010 and 2050 in the Reference Scenario. World employment is 0.9% higher by 2050 in the Decarbonisation Scenario. Inflation is slightly lower, since the world oil price grows less strongly when the demand for oil is lower under decarbonisation.

This higher growth, although hardly noticeable, can be contrasted with the reductions in growth from most modelling studies of deep mitigation in the literature, although not all those of the IEA. The reasons are, first, that we have adopted a post-Keynesian model that does not assume full employment and optimal use of resources in the scenarios, unlike the general equilibrium models widely used by others. This means that the economy can grow faster if there is more demand especially when there are under- or unemployed resources. Second, technological change induced by high carbon prices in the scenarios also reduces the costs of low-carbon equipment and electricity and this also allows for faster growth. Third, we have included higher long-term responses of carbon-intensive demand and production to relative prices, based on cross-section studies, increasing the responses to higher carbon prices. Finally policies of regulation, usually missing in equilibrium models, lead to earlier closures of inefficient coal-fired power plants and additional investments to meet higher standards.

However, to achieve these results we found that a mix of policies is necessary, giving both high carbon prices and strong regulations to promote low-carbon investment. With differentiated action, the average carbon price across all countries and sectors is some US\$29/tCO_2 by 2030 and US\$65/tCO_2 by 2050 in 2012 prices. For the energy-intensive sectors, mainly power generation, in the more developed countries, the prices of permits in emissions trading rises to US\$110/tCO_2 in 2030 and US\$187/tCO_2 in 2050 in 2012 prices. Policies involving a carbon price by itself, even if they rise to very high levels and are extended from the energy-intensive sectors to all other sectors via a carbon tax, are insufficient to achieve the substantial reductions in CO_2 required. This is a feature of our model, but it reflects the fact that some uses of fossil fuels, e.g. aviation, are both rather unresponsive to increases in prices of fuels and have expensive non-fossil fuel alternatives.

These findings are robust to changes in the underlying assumptions about oil prices and the risks that long-term growth may be slower after the Great Recession. Indeed, one finding is that CO_2 emissions may be higher with lower growth, because overall investment is lower and technological change is slower, so there is more use of coal in the energy mix, even although less energy is consumed.

10.5. Transforming the Energy Sector

There are also many ways in which the energy sector can be decarbonised. How it happens will depend on a country's natural resources, policies and trading opportunities. There are three main routes to decarbonise the energy sector: (1) substitution of fossil fuels by renewable sources such as solar, wind, hydro, oceans and biomass; (2) substitution by use of nuclear fuels; and (3) the use of fossil fuels with carbon capture and sequestration (CCS). The combination of biomass as a fuel with CCS yields negative GHG emissions.

However, the available supplies of renewable resources and the costs of bringing them to market will constrain the level and mix of decarbonised electricity, both globally and by region. These supplies are subject to generally increasing unit costs as the most accessible and lowest-cost sources of each type of resource tend to be exploited first. Against these increasing costs, technological progress and economies of specialisation and scale lead to reductions in costs of the equipment and infrastructure necessary to generate the renewable electricity. In addition, emissions trading schemes and subsidies, such as feed-in tariffs, add to the profitability of low-GHG technologies and increase their adoption and hence accelerate the technological learning and further reductions in unit costs.

Solar energy has the highest technical potential for renewable energy sources, with the opportunity costs of land being the main limiting factor. With unit costs falling over the long term and solar energy already cost-effective in niche applications, solar energy is a backstop technology.

However it will require a massive investment in the global electricity grid, given the distribution of the resource in relation to demand. Onshore wind has the highest potential at lower costs, followed by hydro and biomass. Offshore wind and ocean potentials are more limited both by cost and availability. Nuclear and CCS are also more expensive.

The alternatives to conventional fossil fuel power generation are available but will require subsidies to raise their adoption to a scale that will reduce unit costs (solar and wind), and increasing levels of carbon prices to keep them competitive with coal and gas.

10.6. End-use Decarbonisation

Looking at the reduction of GHGs by end-use leads to two related considerations. First, the relevant demand is for services such as warmth, cooling, light, power, mobility and the physical properties of the fossil fuel, as in the use of coke in cement manufacture, rather than for the forms of energy. It is important to consider other ways in which these services can be provided, as well as direct substitutes for the specific energy forms. Thus in transportation the switch in mode, from cars to bicycles, may be as important as a switch from fossil fuels to biofuels in reducing emissions. The second consideration is that the indirect emissions are very important in mitigation policy, e.g. the electricity used to replace gas in domestic heating may be generated from fossil fuels.

There are many ways that end-use emissions can be reduced by different groups and regions over the period to 2050, through demand reduction and new technologies. We have considered these by dividing the global economy into three major end-use sectors, namely, buildings, transport and industry.

In buildings and transport, the emission sources are dispersed depending on the human population, incomes and climate. Those taking decisions about energy use often have many other considerations besides costs and prices, since lifestyles are usually more important. In addition, costs of information are higher in relation to rewards from action. These considerations mean that final demand is likely to be inelastic and regulations imposed on builders and manufactures are likely to be more effective than market-based instruments. Buildings are the sector in which it appears there are the most lowest-cost contributions to GHG mitigation according to cross-sector engineering studies.

There is a difference here between developed and developing countries. In developed countries, retrofitting for low energy use is important. In developing countries, urbanisation will be a critical feature of the next 40 years, so GHG standards for new building are critical for long-term mitigation. Thus different solutions will be required for different countries depending on their stage of development. Transport also has a high mitigation potential, both in improving the efficiencies of vehicles and in the relatively short lifetimes of road passenger vehicles, allowing the vehicle stock to be replaced quickly if there are sufficient price and other incentives.

Mitigation of industrial emissions is highly specific depending on technology and location. Some industries have cost-effective options for CCS. In others there are substantial opportunities for non-destructive recycling.

The role of multinationals and supply chains is also important in that they cut across each of these three sectors and across national borders.

It must be noted that exploitation of opportunities for decarbonisation within any given sector are, alone, insufficient for meeting the required level of global emission reduction. All sectors therefore need to pursue important demand reduction strategies simultaneously. Strategies include the introduction of new technologies that result in lower emissions but also an acceptance that consumption levels of carbon-intensive goods and services must be reduced.

10.7. The Political Feasibility of Decarbonisation

An important contribution of this book to the debate and information available on GHG mitigation is the approach we have taken to the policies modelled. We have modelled those policies that governments have chosen to address the problem, rather than following the conventional more idealised policies adopted in the literature, such as a global carbon tax.

Despite over 20 years of growing scientific concern and global debate, only very weak global policy action has been forthcoming. Under the Durban Platform of 2011, governments have agreed to set up a process for international agreement in 2015 leading to coordinated action from 2020. A gap of 5–6 Gt has been estimated between the expected CO_2 emissions after government pledges and the 44 Gt estimated for 2020 as required for there to be a likely chance of limiting temperatures to the 2°C target. At present there is no convincing set of policy actions by governments to close the gap.

However there have been some successes, with tightened emission standards for energy efficiency in buildings and vehicles in many countries, and the introduction of the EU Emissions Trading Scheme (EU ETS) as well as a number of national and sub-national carbon taxes and trading schemes. We have described and discussed these successes and failures, making observations and drawing lessons for the political feasibility of mitigation.

A preliminary observation is that scientific consensus and the appeal to reason in decision-making has repeatedly faced unprincipled and well-funded campaigns of disinformation, designed to thwart effective climate policies. Despite climate change as an issue being a concern of both right- and left-wing politicians, policies for mitigation have readily been politicised. This has especially been the case when a tax is involved, and has been used by the opposition to gain power. This observation suggests the importance of non-state actors in both opposing action and mobilising support. It is notable that agreements on trading and regulation involve far fewer people and organisations compared to those for tax policies.

Policies are developed in a dynamic evolutionary process, implying path dependence and overlapping policies. Typically successful policies have come about through special circumstances, when they fitted into a broader set of political concerns, e.g. incidents of severe air pollution in large urban areas. Timing is critical. And an institutional history of groups or organisations working together greatly supports collective actions.

The first lesson is that people respond more to prospective losses than to gains. In climate policy, the losses are more immediate, being higher fuel costs and the loss of revenues and profits of producers as soon as demand for fossil fuels is curtailed. The gains are in the reduced risks of climate change and they are much later and experienced largely by those who do not have a current vote. The losses are also more concentrated on those who have the means and ability to lobby successfully to oppose or undermine the policy. The gains are diffused across many more people over a much longer timescale. Even with green tax reform, when the overall tax burden remains the same, those who expect to pay the pollution taxes are usually fewer, better funded and better organised than those who benefit.

Second, there have been lessons for emissions trading versus carbon taxes. In trading schemes, permits can be freely allocated, allowing the political process to compensate the losers. In the EU there has also been a critical advantage for the establishment of the ETS: the agreement required qualified majority voting, whereas an EU carbon tax required a unanimous decision across all of the Member States. Perhaps the smartest feature of the EU ETS is that its phases allow the scheme to evolve, learning lessons from experience. However, a major weakness has been the repeated over-allocation of permits and great difficulties in managing the scheme to remove permits from the market to keep permit prices high enough to signal the profitability of long-term abatement. Proposals for new schemes may be strengthened if they include floors and ceilings for the prices of permits, with intervention or related mechanisms to manage the market.

10.8. Co-benefits of Decarbonisation

While climate-related effects on populations are often in the future (beyond the timescale of political elections) and in foreign places, there are significant co-benefits of decarbonisation that are immediate and local. A particular co-benefit of interest is co-reduction of air toxics when GHGs are reduced. For example, in urban areas in developing countries, the political feasibility of policies for GHG mitigation is greater when there are immediate health benefits from the associated reductions in air pollution.

The economic improvements from these co-benefits can at times be as large as, greater than, or at least a significant fraction of, the apparent costs of GHG mitigation. This is especially true for particulate matter (PM) reduction co-benefits. In addition, the scientific confidence with which co-benefits can be estimated is greater than that for climate-related health benefits, again improving policy acceptance.

A reduction in global CO_2 emissions by 50% would result in a co-benefit from PM reduction equivalent to reducing lifetime risk of mortality by 0.001 (or one in 1000). In other areas of environmental policy, a reduction in risk of this magnitude would be required if feasible. The health care/lost productivity cost savings from this 50% reduction would be on the order of US$60,000,000,000 per year globally, which is not dissimilar to the direct costs of transitioning to a globally decarbonised energy system.

Meeting end goals will also require working with the interdependencies between sectors and realising that energy demand cannot be considered in isolation from energy supply and *vice versa*.

10.9. Conclusion

This book has addressed the feasibility of decarbonising the global economy sufficiently to have a likely chance of keeping the global average temperature below 2°C by 2100. There are many different aspects to the problem. We have largely framed our approach in terms of policies and measures available to governments, taking the view that governments are the only institutional structure to provide the necessary leadership through the coordination of the UN. We have adopted the scenarios approach to assessing the problem and the solution, developing a reference and a decarbonisation scenario and using a formal integrated model of the world economy and energy system to make projections to 2050.

We have judged that decarbonisation is economically and technically feasible, with a combination of regulations for end-use energy-demand sectors and rising carbon prices for the energy-supply sectors. The policy mix we have considered corresponds to that already adopted in practice by governments although strengthened and extended to more countries. However all sectors have roles to play and the energy sector must be transformed.

The political feasibility of decarbonisation is more open to question. The science suggests that urgent action is required; if action is delayed the problem accumulates. The economics suggests that the greater the delay, the higher the cost, because more investment is wasted in extracting fossil fuels and in building coal and gas-fired electricity generation that will have to be retired early. The politics suggests that national action has been repeatedly defeated by opposition to carbon taxation and the vested interests of coal and oil producers and energy-intensive industries. International action, despite over 20 years of scientific consensus, has proven even more difficult.

We conclude that decarbonisation remains feasible in economic and technical terms, but that a wide range of policy instruments is required and that these must be greatly extended in their coverage of the world economy and in their strengths and rates of operation. A well-designed mix of regulation and emissions trading schemes could keep GDP costs negligible and lead to increases in GDP and employment in conditions of unemployed resources. International agreements and coordination of policies will reduce direct costs, and the health co-benefits of reduced air pollution will produce immediate gains.

Index